3

Maths
Teacher's Guide

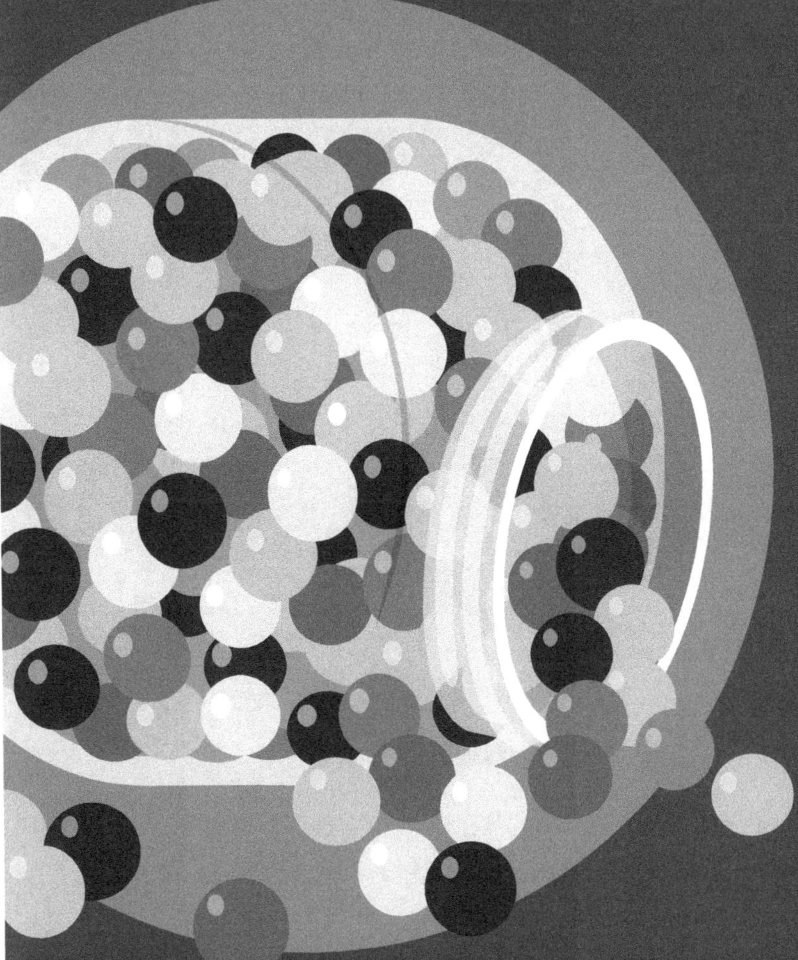

Ray Huntley
Tony Cotton

Caroline Clissold
Linda Glithro
Cherri Moseley
Janet Rees

OXFORD
UNIVERSITY PRESS

UNIVERSITY PRESS

Great Clarendon Street, Oxford, OX2 6DP, United Kingdom

Oxford University Press is a department of the University of Oxford. It furthers the University's objective of excellence in research, scholarship, and education by publishing worldwide. Oxford is a registered trade mark of Oxford University Press in the UK and in certain other countries.

British Library Cataloguing in Publication Data

Data available

ISBN 9781382017282

13

Paper used in the production of this book is a natural, recyclable product made from wood grown in sustainable forests. The manufacturing process conforms to the environmental regulations of the country of origin.

Printed and bound by CPI Group (UK) Ltd, Croydon, CR0 4YY

Acknowledgements

The publisher and authors would like to thank the following for permission to use photographs and other copyright material:

Cover: Artwork by Peskimo. **Photos: pv(t):** foto-bee/Alamy Stock Photo; **pv(m):** Alistair McDonald/Shutterstock; **pv(b):** 2R fotografia/Shutterstock; **px:** Monkey Business Images/ Shutterstock; **pxviii:** monkeybusinessimages/iStockphoto; **p75:** Oxford University Press; **p185(l):** legenda/Shutterstock.

Artwork by Q2A Media Services Pvt. Ltd and OKS.

Every effort has been made to contact copyright holders of material reproduced in this book. Any omissions will be rectified in subsequent printings if notice is given to the publisher.

The manufacturer's authorised representative in the EU for product safety is Oxford University Press España S.A. of El Parque Empresarial San Fernando de Henares, Avenida de Castilla, 2 – 28830 Madrid (www.oup.es/en or product.safety@oup.com). OUP España S.A. also acts as importer into Spain of products made by the manufacturer.

Contents

Introduction

The joy of learning maths

We are living in an ever-changing world, where the way we work, live, learn, communicate and relate to one another is constantly shifting. In this climate, we need to instill in our learners the skills to equip them for every eventuality so they are able to overcome challenges, adapt to change and have the best chance of success. To do this, we need to evolve beyond traditional teaching approaches and foster an environment where students can start to build lifelong learning skills for success. Students need to learn how to learn, how to problem solve, be agile and work flexibly. Going hand-in-hand with this is the development of self-awareness and mindfulness through the promotion of wellbeing to ensure that students learn the socio-emotional skills to succeed.

With *Oxford International Primary Maths*, students develop lifelong learning skills as well as mathematical skills. The course promotes the development of real-world skills including financial literacy. The activities in the Student Books and Practice Books offer numerous opportunities to think creatively and develop interpersonal skills. Fundamentally, *Oxford International Primary Maths* promotes students' self-development as critical thinking and motivation are at the heart of the problem-solving approach in the course.

This series is based on the English National Curriculum Programme of Study for Primary Maths. The *Oxford International Primary Maths* books for each stage meet all the learning objectives from the curriculum. Each lesson includes the learning objectives and a summary of the key teaching points. A full mapping grid identifying the unit and lesson where each objective can be found is available online at www.oxfordowl.co.uk

Oxford International Primary Maths: A problem-solving approach

In this second edition of *Oxford International Primary Maths*, there is a strong focus on using a problem-solving approach. While mathematical facts are important, it is unlikely that simply giving students the information they need will result in them understanding the mathematics and being able to apply their learning in new problem-solving situations. This is often described as a move from 'surface learning' to 'deep learning'.

Many people remember mathematics lessons as places where the teacher stood at the front of the class writing on the board. Students wrote down the information, maybe worked through a couple of examples with the teacher and then proceeded to complete a series of exercises to practise the skill that they had been taught. This can be described as a *didactic* approach and it relies on the idea that direct instruction is the appropriate strategy to adopt. The authors of this series would argue that *heuristic* strategies encourage students to explore the mathematics for themselves supported by the teacher. 'Heuristic' derives from the Greek word meaning to discover, and in mathematics learning, heuristic strategies are ones where students engage in exploration and discovery to solve a problem. Heuristic strategies include making a visual representation of a problem, making a calculated guess or estimate, simplifying a problem or following a known method. This results in a deeper understanding.

When faced with any problem in mathematics, there are recognised stages to go through in order to solve the problem, and these have been developed and agreed by many researchers. One version that summaries the problem-solving process comes from Georg Polya.

1. Understand the problem.

2. Devise a plan.

3. Carry out the plan.

4. Check the reasoning.

In following these stages, students use a number of skills that support problem solving, such as using trial and improvement, working systematically, pattern spotting, visualising, conjecturing and generalising.

Embedding a mastery approach

In recent years, the term 'mastery' has been used in conjunction with mathematics learning. It has been drawn from teaching approaches in countries where mathematics performance is deemed to be very high. The essence of mastery is to produce students who have deep conceptual understanding and procedural fluency through learning in a collaborative and problem-solving context. Mastery learning incorporates use of manipulatives, exposure to different methods of solving a problem, dialogue and explanation.

Following a Concrete Pictorial Abstract (CPA) approach

One of the more successful approaches to learning was provided by Jerome Bruner in his model of enactive, iconic and symbolic modes. This has been developed in recent years to form the CPA approach. CPA stands for concrete, pictorial and abstract, each of which aligns with Bruner's modes. The concrete phase involves students making use of physical manipulatives to help understand the learning, before moving to record the learning in pictorial form as individuals. As the learning develops, students will begin to recognise how to record their learning in a more general and abstract way. The CPA approach is not necessarily sequential, and students might move between the different modes as they work through a problem.

Oxford International Primary Maths and the use of manipulatives

Throughout the series, students are encouraged to use manipulatives, or concrete objects, to model addition, subtraction, multiplication and division. These manipulatives include:

- base-ten equipment (ones-cubes, tens-rods, hundreds-flats and thousands-cubes)

- place-value counters

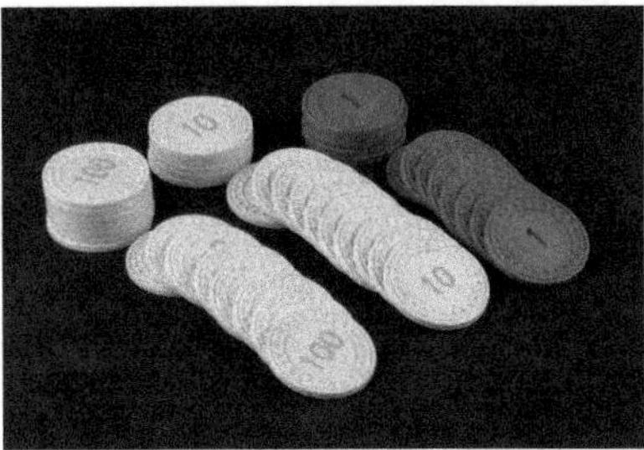

- number rods.

Such manipulatives are used to explain to students how the written methods 'work', for example by modelling exchanging 10 ones-cubes for 1 tens-rod in an addition.

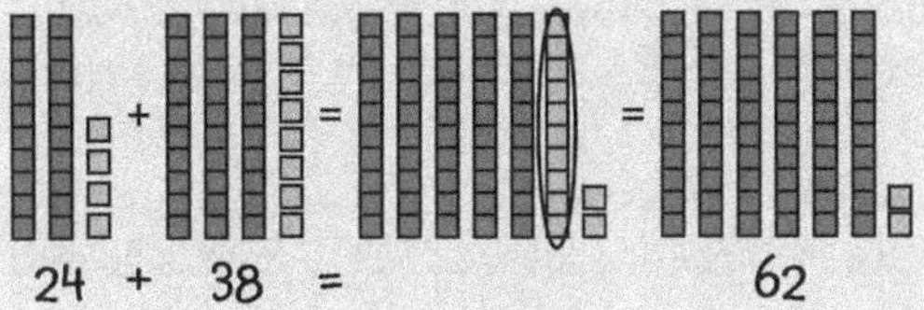

$$24 + 38 = 62$$

Differentiation

There are several ways that you can differentiate learning in the classroom. These include differentiation by:

- task
- outcome
- support
- grouping.

It has been traditional in some schools to offer up to three different levels of tasks for each lesson. This is differentiation by task. It is important that all students are exploring the same area of mathematics as they can collaborate and discuss their mathematics in a way that is not possible if students are engaged on different activities. This approach has been extensively researched and published by Jo Boaler of Stanford University, California. For example, she has outlined projects that gave students in different schools either a differentiated approach in lessons, or lessons where everyone worked on the same task (Boaler, J., 2005. *The Elephant in the Classroom*. Souvenir Press). Where all abilities worked on the same task, every student made and sustained 'better than expected' progress, and performed better on statutory tests and exams. The Education Endowment Foundation teacher's toolkit suggests that collaborative learning can result in a five-month acceleration in students' learning. (See https://educationendowmentfoundation.org.uk/resources/teaching-learning-toolkit.)

The expectation in this series is that all students will be offered the same starting point. The activities are carefully designed to be accessible to all students in your class and the teacher's notes for the activity offer differentiated outcomes for students. It is also important that you offer differentiated support to different students. You will mainly do this through the sort of questioning that you engage in and support you offer. You will ask challenging questions and supporting questions to help all students access the task. For example, when engaging in a simple counting activity with some students you might model the action of counting by placing a finger on each object as you count and emphasise the last number you say to model that the last number you say gives the number of objects in the set. You might ask other students engaged in the same activity to compare two sets, or to find one more or one less than the set they are counting.

Grouping students to promote a growth mindset

When engaging in learning mathematics, it is expected that you will use a variety of student groupings. This may be a change for some teachers who have previously grouped students by prior attainment in their classroom. Research has shown that grouping students 'by ability', which usually means grouping students using test results, can have a negative impact on their future attainment. It is more effective to use a range of ways of grouping students. You will decide on the most appropriate way of grouping depending on the activity. You are also given advice in the teacher's notes. It is important that the teacher is active in deciding which form of grouping is appropriate. It is also important that students learn how to operate in a range of different groups and with a range of different students so that they get used to working in a variety of ways and with different people.

The three main ways of grouping students are based on:

- friendship
- ability/prior experience
- mixed attainment.

Friendship groups, are most appropriate for activities in which the students have been given some element of choice. Perhaps they are carrying out some research for a data handling project or exploring data on animals to develop their understanding of measurement. This grouping is the default if teachers do not actively group students.

Ability groups, or groups based on students' prior experience, may be helpful if the lesson requires a very specific prior knowledge. You can group together the students you know have this knowledge and they can then work with minimal guidance from you, which allows you to focus on groups who need additional support.

Mixed-attainment groups are encouraged for the majority of the activities. This form of grouping is also favoured by those following a mastery approach. Working in collaborative, all-attainment groups also supports students' wellbeing and promotes a growth mindset, as described in research by Carol Dweck. She found that students who were grouped by ability tended to stay in those groupings throughout their school life, and regard themselves as having a fixed ability that could not be changed. This has dire consequences for students in middle or lower sets. When placed in mixed-ability groups, all students can develop a growth mindset which enables them to believe they can learn and improve, whatever their starting point (Dweck, C., 2007. 'The Perils and Promise of Praise'. *Educational Leadership*. October 2007, 65(2), 34–39). A growth mindset is promoted when students do not feel that their future success is predicated on prior achievement. This kind of grouping is particularly helpful for students new to English. Mixed-attainment groups allow students who are less confident in English to hear more-confident peers using mathematical vocabulary. Research has shown that mixed-attainment groups benefit both high attainers, who become more secure in their mathematics knowledge through explaining their thinking to peers, and those less secure in their mathematical knowledge as peer teaching has been shown to be effective.

Whatever form of grouping you choose, it is helpful to assign roles to individuals in the group. Some teachers use 'role cards' to remind members of the group of the role they should play. Here are some examples of roles.

- Leader: You should make sure everyone has a chance to speak and focus the discussion around the task.
- Time keeper: You should encourage the group to stay on task. Announce when the time is half way through and when time is nearly up.
- Recorder: You should write down group members' ideas or draw a collective graphic. You will write on the board during the presentation.
- Presenter: You will present the group's findings to the whole class at the end of the session.
- Resource organiser: You will make sure that group members have all the resources they need during the task.

Assessment is the process of establishing how individual students are progressing and what they have achieved, or a means of measuring their learning. Assessment is usually carried out in two main ways – assessment of learning and assessment for learning.

Assessment of learning is sometimes called summative assessment, and takes place at the end of a lesson, a unit, a term or even a year. It measures what students know at that point as a summary of their learning to that point. In *Oxford International Primary Maths*, summative assessment opportunities are provided in the Review lesson at the end of each unit in the Student Book, while half-termly summative assessment opportunities are provided through printable resources, available online.

Assessment for learning is an approach brought to prominence by Paul Black and Dylan Wiliam and is based on the notion that students have a full, clear sense of what they are learning, where they have reached in their learning and what they need to do to improve further. It is carried out during lessons and gives teachers continuous data on each student's learning, as well as allowing students to track their own learning, which provides greater motivation. (Black, P., Harrison, C., Lee, C., Marshall, B., and Wiliam, D. 2004. 'Inside the Black Box: Assessment for Learning in the Classroom'. *Phi Delta Kappan*. (86)1, 8–21).

It is suggested that there are five key strategies for assessment for learning. These are outlined below with suggestions of how you can do this in your classroom.

1 Being clear about learning objectives and success criteria with the students.

Each activity has at least one learning objective. At the beginning of a lesson, share the activity's learning objective with students. This should be more than simply stating the objective. You should make sure that students understand the objective and how you will measure success. For example, you might say: *I know that you can all count 10 objects* and all count to 10 as a class. Then you point to 20 on a number line and ask: *Does anyone know what this number is?* If a student knows it is 20 praise them, if no-one knows, tell them it is 20 and say: *By the end of the lesson I will be able to listen to you count to 20.*

2 Planning student discussions that give you evidence of their learning.

Every activity plan in the Teacher's Guide offers the opportunity for small-group or whole-class discussion. There are also examples of probing questions that you can ask to assess students' current understanding. For example, if a group has been counting two sets of objects you can ask: *Were there more or less in the second group? How do you know?*

3 Giving students feedback that helps them move forward.

This allows students to know whether or not they are meeting the success criteria and what they can do next to move their learning on. Developing the example above, if a group has been comparing two sets and understands the concept of 'more' and 'less' you could ask them to make sets that are one more and one less, or even two more and two less.

4   Activating students to act as instructional resources for each other.

Collaborative group work in mixed-attainment groups, as described by Jo Boaler in her research (see under Differentiation earlier), gives students the opportunity to operate both as learners and teachers, with peer learning being highly effective. Not only is understanding of the mathematics enhanced, but students can support each other in assessing their progress.

5   Activating students as owners of their own learning.

The key point here is to listen carefully to the students and adapt your questioning to support individual development and to follow individual interests.

Questioning is key

The most skilled mathematics teachers can ask open questions to elicit students' current understandings. Skilful open questioning also allows students to articulate their current understanding carefully and though this process either consolidate their understanding or come to realise where they have made a mistake. The list below offers a series of open questions that can be used whatever mathematics you are teaching.

- *How are these the same/different?*
- *About how many/how long/many more … do you think there will be?*
- *What would happen if …?*
- *How else could you have done that?*
- *Why did you ….?*
- *How did you …?*
- *How do you know that is correct?*

If you want students to check their solutions and consolidate their learning it is helpful to ask them to explain how they reached their solution to a friend. Similarly, to support students in reflecting on their learning you might ask the following.

- *What mathematics did you use to solve the problem?*
- *What new mathematics did you learn?*
- *What key words did you use?*
- *What was the most challenging part of the activity?*
- *What did you do when you got stuck?*
- *What other questions could you ask?*
- *Did this remind you of any other areas of mathematics?*

In *Oxford International Primary Maths*, there is an opportunity to ask these reflective questions, and for students to reflect on their learning, at the end of each unit in the Review lesson of the Practice Book.

Word problems

Word problems are useful as an assessment of children's understanding of the correct mathematics to use in any given situation. In *Oxford International Primary Maths* word problems are included throughout the units and on every Student Book Review page as part of the end-of-unit assessment. Many teachers find teaching word problems a challenge. This area is particularly challenging for students with a limited English vocabulary as word problems are tightly bound to linguistic ability. We have to decode and understand what the problem is asking us to do before we can begin to apply our mathematical knowledge. Some teachers have found the following acronym helpful when working with students on solving word problems.

R: Read the problem carefully.

U: Understand what the problem is asking you to do.

C: Choose the mathematics or arithmetical operations that you need to use to solve the problem.

S: Solve the problem.

A: Answer the problem.

C: Check that the answer is accurate and reasonable.

It is often helpful for students to underline key facts and write down the operations they are going to use before they solve the problem. For example:

> Tony rode his bicycle 7 miles to school with his friend. On his way home he took a short cut which was only 5 miles. How far did he cycle altogether?
>
> *This will be an addition calculation.*

It is a useful activity for students to annotate word problems and write down the operation(s) they will use without carrying out the calculation as this focuses on the skill of understanding the problem and choosing the operations appropriately.

Another activity that helps students to become skilled at solving word problems is asking them to write their own word problems based on a picture or a set of objects. Here is an example.

- How many black cubes are there? (3)
- Two friends took three cubes each. How many were left? (2)
- If I take out the black cubes, how many are left? (5)
- If I share the cubes equally between two people, how many do they each get? (4)

Wellbeing and *Oxford International Primary Maths*

It is thought that students learn more and feel more connected to their learning when they are active in their lessons. *Oxford International Primary Maths* has active learning at its heart. Most lessons start with a whole-class session that usually includes a range of physical or active activites. You will see this signified by a 'star-jump' icon in the Teacher Guide.

Many adults and children have felt anxious about their learning of mathematics at some stage. This anxiety is reduced by working collaboratively in all-attainment groups. There is also a reflective session at the end of each lesson and the formative assessment activity in the Practice Book asks students to reflect on their learning across the unit.

Wellbeing is also supported by effective questioning to support and stretch students and by planning group work carefully. These areas have already been discussed above.

Language support

The challenges

Ministries of Education at both local and national level are increasingly adopting the policy of English Medium Instruction (EMI), for either one or two subjects or across the whole curriculum. The rationale for doing so varies according to the local context, but improving the levels of achievement in English is an important factor.

In international schools an additional reason is likely to be that students do not share a mother tongue with each other or perhaps the teacher. English is, therefore, chosen as the medium for instruction so that all students are in the same position and to provide the opportunity to develop proficiency in an international language.

This does not mean that the mathematics teacher is now being asked to replace the English teacher, or to have the same skills or knowledge of English (though in many primary schools one teacher may indeed teach both). What it does mean, however, is that mathematics teachers have to view their role differently: they have to become much more language aware. It is this recognition of the need to ensure that the delivery of the content is not negatively impacted by the use of the second language that informs the planning and methodology of EMI.

This raises significant challenges, including:

- the teacher's knowledge of English
- students' level of English (which may vary considerably in international schools)
- resources that provide appropriate language support
- assessment tools which ensure that it is the content and not the language that is being tested
- differentiation that acknowledges different levels of proficiency in both language and content.

Meeting the challenges positively

Perhaps lack of confidence in their own English proficiency is one of the most common concerns among teachers. However, while it is a factor, success in EMI is not necessarily linked to teachers' proficiency in English. Teachers who have English as their mother tongue may well lack the sensitivity to, or awareness of, the language that a non-native speaker has acquired through learning and studying the second language. Developing this awareness and demonstrating it in both materials and method is the key to effective EMI.

Classroom language/Teacher Talk

Often non-native-speaker teachers are more concerned about their ability to run and manage the whole class in English than they are about the teaching of the mathematics concepts, as the resources or textbook should help them with the latter. However, this use of English in the class is very important as it provides exposure to the second language, which plays a valuable role in language acquisition. It is also true that the Teacher Talk for purposes such as checking attendance and collecting homework does not have to be totally accurate or accessible to students. When teaching the mathematics concepts, however, it is essential that the Teacher Talk is comprehensible. Some basic strategies to ensure this include:

- simplify your language
- use short, simple sentences and project your voice
- paraphrase (say in a different way) as necessary
- use visuals, write or draw on the board, gestures and body language to clarify meaning
- repeat as necessary
- plan before the lesson
- prepare clear, simple instructions and check understanding.

Creating a language-rich environment

Primary teachers often excel at providing a colourful and engaging physical environment for students. In the EMI classroom, this becomes even more important. Posters, 'word walls', lists of key structures, students' work, English signs and notices all provide a backdrop that provides the opportunity for language exposure and language acquisition.

Planning

When planning, look carefully at each stage of the unit and identify the language demands. This means thinking about what language students will need to understand or produce, and deciding how best to scaffold the learning to ensure that language does not become an obstacle to understanding the concept. This involves providing language support and goes beyond the familiar strategy of identifying key vocabulary.

Support for listening and reading

Listening and reading are receptive skills, requiring understanding rather than production of language. If you are asking students to listen to or read texts in English, ask yourself the following questions when you are planning the unit.

- Do I need to teach any vocabulary before they listen/read?
- How can I prepare them for the content of the text so that they are not listening 'cold'?

- Can I provide visual support to help them understand the key content?
- How many times should I ask them to read/listen?
- What simple question can I set before they listen/read for the first time to focus their attention?
- How can I check more detailed understanding of the text? Can I use a graphic organiser (e.g. tables, charts and diagrams) or gap-fill task to reduce the language demands?
- Do I need to differentiate the task for those students who find reading/listening difficult?
- Could I make the tasks interactive (e.g. jigsaw reading, when students access different information before coming together then share information)?
- How am I going to check their answers and give feedback?

Support for speaking and writing

Speaking and writing are productive skills because students doing these need to produce language. They are different from the receptive skills of listening and reading where students receive language from other sources. These skills may require more input from the teacher.

When you plan to use a task that requires students to *produce* English (speak or write), you need to think about how to help them do this.

This means that you have to think in detail about what language the task requires (Language Demands, LD) and what strategies you will use to help them use English to perform the task (Language Support, LS).

You need to ask yourself the following questions.

- What *vocabulary* does the task require? (LD)
- Do I need to teach this before they start? How? (LS)
- What *phrases/sentences* will they need? Think about the language for learning mathematics (e.g. predicting and comparing). What structures do they need for these language functions? (LD)
- Will they be able to produce these sentences or should I provide some *scaffolding* [e.g. sentence starters/sentence frames/gapped sentences (see below)]? (LS)

  A square has _____ sides.

  A triangle has _____ sides.

  A quadrilateral has _____ sides.

  A pentagon has _____ sides.

- While I am *monitoring* this task is there any way I can provide further support for their use of English (especially for the less-confident students)? (LS)
- What language will students need to use at the *feedback* stage (e.g. when they present their task)? Do I need to scaffold this? (LD, LS)

Teaching vocabulary and structures

Vocabulary

Learning the key mathematics vocabulary is central to EMI and 'learning' means more than simply understanding the meaning. Knowing a word also involves being able to *pronounce* it accurately and *use* it appropriately. Below is a list of strategies that could be useful.

- Avoid writing the list of vocabulary on the board at the start of the unit and 'explaining' it. The vocabulary should be introduced as and when it arises. Word boxes are provided on each page of the Student Books and Practice Books with the key words for the lesson. This helps students associate the word or phrase with the concept and context.
- Before the lesson, check that you are confident with the pronunciation and spelling of the vocabulary that will be used. Write the vocabulary clearly on the board when you first introduce it in the lesson. If you think students may struggle to pronounce words, decide how best to model the pronunciation.
- Give students a chance to say a word once they have understood it. The most efficient way to do this is through repetition drilling.
- Use visuals whenever possible to reinforce students' understanding of the word.
- Ensure that students are recording the vocabulary systematically in their glossaries, at the back of their Student Books, and, if possible, use a word wall that lists the vocabulary under unit or topic headings.
- Remember to use and revise the vocabulary.

Structures

In order for students to talk or write about their mathematics, they will need to go beyond vocabulary: they will also need to use those phrases and sentence frames that a particular task requires.

For example, they may need the following expressions in mathematics.

  X is the same as Y.

  The sides are the same length.

  The next number in the sequence.

  I predict that X will happen.

  If X happens, then Y happens.

  The next step is …

You need to build up banks of common mathematics phrases and encourage students to record them. This is an important part of identifying the language demands and providing the necessary support. You do not have to focus on grammar as the language can be taught as phrases rather than specific grammatical structures.

Every unit of the Teacher's Guide begins with useful background information that includes the following.

The Big idea: The main mathematical concept covered in the unit is outlined.

Look out for: This section focuses on tricky concepts that may need explaining prior to any learning taking place.

Common misconceptions: Common errors that students make, or misunderstandings that students have, are identified. This section offers advice on how to deal with these misconceptions.

Key vocabulary: This is a list of the key mathematical words used in the unit.

Coverage in lessons: The English National Curriculum objectives covered in the unit are listed.

Every lesson in the Student Book and Practice Book has corresponding lesson notes in the Teacher's Guide. These comprehensive lesson notes include the following.

A mini reproduction: This shows the relevant pages from the Student Book.

Global skills: These are the skills that aim to foster a classroom environment where students develop the skills for success. The skills are: *creative skills* where students are problem solving, investigating or exploring new maths content; *real-world skills* where students are taking part in research, or presenting and interpreting information, or if they are dealing with money and developing their financial literacy; *interpersonal skills* where students are practising their teamwork and communication, often through working in pairs or larger groups; and *self-development skills* where students have the opportunity to reflect on their learning and talk about what went well and what they are still uncertain about.

The key vocabulary and resources: Key vocabulary used in the lesson, and the concrete resources required for the activity, are listed.

Language support: This includes a range of strategies, including card sorts and card games, word walls, team games to define or explain words, use of similar words to explain meaning and exploration of the origins of words.

The key principles underpinning the language support are listed below.

Words should be introduced and explained carefully.

Words should be explained in context.

Repetition is vital.

Words should be linked to pictures or actions.

Students should develop their own glossaries.

The learning of mathematics vocabulary should be fun.

Language should not be a barrier to effective learning of mathematics.

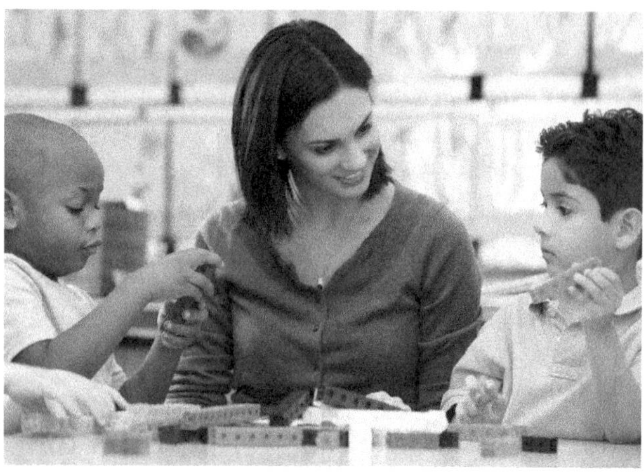

Detailed lesson notes: Comprehensive lesson notes include an introduction activity and main activity. These notes refer to the Student Book and Practice Book, where relevant. The notes include probing questions for formative assessment, which are italicised. Icons are used to suggest the groupings that should be used at each point of the activity (whole class, small group, pairs, individual). A separate 'star-jump' icon indicates that the activities give students an opportunity for physical movement (standing up, jumping, moving around) rather than doing activities sitting down.

Differentiation: The Teacher's Guide offers strategies for you to *support* those students who may have difficulty accessing the task; to *consolidate* the learning for those students who need a little more practice; and to *extend* the learning for those who need more challenge.

The Teacher's Guide also offers differentiated outcomes. These outcomes are listed in the form of:

All students

Most students

Some students

Stretch zone: Each activity in the Student Book and the Practice Book has a stretch zone question to support deeper learning. The Teacher's Guide provides additional notes on these activities.

Reflection time: Suggestions are made on how to bring the class back together to reflect on the learning and share ideas.

Answers: Answers to all the Student Book and Practice Book activities are provided.

Review pages: The Teacher's Guide has notes on the Review pages of the Student Book (summative assessment), with answers to the assessment questions, and the Practice Book (a formative, reflective review).

Digital resources: Where it is appropriate to use digital resources in a lesson, such as sharing the interactive Student eBook page on an interactive whiteboard (IWB), suggestions are embedded in the lesson plan.

Resources sheets: These photocopiable resources can be used with some of the main activities. They are referenced in the resources section of the lesson plan and are available on the Oxford Owl website (www.oxfordowl.co.uk).

Tour of a typical unit

The 'Big question' provides a discussion stimulus about the key idea of the unit.

1 Numbers and counting

?

How do we use numbers?

In this unit you will:

- count, read and write numbers to 100
- count in twos, fives and tens
- know and make numbers using objects and pictures
- use words such as equal to, more than, less than (fewer), most, least
- read and write numbers from 1 to 20 in words.

Learning objectives are stated clearly at the beginning of every unit.

Engage

Which numbers can you see in the classroom?

Which numbers can you see on your way to school?

What is the biggest number you have ever seen?

Further questions allow students to develop communication skills.

6

The Engage spread is bright and colourful, with artwork or photos to spark interest in young students and provide discussion points.

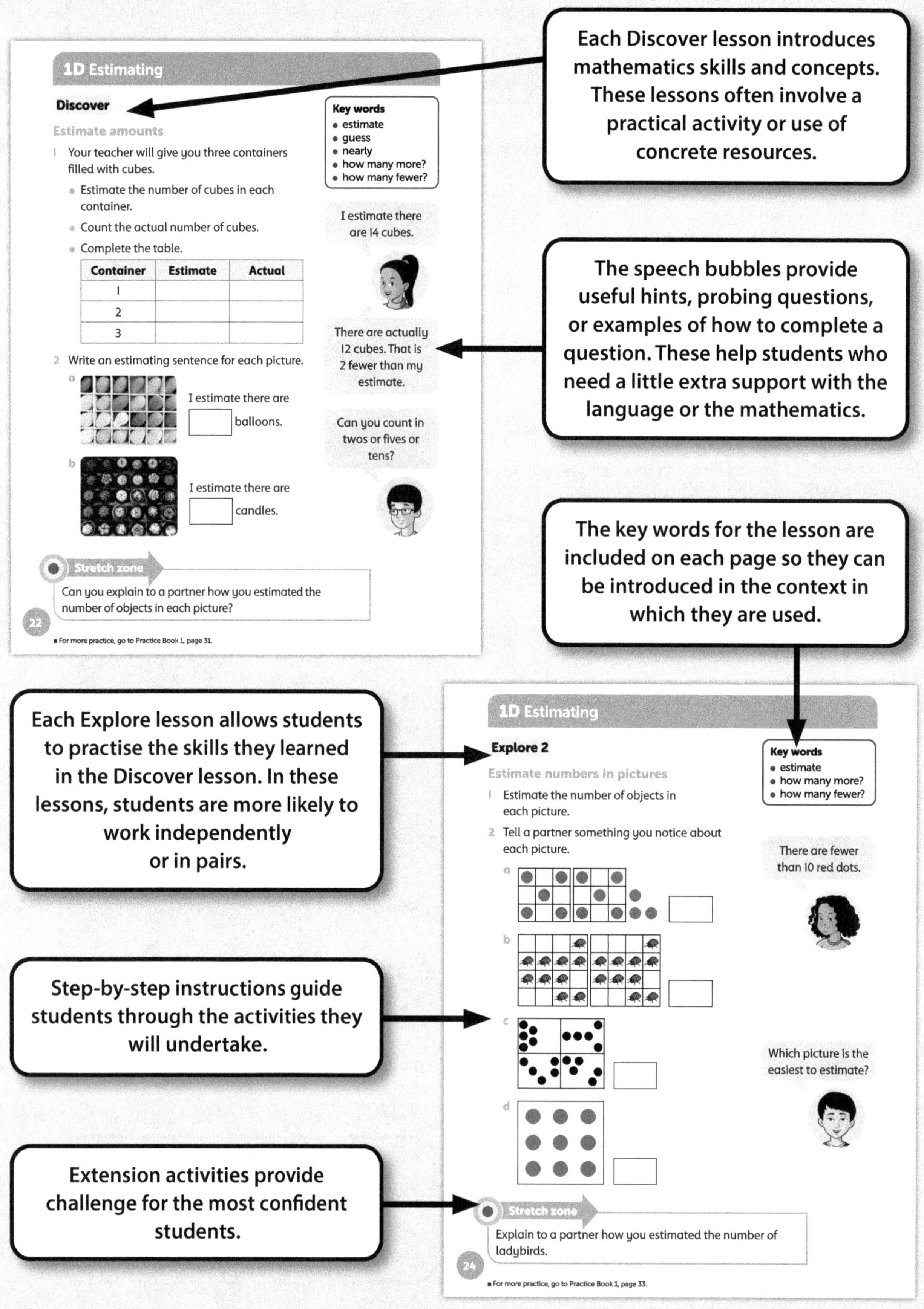

Each Discover lesson introduces mathematics skills and concepts. These lessons often involve a practical activity or use of concrete resources.

The speech bubbles provide useful hints, probing questions, or examples of how to complete a question. These help students who need a little extra support with the language or the mathematics.

The key words for the lesson are included on each page so they can be introduced in the context in which they are used.

Each Explore lesson allows students to practise the skills they learned in the Discover lesson. In these lessons, students are more likely to work independently or in pairs.

Step-by-step instructions guide students through the activities they will undertake.

Extension activities provide challenge for the most confident students.

1 Numbers and counting

Connect

Make a number poster

Work as a group.

1 Collect some magazines. Talk about which magazines might have numbers in them. What do the numbers tell us?

2 Cut out pictures that have numbers.

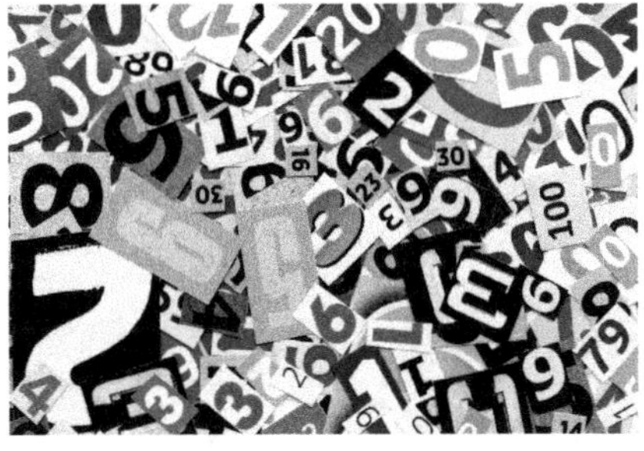

3 Make a poster to display in class.

4 Talk in your group about the numbers you have found.

Take photographs of numbers on the way home from school. What job are the numbers doing? Explain your ideas to a partner.

We use numbers to count or to say how many of something there are.

The 'Big idea' sums up what students have discovered in the unit. It answers the Big question on the Engage page.

What is the biggest number on your poster?

What is the smallest number on your poster?

A further extension activity provides a challenge for the most confident students.

1 Numbers and counting

Review

Students' progress is assessed through the questions and tasks at the end of each unit. In Student Books 2 and 6, these questions reflect the style of the SATs (national Standard Assessment Tests).

1 Draw the beads and write the numbers in the spaces.

Beads	Numbers	Words
	5	
		sixteen
		three
	12	
		nineteen
	1	
		four
	14	
		twenty

2 Samir has a bracelet with 19 beads. Lina's bracelet has one more bead than Samir's. How many beads are on Lina's bracelet?

Celine's bracelet has 10 more beads than Lina's. How many beads are on Celine's bracelet?

26

A word problem is always included on the Review page.

Practice Book Discover and Explore

Practice Book activities can be completed in the school lesson or as homework.

Most of the Discover and Explore lessons in the Student Book have a corresponding page in the Practice Book. These activities provide opportunities for students to consolidate and deepen their learning.

Step-by-step instructions guide students through the activities they will undertake.

If students require concrete resources, these are listed in a box at the top of the page. This is particularly useful if students are completing the activities at home.

Extension activities provide challenge for the most confident students.

1D Estimating

Discover — Student Book 1, page 22

- four different sorts of small objects that you can hold in your hands

How many of each object do you think you can hold?

Write your estimate and then take a handful to find out.

Was your estimate more or less than the actual number? How many more or less?

An example is shown in the table.

Object	Estimate	Actual number	More or less
cherries	11	8	My estimate was 3 more.

Stretch zone

Did your estimates get better each time? _______________

If they did get better, can you explain why?

____

31

1 Numbers and counting

1D Estimating

Explore 2 — Student Book 1, page 24

Estimate the number of dots. Do not count them!

Draw a circle around the number that you think is a good estimate.

1. 2   5   9

2. 9   12   14

3. 3   4   6

4. 6   9   12

5. 10   13   16

6. 5   8   10

Stretch zone

How many sweets do you think there are in this jar? ☐

How did you make your estimate?

____

33

1 Numbers and counting

Each Review page in the Practice Book includes a reminder of all the topics learned in the unit.

1 Numbers and counting

Review

1 Draw a face next to each bubble to show how you feel about your learning.

counting objects

reading and writing numbers

counting in twos, fives and tens

estimating quantities

2 Tell a partner about one thing you did really well in this unit.

3 Draw or write about things you found easy, challenging or really hard.

What work did you feel confident doing?

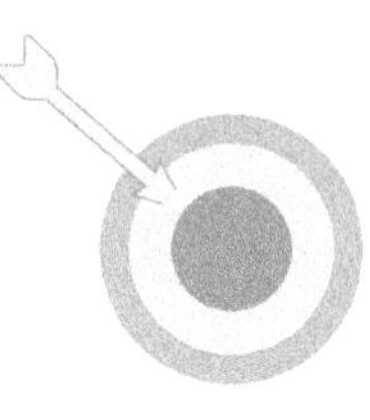

What work was challenging?

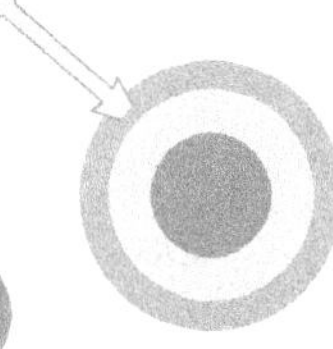

Is there any work you might need some extra help with?

34

Self-assessment activities help students to reflect on their learning.

The Student Books

The Student Books are write-in textbooks for students to read and use. There are six Student Books: one for each school year at primary school. The Student Books introduce learning through a mixture of practical, discussion and independent activities.

Student Book	Typical student age range
Student Book 1	Age 5–6
Student Book 2	Age 6–7
Student Book 3	Age 7–8
Student Book 4	Age 8–9
Student Book 5	Age 9–10
Student Book 6	Age 10–11

  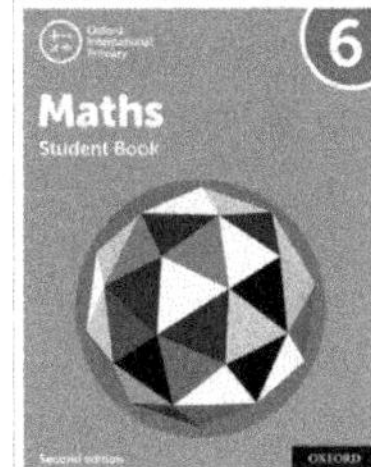

The Practice Books

The Practice Books are write-in workbooks for students to read and use. There are six Practice Books: one for each school year at primary school. The Practice Books provide deeper learning opportunities through a range of independent activities, which can be completed in school or at home.

Practice Book	Typical student age range
Practice Book 1	Age 5–6
Practice Book 2	Age 6–7
Practice Book 3	Age 7–8
Practice Book 4	Age 8–9
Practice Book 5	Age 9–10
Practice Book 6	Age 10–11

 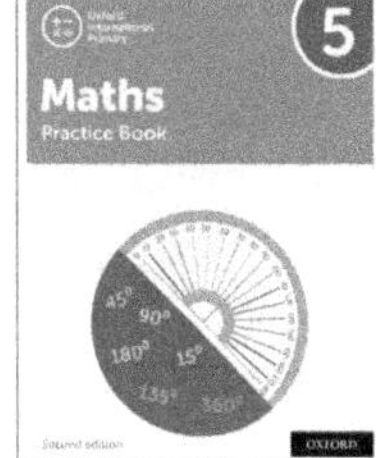 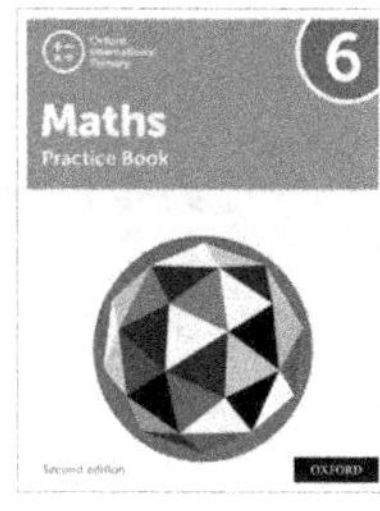

The Teacher's Guides

There are six Teacher's Guides: one for each school year at primary school. Each Teacher's Guide includes:

- an introduction with advice about delivering mathematics in primary schools using *Oxford International Primary Mathematics*
- a unit overview, giving advice on teaching each unit, including common misconceptions and how to deal with them
- a lesson plan for every lesson in the Student Book and corresponding pages in the Practice Book
- model answers to each question in the Student Book and Practice Book.

  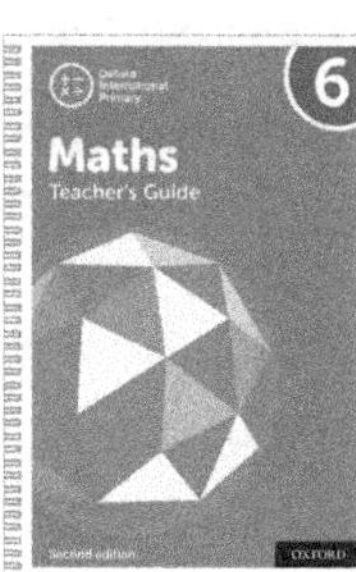

Digital resources

Interactive eBooks

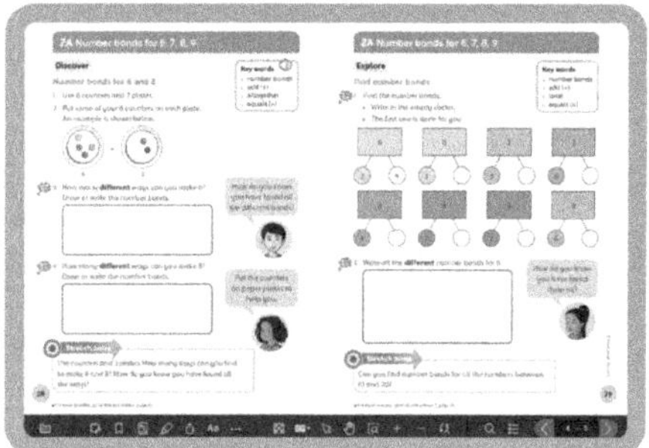

For the teacher

Teachers can access the Student Books, Practice Books and Teacher's Guides online in eBook format, on the Oxford Owl website (www.oxfordowl.co.uk).

The enhanced eBooks show the course content on screen, making it easier for teachers to deliver engaging lessons.

For the students

Teachers can allocate an eBook version of the Student Books to students for use at home. The Student eBooks include interactive activities, worksheets and audio of all the key vocabulary,

Assessment resources

The downloadable assessment materials offer you additional opportunities to assess students' progress. The materials include:

- end-of-unit summative assessment

- end-of-year summative assessment.

Every test comes with everything you need to assess and record progress including:

- answers

- mark schemes and guidance on assessment.

Oxford Primary Illustrated Maths Dictionary

The *Oxford Primary Illustrated Maths Dictionary* gives comprehensive coverage of the key maths terminology students use in the course. Entries are in alphabetical order, and each includes a clear and straightforward definition along with a fun and informative colour illustration or diagram to help explain the meaning. The dictionary is suitable for Students with English as an Additional Language.

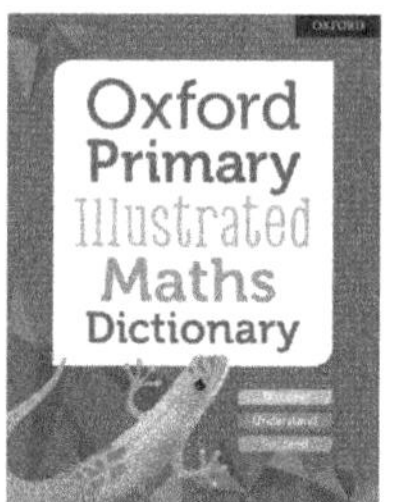

The curriculum

The Oxford International Curriculum offers a new approach to teaching and learning focused on wellbeing, which places joy at the heart of the curriculum and develops the global skills students need for their future academic, personal and career success.

Through six subjects – English, Maths, Science, Computing, Wellbeing and Global Skills Projects – the Oxford International Curriculum offers a coherent and holistic approach to ensure continuity and progression across every student's educational journey, equipping them with the skills to shape their own future. Through this approach, we can help your students discover the joy of learning and develop the global skills they need to thrive in a changing world.

Overview

Big idea

Students have previously been introduced to numbers up to 100. In this unit, students draw on their knowledge of place value up to 100 to build their understanding of numbers up to 1000. The key knowledge is to understand what each digit in a number represents and to realise that the value of a digit is 10 times greater than the same digit one place to its right. This will help students understand the effect of multiplying by 10, which they will do in later units.

Students need to continue to use 100-squares and number lines to develop their mental images of the number system. Place-value cards and base-10 equipment will also help them come to understand how they can partition numbers up to one thousand.

Look out for

- **Students who read numbers as a series of digits. For example, they read 503 as five, zero, three (or, as five, 'oh', three), rather than five hundred and three.** Encourage students to say the place-value names, so this would be 'five hundred and three'. Likewise, students might read 348 as 'three, four, eight'. Encourage them first to read this as 'three hundred', 'forty' and 'eight' and to then combine these to say 'three hundred and forty-eight'.
- **Students who use a 100-square incorrectly. When they come to the end of a row, they move directly down to the next row and read from right to left.** Cut out the individual rows of a 100-square and lay the rows end-to-end to model how a 100-square can be seen as one long number track. Point out that, on a 100-square, each new row is therefore read from left to right.

Possible misconceptions

- **Students think of numbers as a sequence of digits, and so do not realise the value of each digit.** Much like reading numbers as a series of digits, this is a misunderstanding of the individual value of each digit within a number. To overcome this, model numbers using base-10 equipment and by writing numbers in a place-value grid, including using 0 as a place holder where necessary.
- **Students write numbers as they say them. For example, they write 423 as 400203.** Model how 0 is used as a place holder in numbers. Use base-10 equipment to make the numbers and then write out the number to represent each part of the base-10 model. Demonstrate the 0 is only used when there are no hundreds, or tens or ones.

Key vocabulary

- number names from zero to one thousand
- count to 100 in ones, twos, threes, fours, fives, count on/back
- How many? How many more/less/fewer?
- place value, ones, tens, hundreds, thousands
- exact, approximation, estimate, group
- digit, 1-, 2-, 3-, and 4-digit number, what does that digit represent?
- equal to, greater than, more than, smaller than, less than, biggest, smallest, closest to, 1 more, 10 more, 100 more, 1 less, 10 less, 100 less, compare, order, size
- number line, 100-square
- round, to the nearest, multiple of 10, multiple of 100
- first, second, third, … twenty-first, twenty-second, …

Coverage in lessons

Learning objective	Engage	1A	1B	1C	1D	Connect	Review
Count from 0 in multiples of 4, 8, 50 and 100; find 10 or 100 more or less than a given number.		✓			✓		
Recognise the place value of each digit in a 3-digit number (hundreds, tens, ones).		✓	✓	✓	✓	✓	✓
Compare and order numbers up to 1000.	✓	✓	✓	✓	✓		✓
Identify, represent and estimate numbers using different representations.		✓	✓	✓	✓	✓	✓
Read and write numbers up to 1000 in numerals and in words.	✓	✓	✓				✓
Solve number problems and practical problems involving these ideas.	✓	✓	✓	✓	✓	✓	✓

1 Number and place value

Big question

- How can I count large numbers of objects?

Global skills

- **Creative skills:** problem solving
- **Real-world skills:** interpreting information
- **Interpersonal skills:** communication

Key vocabulary

- ones, tens, hundreds, estimate, compare, exact, approximate

Resources

- beans (dried beans or pulses), rulers, metre rules, calculators

Language support

Ask students whether they remember what 'estimate' means and whether they can think of familiar situations to put it into context, for example: *Estimate how many people were at the park on Saturday. How did you estimate? Did you visualise the people in groups?* For example, there were 16 playing football, 4 pairs flying a kite, around 12 people walking their dogs.

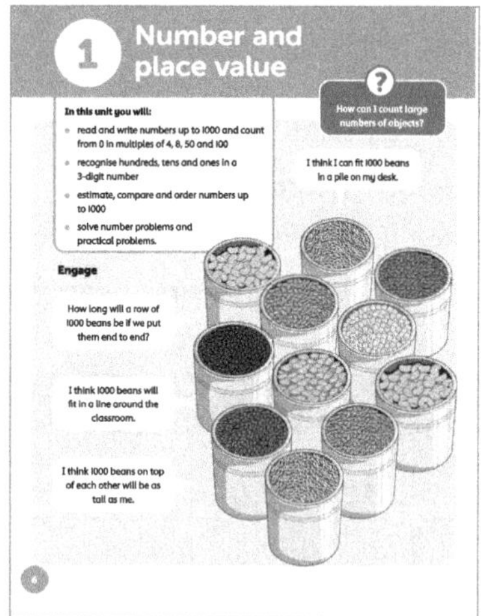

Introductory activity

Give each student a small packet of beans (or raisins) either in a packet or as a handful.

Ask them to **estimate** how many there are without counting. Students feed back to you how many beans they estimate are in the packet. Then ask them to count the beans. Notice those students who count by 'grouping'. When everyone has counted their beans, ask who grouped to count and discuss the different ways of grouping. Ask all students to group in **tens** and **ones**. Use this to calculate how many beans there are, altogether, in the class.

Main activity

Look together at page 6 of the Student Book. If you have access to an IWB you could use this. The groups should now look at the speech bubbles in the Student Book and solve the problems or work out whether the statements are correct. Groups may need support in realising that they need to break down the problem. For example, they first work out how much space 100 beans will take up, and then use this to calculate how many beans will fit on the desk altogether.

Students can make a template of the area covered by 100 beans and see how many times this fits on the desk. Students can use similar methods for the other questions. Students can either all work on the problems, or you could ask different groups to work on different problems.

Differentiation

Students should work in mixed-attainment groups so that the less-confident learners can see grouping and counting modelled by their peers.

Supporting: Count with students in ones to 100. Model counting in tens and **hundreds**.

Consolidating: Ask students to justify their estimates to you. *How do you know that this is a sensible estimate?* Encourage them to count in hundreds to 1000.

Extending: Ask students to share their strategies with the rest of the group.

Stretch zone: Ask students to write some tricky problems that they can now work out, using the information they have collected in the main activity. For example, once they know how many beans tall they are, can they work out how high the school is, or how high their house is?

Reflection time

Share answers to each of the problems. Introduce the language of estimation, for example: *Are your answers '**exact**' or an '**approximation**'?*

Ask students for other estimations: *How many beans could you hold in your hands? How many beans would fit in your school bag? How many beans would fit on the teacher's desk?* Ask those students who completed the Stretch zone activity to read out some of their problems.

1A Place value

Discover 1 — Student Book page 7 · Practice Book page 14

Specific learning focus

- Read and write numbers to at least 1000.
- Understand what each digit represents in 3-digit numbers.
- Partition 3-digit numbers into hundreds, tens, and ones.

Global skills

- **Creative skills:** exploring
- **Interpersonal skills:** communication

Key vocabulary

- hundreds, tens, ones, order, digit

Resources

- large place-value cards
- large 0–9 digit cards and smaller sets of 0–9 digit cards for each pair

Language support

Encourage students to say the number names rather than just speak the names of the digits. For example: they should say 'forty-five' and not 'four five'. Use the language of place value, asking: *How many tens are there in 60? How many hundreds are there in 500?*

Introductory activity

Ask three students to come to the front of the class. They should pick a **digit** card each and stand in line in the **order** they came out. For example, if the three students picked 4, 2, 9, the number would be 429. Ask the class to say the number. Make sure that they say 'four hundred and twenty-nine' and not 'four two nine'. Model this number using the place-value cards on the board:

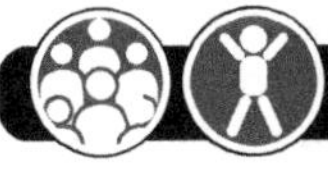

Overlap these cards on the board to form the number 429, and model the language: *the 4 represents four hundreds, the 2 represents two tens, and the 9 represents nine ones.*

Repeat, with three different students picking the digit cards each time.

Main activity

Keep the last set of digit cards visible and now ask students (in their pairs) to rearrange the cards so that they make the largest number possible. Repeat the modelling from the introductory activity, with the place-value cards and the language needed. For example, using the digit cards 4, 2 and 9 say: *In 942, the 9 represents nine hundreds, the 4 represents four tens and the 2 represents two ones.* Finally, ask students to rearrange the cards to make the smallest number possible. Again, repeat the modelling and language using the place-value cards.

Ask students to complete the activities on Student Book page 7. They have their own sets of digit cards and select their own digits to generate 3-digit numbers. Refer students to the second speech bubble: *How can we make a 3-digit number only using two digit cards? What do we need to include?* Discuss the use of zero as a place holder in numbers such as 205 or 230.

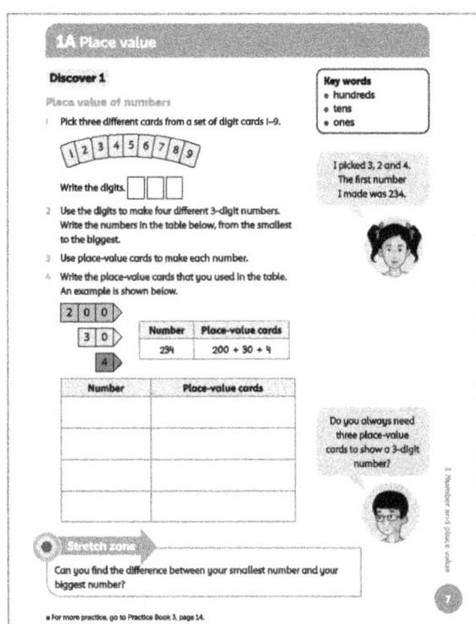

Differentiation

Supporting: Model the number names with students. Say them and ask students to repeat them. Make deliberate mistakes and ask students to spot them.

Consolidating: Ask open questions that encourage students to develop extended responses: *How do you know that is the largest number? Why did you choose to place that digit there?*

Extending: Ask students to think about rounding the number, for example to the nearest hundred. *How close is each number to 500? Which number is closest to 500?*

Stretch zone: *Can you find the difference between your smallest number and your biggest number?*

Check that students have correctly calculated the difference for the numbers they used. *How did you find the difference? What strategy did you use?*

Reflection time

Ask students to share some of the numbers they made and ask them to use the same digits to make the largest/smallest numbers. *How do you know that is the largest/smallest number that you can make?* Ask one student to replace one of the digits in their number with a zero. *How does this change the number? How many hundreds/tens/ones do we have now?* This will allow you to ensure that all students understand the use of zero as a place holder.

Practice Book: Students complete Practice Book page 14. They can do this directly after the main activity, as homework, or as the focus of a separate mathematics session to help students consolidate their learning and build fluency.

Students write the number in words as well as numerals. Use evidence from this work to determine which students need extra support with writing the numbers in words.

Differentiated outcomes	
All students	should understand the effect of place value when making large and small numbers.
Most students	will be able to order a small set of 3-digit numbers from largest to smallest.
Some students	may be able to round a number to the nearest 100.

Answers

Student Book page 7

Observe students while they do the activity and note who can read and say the numbers they make correctly. In the activity on page 7, students complete the table by looking at all the numbers that they made. Check that they have used the correct place-value cards to make the 3-digit numbers they formed.

For example, if they had chosen the digits 2, 3, 7 they could make any of the following numbers: 237, 273, 327, 372, 723, 732. Check that they have four of these in the correct order.

With place-value cards, they would be:

$237 = 200 + 30 + 7$

$273 = 200 + 70 + 3$

$327 = 300 + 20 + 7$

$372 = 300 + 70 + 2$

$723 = 700 + 20 + 3$

$732 = 700 + 30 = 2$

Practice Book page 14

Observe students as they complete the activity. Check that the digits they pick each time are being written correctly in place-value columns and in words. Check that they have completed the sentences about their numbers correctly.

Stretch zone: Check that students have correctly ordered their numbers and have used the < and > signs correctly.

1A Place value

Discover 2 · Student Book pages 8–9 · Practice Book page 15

Specific learning focus

- Read and write numbers to at least 1000.
- Understand what each digit represents in 3-digit numbers.
- Partition 3-digit numbers into hundreds, tens, and ones.

Global skills

- **Creative skills:** exploring

Key vocabulary

- hundreds, tens, ones, biggest, smallest

Resources

- large place-value cards
- large 0–9 digit cards and smaller sets of 0–9 digit cards for each pair
- base-10 equipment

Language support

Continue to develop the language of place value by asking: *What is that digit worth? What does that digit represent?* Encourage students to say the worth of each digit aloud and say the whole number.

Look together at page 8 of the Student Book. If you have access to an IWB you could use that. Refer students to the Think back, which reminds them of the **place value** of the digits in a 3-digit number.

Write on the board the number 428. Ask a student to say the number aloud. *How many ones in the number? How many tens in the number? How many hundreds in the number?* Then ask for each digit: *What is this digit worth?*

Use base-10 equipment to show what each digit represents. For example, when saying there are four hundreds in 428, show students four of the hundreds-blocks. You can also hold the wrong equipment on purpose, for example holding four of the tens-rods and asking: *Is this 400?*

 **Main activity**

Students should work individually on the activities in the Student Book on pages 8–9. As they work, check that they are writing their 3-digit numbers in words correctly. Ask students to say their numbers as well, so that you can be sure that they are recognising the 3-digit numbers by how they appear in digits, how they are written in words and how they are spoken. Refer students to the question in the speech bubble on page 9. Can they say how many hundreds, how many tens and how many ones they would need to get to 1000?

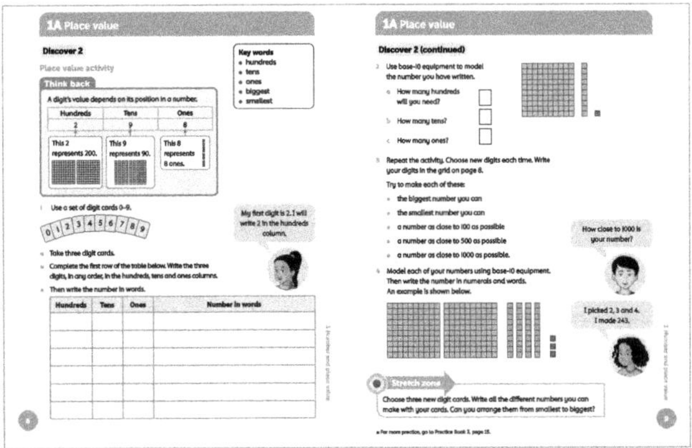

Differentiation

Supporting: Model the number names with students. Say them and ask students to repeat. Make deliberate mistakes and ask students to spot your mistake.

Consolidating: Ask questions that encourage students to develop extended responses: *How do you know how many hundreds there are in that number? Tell me all the numbers you can make using those three digits. How do you know you have found them all?*

Extending: Encourage students to make numbers that have a zero in them. *What does the digit 0 tell you about the value in a column? How do you say that number name? How do you write that number?*

Stretch zone: *Choose three new digit cards. Write all the different numbers you can make with your cards. Can you arrange them from smallest to biggest?*

Check that students have correctly found all six of the possible numbers, and ordered them correctly.

 **Reflection time**

Ask students to share one of their choices of three digits and explain how they made a number with base-10 equipment, and how they wrote it in words. Ask a student who has used zeros to come to the front of the class and share their answers. This will allow you to ensure that all students understand the use of zero as a place holder.

Practice Book: Students complete Practice Book page 15. They can do this directly after the main activity, as homework, or as the focus of a separate mathematics session to help students consolidate their learning and build fluency.

Students consolidate reading and writing numbers in numerals and in words. If students are not confident writing the words for the numbers, ask them to say the numbers to you and check that they say the full number names. For example, for 245, they should say 'two hundred and forty-five' rather than 'two four five'.

Differentiated outcomes	
All students	should understand place value when making large and small numbers.
Most students	will accurately use number names for numbers with up to three digits.
Some students	may be able to say how many hundreds, tens and ones are needed to reach 1000.

Answers

Student Book pages 8–9

1–2 In the activity on pages 8–9, students complete the table by looking at all the numbers they made. Check that they have identified the correct numbers to meet the description in each row of the table.

3 For example, if they had made the numbers 392, 648, 257, 820, 504, then:

- their biggest number was 820
- their smallest number was 257
- their number closest to 100 was 257
- their number closest to 500 was 504
- their number closest to 1000 was 820

Practice Book page 15

Check that students complete the rows of the table correctly for their chosen digits.

Stretch zone: If students chose, for example, 2, 4, 7, 8, then their sequence of numbers in order could be: 847 > 782 > 724 > 482 > 428 > 278 > 248 > 247

1A Place value

Explore 1
Student Book page 10–11 • Practice Book page 16

Specific learning focus
- Find 1, 10, or 100 more or less than any 2- or 3-digit number.

Global skills
- **Creative skills:** investigating

Key vocabulary
- 1 more, 1 less, 10 more, 100 more, 10 less, compare, more than

Resources
- set of 0–9 digit cards per student
- coloured crayons or pens
- a counting stick (these are easily made by using coloured tape and a piece of dowelling or a metre stick), the numbers 8, 57, 85, and 130 on sticky labels that you can attach to the counting stick
- 100-square and a cross of paper to sit over five squares on the grid

Language support
Encourage students to use the language of **1 more**, **1 less**, **10 more** and **10 less** as much as possible. As they make numbers from the digit cards, ask them to say the numbers aloud, and support them to say the names correctly. Encourage students to use their knowledge of place value to say related numbers, for example 'one more than forty-four is forty-five'.

Introductory activity
Stick the label '8' at one end of the counting stick. Ask students to count up in ones from 8 to 18, pointing to each number as you count: eight, nine, ten, eleven, twelve, thirteen, fourteen, fifteen, sixteen, seventeen, eighteen. *What do you notice about the finishing number?* (It is 10 **more than** the starting number.) *What do you notice about the digits in the starting number and finishing number?* (The ones are both 8.) Repeat this activity, with starting numbers 57, 85, and 130, counting on ten, in ones, each time. Ask students to look at each starting number and corresponding finishing number. *What is the same? What is different?*

Next, repeat the activity, counting on in tens instead of ones, from each starting number, to the number that is **100 more**. *What to you notice about the digits in each starting number compared to each finishing number?*

Finally, repeat the activity, counting up in hundreds (the finishing number will be one thousand more than the starting number). *What to you notice this time about the digits in each starting number compared to each finishing number? What is the same? What is different?*

Main activity
Students should complete the tasks in the Student Book individually. As you move around the class, ask students to talk to each other about the patterns they are noticing. Encourage students to talk about moving up and down the rows of the 100-square or about moving to the left or right across the columns. Ask students whether they can spot any patterns in the numbers as they move up and down or left and right.

Direct students to look at the second speech bubble in the Student Book on page 10, which asks if they always move one square to the right to find 1 more. Use this to start a discussion about what happens at the end of a row.

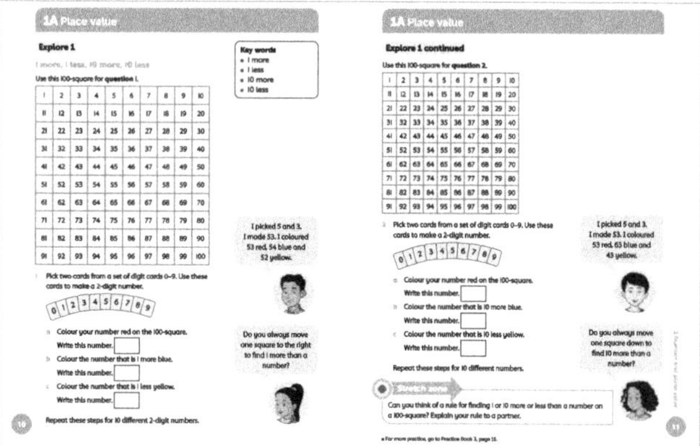

Differentiation
Supporting: Ask students to read the numbers aloud as they colour them in on the 100-square.

Consolidating: Ask students to predict the outcome before they look at the 100-square.

Extending: Ask students to extend the activity by making 3-digit and then 4-digit numbers using digit cards and to say the number that is 1, 10 or 100 more or less than each of their numbers.

Stretch zone: *Can you think of a rule for finding 1 or 10 more or less than a number on a 100-square? Explain your rule to a partner.*

Ask students to explain their rule.

Reflection time
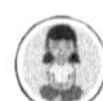
Use the 100-square and a cross of paper that exactly fits over 5 numbers in the grid. Students describe the relationship between the covered numbers when the cross is placed in different positions on the grid. You can repeat this for different-shaped templates.

Practice Book: Students complete Practice Book page 16. They can do this directly after the main activity, as homework, or as the focus of a separate mathematics session to help students consolidate their learning and build fluency.

Students say a number 10 or 100 more or less than any 3-digit number. They then use the < and > symbols to say which numbers are greater or smaller than others.

Differentiated outcomes	
All students	should understand the effect of moving up, down, left and right on the 100-square.
Most students	will be able to predict the outcome of adding or subtracting 1, 10 or 100, and check using the 100-square.
Some students	may be able to extend the activity to using 3- and 4-digit numbers.

Student Book pages 10–11

1   Check that students have coloured the correct square on the 100-square and found the numbers 1 more and 1 less.

2   Check that students have coloured the correct square on the 100-square and found the numbers 10 more and 10 less.

Stretch zone: Students may say 'to find 1 more on a 100-square, move 1 number to the right' or similar for 1 less, 10 more and 10 less.

Practice Book page 16

Check that students have written the correct numbers in each row of the table for their starting numbers. Ask students to say which digit changes when looking for the number that is 1/10/100 more/less than 'My number'.

Stretch zone: The tens and ones are the same; the hundreds are different: 1065 has one thousand and no hundreds; 965 has nine hundreds.

1A Place value

Explore 2   Student Book page 12 · Practice Book page 17

Specific learning focus

- Find 1, 10, or 100 more or less than any 2- or 3-digit number.

Global skills

- **Creative skills:** investigating
- **Interpersonal skills:** communication

Key vocabulary

- count on, count back, 100 more, 100 less

Resources

- set of 0–9 digit cards per student
- base-10 equipment

Language support

Support students by helping them read numbers correctly when they have added one ten or hundred. Use a place-value chart on display and refer to it to help students select the right number names.

 Introductory activity

Write a 3-digit number on the board (e.g. 723). Ask a student to come forward and make that number using base-10 equipment. Ask another student to check that it is correct and say how each digit is being represented.

Look together at page 12 of the Student Book. If you have access to an IWB you could use this. Discuss the worked example, which shows 427 made up of base-10 equipment, and then the new number when adding 1, 10 and 100 to the number.

Refer back to the number on the board: 723. *How can I make this number 10 more? What piece of base-10 equipment should I use to make this number 10 more?* If a student responds with an extra tens-rod, ask them to come to the front and add the extra piece. Then ask them to say what the number is now and to write it on the board.

Repeat this for increasing or decreasing the number by 1 and 100.

 Main activity

Students should work in pairs to complete the tasks in the Student Book on page 12. As you move round the class, ask them to talk to each other about what they notice as they add or remove base-10 pieces from their numbers. Refer students to the third speech bubble. *What stays the same and what changes when you add 1 to a number? What about when you add 10, or 100?*

For question 2, encourage students to discuss and write the rules they spotted from the completed table in question 1. This is an opportunity for them to practise the new language they have learned. For example, encourage them to talk about 'increasing' or 'decreasing' their numbers by 1, 10 or 100 each time, and to explain how they know a number is greater or smaller than another number in terms of the place value of the digits.

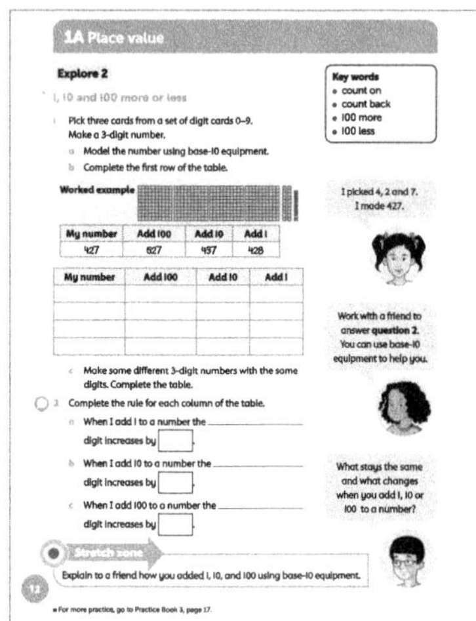

Differentiation

Supporting: Ask students to read the numbers aloud as they make them.

Consolidating: Ask students to predict the outcome before they change the base-10 equipment and find the new number.

Extending: Ask students to extend the activity by making 3-digit and then 4-digit numbers, first choosing digit cards and making the numbers with base-10 equipment, then increasing or decreasing the number by 1, 10, 100 or 1000.

Stretch zone: *Explain to a friend how you added 1, 10, and 100 using base-10 equipment.*

Listen to students' explanations to check that they are accurate.

 ## Reflection time

Ask some students to share the numbers they made and how they represented them with base-10 equipment. Ask them to explain how they know what happens to a number when it is made greater or smaller by 1, 10 or 100.

Practice Book: Students complete Practice Book page 17. They can do this directly after the main activity, as homework, or as the focus of a separate mathematics session to help students consolidate their learning and build fluency.

Students complete and write number sequences that increase or decrease by 1, 10 or 100 each time.

Differentiated outcomes	
All students	should understand the effect of adding or taking away one piece of base-10 equipment.
Most students	will be able to predict the outcome and check using base-100 equipment.
Some students	may be able to extend the activity to using 4-digit numbers.

Answers

Student Book page 12

1 Answers will vary according to the digits chosen. Check that the table is completed correctly for the digits used.

2 **a** ones, 1   **b** tens, 1   **c** hundreds, 1

Stretch zone: Students may say that they added 1 to a number by adding one more base-10 ones-cube to their number, for example.

Practice Book page 17

1	30	40	**50**	60	**70**	80	**90**
2	35	45	**55**	65	**75**	85	**95**
3	85	75	**65**	55	**45**	35	**25**
4	**48**	58	68	**78**	88	98	**108**
5	**91**	**81**	71	**61**	51	41	**31**
6	350	450	**550**	650	**750**	850	**950**
7	990	890	**790**	690	**590**	490	**390**
8	50	**150**	250	**350**	450	550	**650**
9	**645**	545	445	345	**245**	**145**	45
10	873	773	673	**573**	473	**373**	273

11   10 less: 10 more: 100 less: 100 more:
    3, 5   1, 2, 4   7, 9, 10   6, 8

1B Comparing 3-digit numbers

Specific learning focus

- Compare 3-digit numbers, using a number line, and use < and > signs.

Global skills

- **Creative skills:** investigating
- **Interpersonal skills:** communication

Key vocabulary

- number line, greater than (>), smaller than (<)

Resources

- large 0–9 digit cards and smaller sets of 0–9 digit cards for each pair
- counting stick, small coloured sticky dots

Language support

Students will need to be encouraged to speak aloud. Make sure that they read the numbers to you. You will need to model the pronunciation for some students. Remind them that the first digit is always 'hundreds', then 'tens' and finally 'ones'. A poster with all the number names would be helpful.

 Introductory activity

Ask three students to come to the front of the class. They should each take a large digit card and stand in a row facing the rest of the class, so that their three digits make a 3-digit number. Ask students to turn to a partner and agree how to say the number. Ask a student to say the number and then repeat this as a class.

Draw a 0–1000 **number line** on the board marked off in hundreds. Ask another student to come to the front of the class and mark the number on the number line. For example, if the number the students have made is 628, they would mark the following:

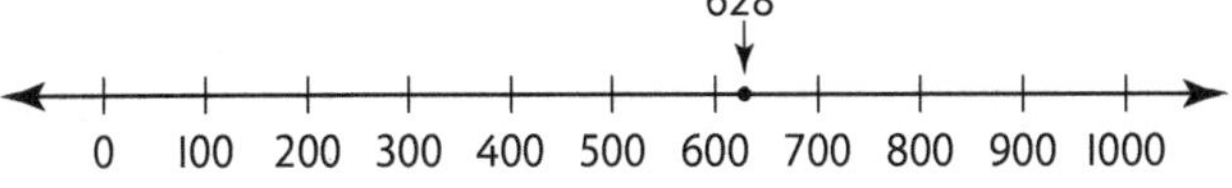

Ask the three students to rearrange their digit cards to make a number that is smaller than the number they first made (or larger if a smaller number is not possible). Repeat the process above and then introduce the **smaller than** '<' sign, writing underneath the number line, for example:

286 < 628

Repeat the process with students rearranging themselves to form a different 3-digit number, using the same digits as before. Introduce the **greater than** '>' sign, writing:

826 > 286   and   826 > 628 (for example)

 Main activity

Give each table sets of digit cards 0–9. Students take it in turns to pick three cards and so don't need a set of cards each. As a group, they make all the different 3-digit numbers possible using this set of three cards. Refer students to the first speech bubble on Student Book page 13 as an example of the numbers they can make using three digit cards.

Students work as a group to complete the activity on page 13 of the Student Book. Encourage them to say the names of the numbers aloud. Emphasise the 'hundreds' aspect. If students pick two digits the same, ask them to pick again so that they are always using three different digits to make their 3-digit numbers.

As students try to place the numbers on a number line, ask questions such as: *What hundreds number is that number between? Is that number closer to 400 or 500? So where would it go on the number line?* Explain that the positions on the line will be approximate positions.

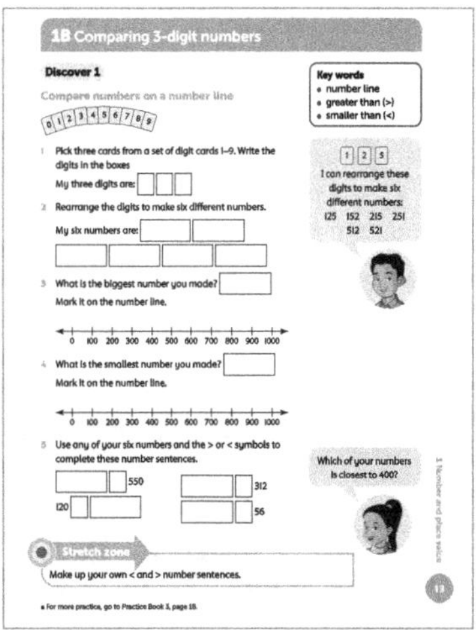

Differentiation

Supporting: Ask students to say the number names aloud as they write them down.

Consolidating: Ask students to explain their thinking when they place numbers on a number line.

Extending: Ask students to choose four digits and to draw a 0–10 000 number line marked in 1000s.

Stretch zone: *Make up your own < and > number sentences.*

Check that students have made correct statements using < and >. Ask students, for example, *Which digits in the numbers did you look at first? Then which? How can you be certain you have used the correct symbol?*

 Reflection time

Use a counting stick. Tell the class that one end is zero and the other end is 1000. Ask what the divisions represent. (hundreds) As a class, count along the stick. Choose one student and ask them to pick the largest number they made. They should come up to the front and put a sticky dot on the counting stick where their number would be. They should also write their number on the board and say it aloud. Repeat this, asking for a smallest number. This time, write the two numbers on the board using the > and < signs. For example, write 652 > 256 and 256 < 652.

Practice Book: Students complete Practice Book page 18. They can do this directly after the main activity, as homework, or as the focus of a separate mathematics session to help students consolidate their learning and build fluency.

Students practise ordering and comparing 3-digit numbers. If they aren't confident writing the words for the numbers, ask them to say the numbers to you and check that they say the full number names. For example, for 5245, they should say 'five thousand, two hundred and forty-five' rather than 'five two four five'.

Differentiated outcomes	
All students	should write and name 3-digit numbers accurately.
Most students	will use < and > correctly to compare 3-digit numbers.
Some students	may extend the activity to 4-digit numbers.

1B Comparing 3-digit numbers

Discover 2   Student Book page 14 · Practice Book page 19

Specific learning focus
- Place a 3-digit number on a number line marked off in multiples of 100.
- Compare 3-digit numbers, use < and > signs.

Global skills
- **Creative skills:** problem solving

Key vocabulary
- number line, greater than (>), smaller than (<)

Resources
- large digit cards 0–9
- base-10 equipment
- number lines marked in tens to 100 and in hundreds to 1000 (these can be drawn on the board or use Resource sheet 1.1)

Language support

Use marked number lines to support students' placing of numbers. Listen to how each student says the numbers out loud when counting on or back in tens or hundreds. Encourage the correct use of 'greater than' and 'less than'.

 ### Introductory activity

Count together as a class from 0 to 1000, in hundreds. Use the number line to help. Then start from any 2-digit number and count on in hundreds from that number. Choose different starting numbers and repeat.

Using the number line to help, count back from 1000 to 0 in hundreds, and then from different 3-digit numbers. For example, count back in hundreds on the number line from 840.

 ### Main activity

Show students a number line between two hundreds numbers (e.g. 500 and 600) marked in tens. Start at one of the marked numbers (e.g. 520) and count on one jump of ten (530). Now jump a further two jumps of 10 from 530 (550).

Start at 580 and jump back two jumps of 10 (560). Now jump back a further three jumps of 10 (530).

Ask students to make a series of jumps of 10, on or back, from different numbers. For example, four jumps of 10 starting from 520 (560), three jumps of 10 back from 580 (550) and so on.

Once students are confident with what each mark on a number line represents, ask them to place a given 3-digit number on a number line. For example, using the number line from 500–600, ask them to tell you where 542 is on the line. Repeat with other numbers between 500 and 600.

Show students a different 3-digit number that does not go on this number line (e.g. 729). Draw an empty number line. *Which number should I write at the start of the line? And at the end?* Point to the middle of the line. *Which number would go here? So where would 729 go?*

Ask students to complete the activity on page 14 of the Student Book individually.

Once they finish the activity, refer students to the second speech bubble: *Can you write your numbers in words instead of numerals?* Challenge them to write as many of the numbers as they can. *When you write 3-digit numbers in words, what is always the same? What is different?*

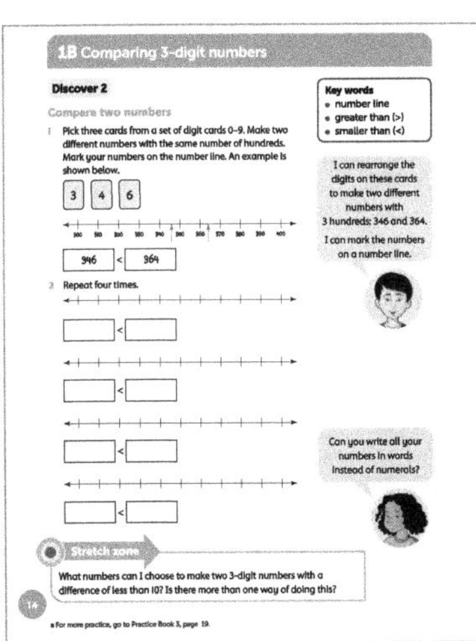

Differentiation

Supporting: Help students to keep their place on the number line as they count on or back in tens. Help them locate where 3-digit numbers should be placed on the number line.

Consolidating: Encourage students to count on and back from a variety of 3-digit numbers and say where each number will go on a number line.

Extending: Encourage students to count on and back from any number, on a number line. Ask them to write inequality sentences comparing two 3-digit numbers.

Stretch zone: *What numbers can I choose to make two 3-digit numbers with a difference of less than 10? Is there more than one way of doing this?*

Students should notice that they can use, for example, 2, 6 and 7 to form 267 and 276. If two of their digits have a difference of 1, then they can make a 3-digit number with a difference of less than 10.

Choose students to share two different 3-digit numbers that they formed from the digit cards and explain which is greater and why. Choose the higher number of the two. Ask the class to tell you what number is 10 more than the number and 10 less than the number. For example, if the student made 237, then 10 more is 247 and 10 less is 227. Repeat with examples given by two more students.

Practice Book: Students complete Practice Book page 19. They can do this directly after the main activity, as homework, or as the focus of a separate mathematics session to help students consolidate their learning and build fluency.

Students write 3-digit numbers and position them on number lines. The number lines are labelled at either end so students must choose numbers that will fit between these two numbers. *In the first number line, what number must be in the hundreds place in your three numbers? What does each division represent in the second number line?*

Differentiated outcomes	
All students	should be able to say the number names counting in tens from a multiple of 100 and say where they go on the number line.
Most students	will count on or back from any number in tens on a number line and say where they go on the number line.
Some students	may be able to count on or back in tens or hundreds on the number line and say where they go on the number line.

Answers

Student Book page 14

Answers will vary depending on the digits chosen and the numbers formed. Check that students have marked and ordered the numbers correctly.

Stretch zone: For example, 145 and 154, 245 and 254, 345 and 354, and so on.

Practice Book page 19

Answers will vary depending on the numbers chosen. Check that students have marked and ordered the numbers correctly on the number line. For example:

1 417, 438, 492

2 163, 167, 168

3 708, 748, 778

Stretch zone: There are 100 numbers less than 1000 with a 5 in the tens place: 50–59, 150–159, 250–259 and so on to 950–959.

1B Comparing 3-digit numbers

Explore 1   Student Book page 15 • Practice Book page 20

Specific learning focus

- Make 3-digit numbers using base-10 equipment.

Global skills

- **Creative skills:** exploring
- **Interpersonal skills:** communication

Key vocabulary

- hundreds, tens, ones

Resources

- sticky labels with a range of 3-digit numbers on them to stick on students' backs or foreheads (make sure that there are at least three numbers for each digit in the hundreds column)
- counting stick
- sticky dots
- base-10 equipment

Language support

- Continue to encourage students to speak the numbers aloud. Make sure that they read the numbers to you. You will need to model the pronunciation for some students. Focus on the phrases 'is less than' and 'is greater than' in this lesson.

Introductory activity

Give each student a sticky label with a number on it. They should stick this label onto another student's back or forehead. Students have to find out what their number is by asking other students questions, to which they can only answer 'yes' or 'no'. When all students have found out their number, ask them to stand in groups that share the same hundreds digit in their numbers.

Pick one group and ask them to come and stand at the front. Put a piece of paper on the floor to represent the 'hundred' below and another, a distance away, to represent the 'hundred' above. The students should organise themselves, in order, so that they represent a number line accurately. So, for example, if the students' numbers are 423, 465 and 482, they would stand a little more than two steps from the 400 label for 423, another four steps away for 465 and another two steps away for 482. Allow students to make their own decisions as they decide how to represent the number line.

Main activity

Choose a 3-digit number (e.g. 372). Ask students whether they can say the value of each digit. For each digit, model the base-10 equipment that represents the digit and finish with the base-10 representation of 372. Now choose another 3-digit number to make (e.g. 185). *How close is this number to 372?* Help them to see that the 'closeness' of the numbers means finding how far apart they are, or their difference. Model with the base-10 equipment to show them how to compare the two 3-digit numbers, by looking at the number of hundreds-blocks, tens-rods and ones-cubes.

On the board, model how to write an equality sentence comparing the two numbers. For example, write 185 < 372 or 372 > 185.

Students work, in pairs, on the activity in the Student Book on page 15. Ask them to check each other's work. As with the previous activity, encourage students to say the names of the numbers aloud, emphasising the 'hundreds'.

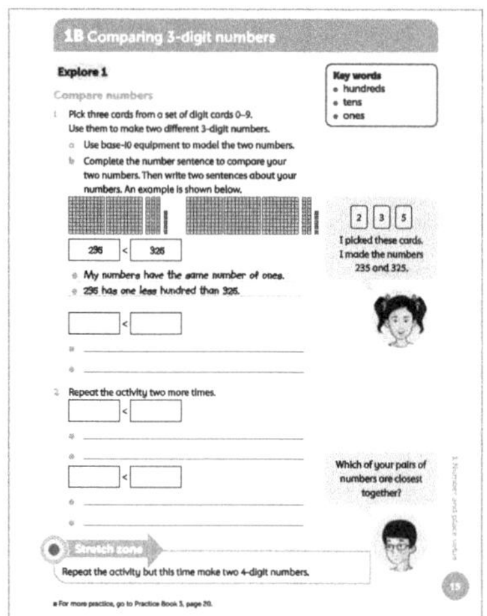

Differentiation

Supporting: Ask students to say the number names aloud as they write them down.

Consolidating: Ask students to explain their thinking when they compare 3-digit numbers using base-10 equipment.

Extending: Ask students to make 4-digit numbers from digit cards and then make them using base-10 equipment.

Stretch zone: *Repeat the activity but this time make two 4-digit numbers.*

Check that students are describing and ordering the numbers correctly.

 Reflection time

Use a counting stick. Tell the class that one end is 400 and the other end is 500. Ask what the divisions represent. As a class, count along the stick. Choose any two students who have made numbers with a 4 in the hundreds column and ask them to put a sticky dot on the counting stick where their numbers will be. Now ask students to write inequality sentences, or use the language of 'greater than' and 'less than' to compare the two numbers. Choose another number with 4 in the hundreds column and ask students to decide where to place a dot for this number. Write an equality sentence to compare the three numbers. Repeat with three different numbers that have the same number in the hundreds column.

Practice Book: Students complete Practice Book page 20. They can do this directly after the main activity, as homework, or as the focus of a separate mathematics session to help students consolidate their learning and build fluency.

Students compare two 3-digit numbers and use the < and > symbols to compare them.

Differentiated outcomes	
All students	should make a range of 3-digit numbers and say their names.
Most students	will make 3-digit numbers correctly using base-10 equipment.
Some students	may extend the activity to 4-digits.

1B Comparing 3-digit numbers

Explore 2 Student Book page 16 · Practice Book page 21

Specific learning focus

- Place a 3-digit number on a number line marked off in multiples of 10 or 100.

Global skills

- **Creative skills:** problem solving

Key vocabulary

- number line, multiple of 10, multiple of 100

Resources

- number lines marked in hundreds to 1000 (these can be drawn on the board or use Resource sheet 1.1)
- digit cards 0–9

Language support

Give students opportunities to say 3-digit numbers aloud as they form them, using the digit cards. Emphasise the comparison of numbers as being 'greater than' or 'less than' as students sequence their set of numbers.

Student Book page 15

Check that students have made two different 3-digit numbers and that these are written in the correct order (from smallest to largest, e.g. 437 < 743).

Check that students compared their numbers using appropriate sentences.

Practice Book page 20

Check that students have written the correct numbers in each row of the table, using the clues given.

Stretch zone: Students may say that two of the numbers have 2 hundreds while the other does not, or that two of them have 5 tens and one does not, or that two of them have no ones but the other has 5 ones.

 **Introductory activity**

Ask a student to come to the front of the class and choose three digits at random from a set of digit cards 0–9. Write the digits on the board for the class to see. Ask them to explain how they would make the smallest possible number using the three digits, encouraging them to focus on having the smallest number of hundreds first.

Now ask students to make the largest possible number using the same three digits. *How did you do this?* Ask students to look at the smallest and largest 3-digit numbers. *What do you notice?* Repeat with three new digit cards, creating other pairs of 3-digit numbers. *What do you notice about each set of numbers?* (The digits are reversed – e.g. 249 and 942).

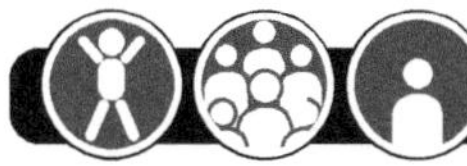 **Main activity**

Ask a student to come to the front and to choose three digit cards to make a 3-digit number. Ask two other students to make different numbers using the same three digits. Write the three numbers on the board and ask students to put them in order from smallest to largest. For example, if the three digits are 1, 4 and 8, the numbers could be 184, 418 and 841.

Draw an empty number line on the board with 11 equally spaced marks. Ask students how they would mark the number line so it could show all three of the numbers made. They should suggest that the line goes from 0 to 1000 labelled in **multiples of 100**. Mark these on the number line and then ask students, in turn, to come and point to approximately where each number will be on the line. Now ask students to make other numbers using the same three digits and place them on the number line.

Ask students to complete the activities on page 16 of the Student Book individually. They need to order ten numbers from smallest to largest and mark numbers on number lines.

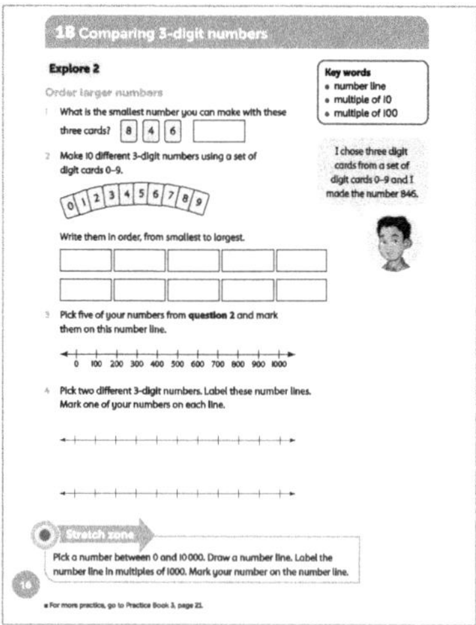

Differentiation

Supporting: Start by asking students to order 2-digit numbers on a 0–100 number line marked in multiples of 10. Then help students with ordering their 3-digit numbers using the hundreds digits, then the tens and then the ones.

Consolidating: Encourage students to use the tens digits to order the numbers more accurately and locate them on the number line.

Extending: Ask students to draw and label a number line to place 4-digit numbers.

Stretch zone: *Pick a number between 0 and 10 000. Draw a number line. Label the number line in multiples of 1000. Mark your number on the number line.*

Students should use their understanding of place value to locate their numbers as accurately as possible.

 ## Reflection time

Choose some students to share the numbers they made with their digits and explain how they sequenced them and placed them on a number line. Ask whether other students can make any other numbers using those three digits and say between which numbers on the line they would place these.

Practice Book: Students complete Practice Book page 21. They can do this directly after the main activity, as homework, or as the focus of a separate mathematics session to help students consolidate their learning and build fluency.

Students make all the different 3-digit numbers possible from a given set of digits. They then write and draw different representations of a number. Encourage them to draw base-10 equipment, mark the number on a number line and write the number in numerals and in words.

Differentiated outcomes	
All students	should make some 2- and 3-digit numbers using two or three digit cards and order them.
Most students	will place 3-digit numbers on a 0–1000 number line.
Some students	may be able to make 4-digit numbers and place them on a number line.

Answers

Student Book page 16

1 and **2** Check that students have made ten different 3-digit numbers and that these are written in ascending order (from smallest to largest). For example:

168	279	290	401	502	523	734
835	856	967				

3 and **4** Check that students have labelled the ends of the number lines appropriately for the numbers they have chosen, and that they have placed their chosen numbers in the correct places on the number lines.

Practice Book page 21

1 123, 124, 132, 134, 142, 143, 213, 214, 231, 234, 241, 243

312, 314, 321, 324, 341, 342, 412, 413, 421, 423, 431, 432

The largest number is 432. Check that students have represented it correctly in different ways, for example, using base-10 equipment, on a number line, in numerals and in words.

Stretch zone: Students may suggest choosing a hundreds digit first and making all the possible numbers using the other three digits, then choose a different hundreds digit and repeat until all the different hundreds digits have been used.

1C Ordering, rounding and estimating

Discover Student Book page 17–18 • Practice Book page 22

Specific learning focus

- Understand what each digit represents in a 3-digit number.
- Compare and round 3-digit numbers.

Global skills

- **Creative skills:** problem solving
- **Interpersonal skills:** communication

Key vocabulary

- tens, rounding, round to the nearest 10/100, greater than (>), less than (<)

Resources

- set of large 0–9 digit cards
- mini whiteboards and markers

Language support

Continue to make sure that students say the number rather than the individual digits, for example 'forty-three' rather than 'four three'. In this way, they will come to 'hear' the place value within a number. Revise some of the vocabulary associated with rounding, such as 'round', 'to the nearest 10/100' and so on.

Introductory activity

Select three digit cards at random and make a 3-digit number. Write this on the board. Ask students, in pairs, to write down as many facts as they can about the number on their whiteboards. They take it in turns to think of facts and to write them down. Each of the pair then reports a fact that their partner expressed. List all these facts on the board around the starting number. Make sure that they have included facts about place value, such as the following.

- 692 has a 9 in the tens column worth 90.
- You can partition 692 into 6 hundreds, 9 tens, and 2 ones.

Students should also include facts about **rounding** and ordering, for example these.

- 692 is 690 rounded to the nearest ten.
- 692 is 700 rounded to the nearest hundred.
- 692 is between 600 and 700.
- 692 is between 690 and 700.

Repeat this four times with different 3-digit numbers. Then ask students to arrange the four numbers in ascending order, using < and > signs to separate them.

 Main activity

Look together at page 17 of the Student Book. Display on an IWB, if possible. Talk about the clues (facts) written in question 1, which describe a 2-digit number. Cover up the number 48 and then, as a class, go through the clues discussing how they can only link back to the number 48. *Which clue gave the biggest clue to the number? When did you work out what the number was?* Individually, students complete the activities in the Student Book on pages 17–18. As they complete their lists of facts in questions 1, 2 and 3, ask questions such as: *What is that **rounded to the nearest** ten/hundred? What does that digit represent? How can you partition that number?* Students can then challenge a partner to guess their number based on their clues. Refer them to the second speech bubble on page 17. Ask them to make a note of how many clues it takes for their partner to guess this number.

Direct students to look at question 4, which gives a real-life rounding context. Say that we don't always need to give exact numbers; sometimes approximate numbers are acceptable and often more useful. *Can you think of other situations where rounding a number is useful?*

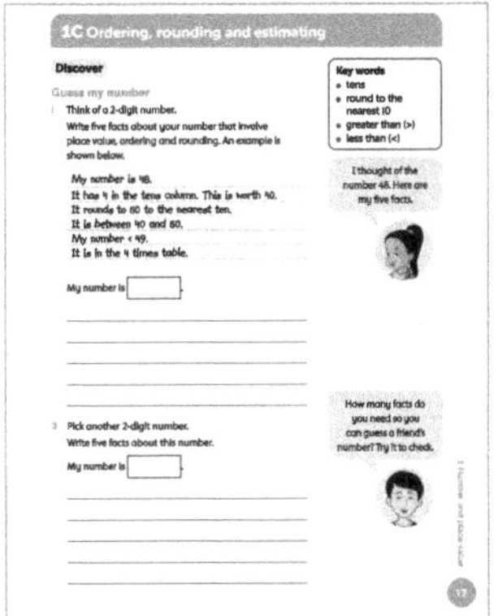 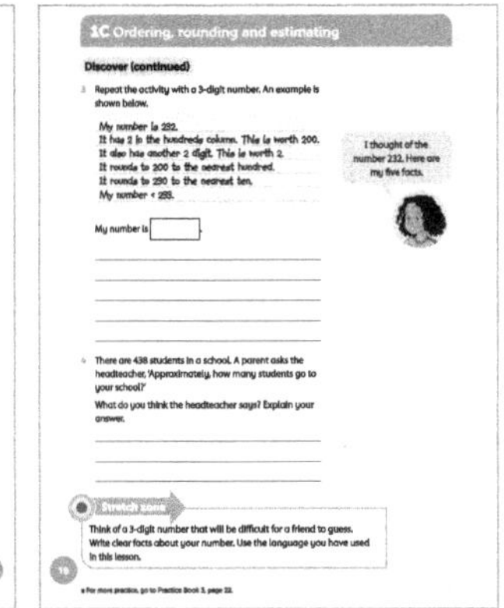

Differentiation

Supporting: Prompt a range of facts by asking questions such as: *What digit is in the tens column? Which numbers are smaller than this number?*

Consolidating: Remind students to use a range of facts like those in the lists on the board from the introductory activity.

Extending: Encourage students to think beyond the facts that you have listed.

Stretch zone: *Think of a 3-digit number that will be difficult for a friend to guess. Write clear facts about your number. Use the language you have used in this lesson.*

Check that students have described their number accurately. *Did your facts challenge your partner? What made your number difficult to guess?*

 Reflection time

Play a game of Guess My Number. Write a 3-digit or 4-digit number on a piece of paper. Students have to guess the number by asking questions to which you can reply only 'yes' or 'no'. Students can take the role of picking a number for the rest of the class so that you can model a range of questions covering all the properties you want to explore.

Practice Book: Students complete Practice Book page 22. They can do this directly after the main activity, as homework, or as the focus of a separate mathematics session to help students consolidate their learning and build fluency.

Students round numbers, following some given rules. Remind students that numbers with a 5 in the ones, tens or hundreds place need to round up to the next nearest multiple.

Differentiated outcomes	
All students	should write down some facts about each number.
Most students	will know a wide range of facts about each number including rounding successfully.
Some students	may make up interesting facts about the numbers.

Answers

Student Book pages 17–18

1, 2 and **3** Check that students have made 2-digit and 3-digit numbers as requested, and that the facts written for each number are correct. In particular, check that the rounding is correct. For example:

325 rounds to 330 to the nearest 10 (the 5 ones mean that we round up to 3 tens)

325 rounds to 300 to the nearest 100 (the 2 tens mean that we round down to 3 hundreds).

4 Students could have rounded to the nearest 10 and said 440, or to the nearest hundred and said 400. The answer should include an explanation about rounding to either the nearest 10 (because it is more accurate) or the nearest 100 (because the headteacher was only asked for an approximate number).

Practice Book page 22

Answers will vary because students choose the numbers. For example:

1	Numbers that round to 500 to the nearest 100	456, 478, 503, 510, 542
2	Numbers that round to 360 to the nearest 10	355, 358, 359, 361, 364
3	Numbers that round to 700 to the nearest 100	654, 687, 719, 734, 749
4	Numbers that round to 1000 to the nearest 100	950, 969, 999, 1025, 1049
5	Numbers that round to 200 to the nearest 10	157, 168, 183, 206, 239
6	Numbers that round to 990 to the nearest 10	985, 987, 991, 993, 994
7	Numbers that round to 1000 to the nearest 10	995, 996, 999, 1001, 1004
8	Numbers that round to 750 to the nearest 10	745, 748, 751, 752, 753
9	Numbers that round to 500 to the nearest 10	495, 497, 499, 502, 504

Stretch zone: 450, 451, 452, 453, 454.

1C Ordering, rounding and estimating

Explore 1 Student Book page 19 • Practice Book page 23

Specific learning focus

- Round 3-digit numbers to the nearest ten or hundred.
- Place a 3-digit number on a number line marked in multiples of 10 or 100.

Global skills

- **Creative skills:** exploring
- **Interpersonal skills:** communication

Key vocabulary

- round, to the nearest 10, to the nearest 100

Resources

- set of large 0–9 digit cards
- mini whiteboards and markers
- counting stick; labels for counting stick

Language support

Particular language to model during this lesson includes the phrases, 'to the nearest ten' and 'to the nearest hundred'. Use the counting stick to give a clear mental image of rounding meaning 'to the nearest'.

 **Introductory activity**

Ask four students to come to the front of the class. One at a time, each student picks three digit cards and uses them to make a 3-digit number. They write their number on a whiteboard and then replace the cards. When you have four different 3-digit numbers, the students who picked the numbers arrange themselves, with their whiteboards, in ascending order, according to their numbers. Write the numbers on the board, using the < sign between each number.

The four students should then rearrange themselves in descending order, according to their numbers. Rewrite the numbers on the board using > sign.

Main activity

Use the counting stick. Explain to students that one end is zero and the other end is 1000. Use the students and their numbers from the introductory activity. Each student should point to the position of their number on the counting stick and explain why they are placing it at that point.

Repeat, using divisions of ten on the counting stick. So, for example, if one student has 437, one end of the counting stick should be 400 and the other end 500. Change the end numbers as appropriate and ask students, one at a time, to position their numbers.

Students complete the activities in the Student Book page 19 individually. They should make six numbers from their three digits and mark them on the number line. They then order them from smallest to largest, and round them to the nearest 10 and 100. While students are working on question 3, ask questions such as, *How did you decide how to put your numbers in order from smallest to largest? How does using the number line help you to round to the nearest 100? Which 'rule' did you use to decide whether to round up or down?* Encourage pairs to check each other's work and explain their answers to each other.

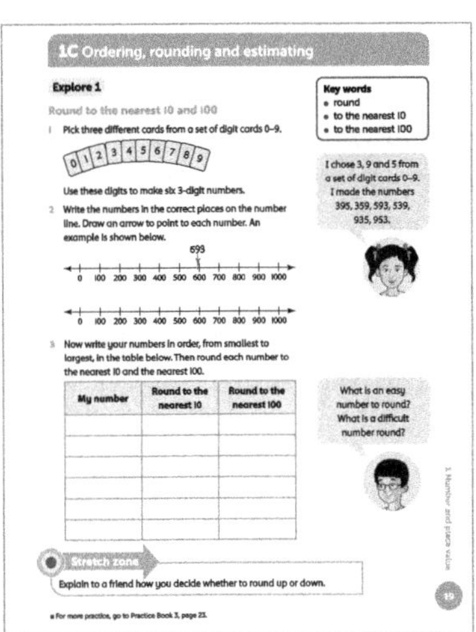

Differentiation

Supporting: Ask students to say the names of the numbers. Rearrange the digits and ask for the new number.

Consolidating: Ask for the smallest and largest possible numbers that will round to a particular 10 or 100.

Extending: Refer students to the second speech bubble in the Student Book on page 19 and ask them to discuss this in pairs: *What makes a number easy to round to the nearest 10 or 100? What makes it more difficult to decide how to round a number?*

Stretch zone: *Explain to a friend how you decide whether to round up or down.*

Check that students understand that numbers round up when halfway or more to the next multiple.

 **Reflection time**

Write the number 480 on the board. Ask pairs of students to write on their whiteboards any number that would be 480 when rounded to the nearest ten. As a class, decide what the smallest answer could be and what the largest answer could be. Repeat for numbers that would be 600 when rounded to the nearest hundred.

Practice Book: Students complete Practice Book page 23. They can do this directly after the main activity, as homework, or as the focus of a separate mathematics session to help students consolidate their learning and build fluency.

Students round numbers to the nearest 10 and 100, following the rules they have learned in the unit so far.

Encourage students to justify the order in which they have written the numbers from smallest to largest. *Why does this number come before this number? How do you know it is smaller than this number?*

Differentiated outcomes	
All students	should understand place value in 3-digit numbers.
Most students	will round 3-digit numbers accurately and place them on a number line.
Some students	may be able to explain how to round and to be able to spot easy and difficult numbers to round.

Answers

Student Book page 19

1 and **2** Check that students have made six 3-digit numbers and have placed them in their correct positions on the number line.

3 Check that students have written their numbers in the correct order in the table and rounded them to the nearest 10 and 100.

Stretch zone: Students may suggest they look at the ones digit when rounding to the nearest 10, or the tens digit when rounding to the nearest 100, and so on.

Practice Book page 23

Check that students have made six 3-digit numbers and have rounded each of them correctly to the nearest 10 and 100.

Stretch zone: 3000 – students' pictures should show how they arrived at this answer.

1C Ordering, rounding and estimating

Explore 2 Student Book page 20 • Practice Book page 24

Specific learning focus

- Count a number of objects by estimating or by grouping in tens

Global skills

- **Real-world skills:** presenting/interpreting information
- **Interpersonal skills:** teamwork

Key vocabulary

- estimate, group in tens, count in tens

Resources

- collections of small items to count, for example, beans or pulses, small buttons, paper clips and so on

Language support

Model the language of estimating by using, 'to the nearest ten' and 'to the nearest hundred'. Use number lines to give a clear image of rounding meaning 'to the nearest' and encourage students to use this language when estimating a number of small objects.

Introductory activity

Tip onto a table a pile of small objects, perhaps paper clips. Tell students that they are going to find out how many paper clips are in the pile, first by estimating and then by counting. Remind students that estimating means 'making a sensible guess' and, in this case, making a sensible guess of how many of something there are. Ask how they might go about it.

Listen to students' suggestions about estimating and ask them why they think their number will be close to the real total. When thinking about counting the paper clips, encourage students to consider counting the paper clips into piles of 10, as this will help them keep count and the total can then be counted up quickly at the end by **counting in tens**.

Main activity

Explain to students that they are going to be doing some more estimating and counting of numbers of small objects. They should estimate the whole amount first, then count the objects into groups of ten to help improve their estimate. Finally, they will be able to count the groups of ten and any remaining objects to get an actual total.

Students work in groups of four to complete the activities in the Student Book on page 20. As they work through the activities, check that students have used accurate methods of grouping and counting to find the correct number of beans and stars.

Refer students to the second speech bubble. *What would make your estimate even more accurate than grouping the objects into groups of ten? When is it better to group objects into smaller groups? When is it better to have larger groups?* You could discuss as a class that with a larger number of objects it might be better to group them into larger groups (of say 10, 20, or more) but with a smaller number of objects, they could group and count the objects in twos or fives.

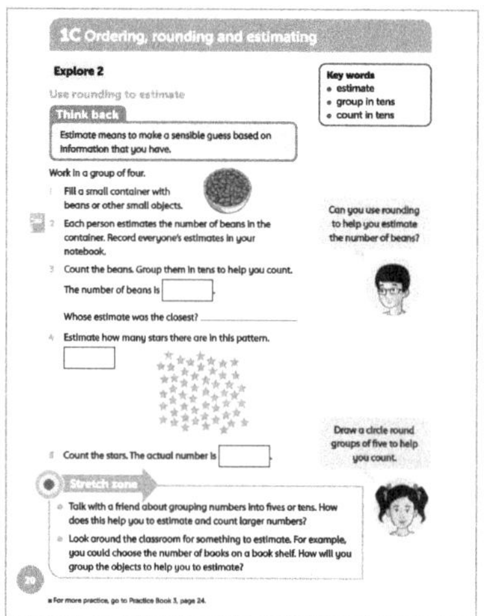

Differentiation

Supporting: Ask students to work in mixed-attainment groups, to group the objects and count them. Students can help each other by modelling the language and the grouping and counting process.

Stretch zone: *Talk with a friend about grouping numbers into fives or tens. How does this help you to estimate and count larger numbers?*

Look around the classroom for something to estimate. For example, you could choose the number of books on a bookshelf. How will you group the objects to help you to estimate?

Check that students can group objects appropriately to help them estimate or find an exact total.

Reflection time

Ask students in each group to explain to the class how they estimated their objects and how they grouped to count them. Ask them to say how close their estimates were to the actual total.

Practice Book: Students complete Practice Book page 24. They can do this directly after the main activity, as homework, or as the focus of a separate mathematics session to help students consolidate their learning and build fluency.

Students estimate the number of objects in four different-sized containers. Encourage students to use their estimates for one container to make more accurate estimates of the number of objects in the other containers. *Did your estimates improve each time? Can you explain why?*

Differentiated outcomes	
All students	should count objects in groups of ten to help find the total.
Most students	will estimate using groups of ten.
Some students	may estimate accurately by sight and then count for accuracy using appropriate grouping.

Student Book page 20

1–4 Estimates can vary but check that students have used accurate methods of grouping and counting to find the correct number of beans and stars.

5 There are 53 stars.

Practice Book page 24

Estimates will vary but check that students have used accurate methods of grouping and counting to find the correct numbers of beans in the different containers.

1D Number sequences

Discover 1
Student Book pages 21–22 • Practice Book page 25

Specific learning focus
- Count on in fours and eights to at least 50.

Global skills
- **Creative skills:** investigating
- **Interpersonal skills:** communication

Key vocabulary
- count on, count back

Resources
- mini whiteboards and markers
- coloured crayons/pens

Language support

Encourage the use of the words 'one more than' and 'one less than'. Point to a number and ask: *Can you show me the number that is 1/10 more/less than this number?* Revise the vocabulary associated with a 100-square (e.g. row, column, up, down, left, right).

Introductory activity

This is a game students may be familiar with. Once you have played it, you can play it on many occasions with different number patterns. It is sometimes called a 'lowest unique number'. Tell students that you will start counting from 1 and will count on in 'ones' (so you will say one, two, three and so on). Each student is to write a number on their whiteboard. Their aim is to write down the smallest number that no-one else in the class writes down. As you count, students put up their hands when you say their number. If more than one student has chosen this number, they all put their hands down and you continue your count. The winner is the first person who is the only one with their hand up.

Play the game two or three times so that students understand the rules. Then tell the class that you will start counting at 3 and count on in twos (so the sequence will be 3, 5, 7, 9, 11, 13, 15 and so on). Repeat the game with different starting numbers and count on and back in twos, threes, fours and fives.

Main activity

Students should complete the activities on Student Book pages 21–22 individually. When they come to describe the patterns, ask them to talk to a partner before they write their answer down. They shouldn't write anything down until their partner understands their description of the pattern. Ask questions to encourage students to spot patterns that counting in fours and eights have in common. *Are any of the numbers the same if you count in fours and count in eights? What is the same and what is different about the patterns created? Why do you think this is? Would any multiples of 3 be in this pattern of eights? What about multiples of 2? What other multiples would be in the pattern?*

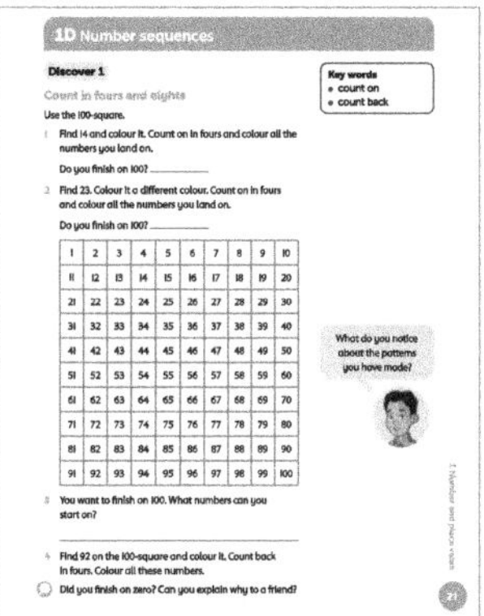

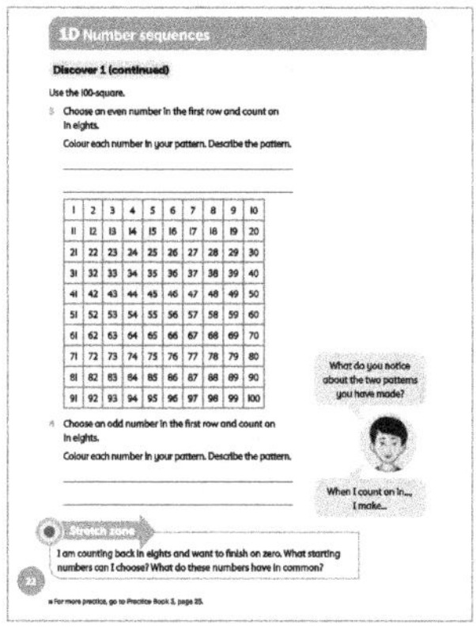

Differentiation

Supporting: Check the patterns that students colour and ask them to check by counting carefully.

Consolidating: Ask students to predict the form that the patterns will take.

Extending: Ask students to explore patterns for other multiples including 2, 3 and 12.

Stretch zone: *I am counting back in eights and want to finish on zero. What starting numbers can I choose? What do these numbers have in common?*

Check that students understand that if they start on a multiple of 8, when they count back they end on 0.

 **Reflection time**

Ask students to describe the pattern of numbers produced on the 100-square if they are counting in multiples of 4 or 8. How did they know their number would be in that count?

Play the game from the introductory activity again. This time, ask a student to be the teacher and have them count in fours or eights on or back. *Who will have the lowest (or highest) unique number?*

Practice Book: Students complete Practice Book page 25. They can do this directly after the main activity, as homework, or as the focus of a separate mathematics session to help students consolidate their learning and build fluency.

Students write number sequences for 4 and 8, counting on and back from different starting numbers.

Differentiated outcomes	
All students	should complete number patterns accurately with support.
Most students	will complete number patterns and describe patterns.
Some students	may complete and describe a wider range of number patterns.

Answers

Student Book pages 21–22

1

1	2	3	4	5	6	7	8	9	10
11	12	13	14	15	16	17	18	19	20
21	22	23	24	25	26	27	28	29	30
31	32	33	34	35	36	37	38	39	40
41	42	43	44	45	46	47	48	49	50
51	52	53	54	55	56	57	58	59	60
61	62	63	64	65	66	67	68	69	70
71	72	73	74	75	76	77	78	79	80
81	82	83	84	85	86	87	88	89	90
91	92	93	94	95	96	97	98	99	100

You do not finish on 100.

2

1	2	3	4	5	6	7	8	9	10
11	12	13	14	15	16	17	18	19	20
21	22	23	24	25	26	27	28	29	30
31	32	33	34	35	36	37	38	39	40
41	42	43	44	45	46	47	48	49	50
51	52	53	54	55	56	57	58	59	60
61	62	63	64	65	66	67	68	69	70
71	72	73	74	75	76	77	78	79	80
81	82	83	84	85	86	87	88	89	90
91	92	93	94	95	96	97	98	99	100

You do not finish on 100.

3 Answers will vary but could include the following.

- 12, 24, 32, 48.
- You have to start on a multiple of 4 to finish on 100.

4

1	2	3	4	5	6	7	8	9	10
11	12	13	14	15	16	17	18	19	20
21	22	23	24	25	26	27	28	29	30
31	32	33	34	35	36	37	38	39	40
41	42	43	44	45	46	47	48	49	50
51	52	53	54	55	56	57	58	59	60
61	62	63	64	65	66	67	68	69	70
71	72	73	74	75	76	77	78	79	80
81	82	83	84	85	86	87	88	89	90
91	92	93	94	95	96	97	98	99	100

You do finish on 0 if you continue the count. You can start on any multiple of 4 to finish on 0.

5

1	2	3	4	5	6	7	8	9	10
11	12	13	14	15	16	17	18	19	20
21	22	23	24	25	26	27	28	29	30
31	32	33	34	35	36	37	38	39	40
41	42	43	44	45	46	47	48	49	50
51	52	53	54	55	56	57	58	59	60
61	62	63	64	65	66	67	68	69	70
71	72	73	74	75	76	77	78	79	80
81	82	83	84	85	86	87	88	89	90
91	92	93	94	95	96	97	98	99	100

The shaded numbers form diagonal patterns on the 100-square. As you move down the 100-square, the ones digit is 2 less than the previous ones digit and the tens digit is 1 more, with the exception of 30 to 38 and 70 to 78.

6 The pattern when you start with an odd number is the same as when you start with an even number.

Stretch zone: Any multiple of 8 (e.g. 8, 16, 24, 32, 40).

Practice Book page 25

Count back in fours from 48	48, 44, 40, 36, 32, 28, 24, 20, 16, 12
Count on in fours from 39	39, 43, 47, 51, 55, 59, 63, 67, 71, 75
Count back in fours from 59	59, 55, 51, 47, 43, 39, 35, 31, 27, 23
Count on in eights from 2	2, 10, 18, 26, 34, 42, 50, 58, 66, 74
Count back in eights from 88	88, 80, 72, 64, 56, 48, 40, 32, 24, 16
Count on in eights from 10	10, 18, 26, 34, 42, 50, 58, 66, 74, 82
Count back in eights from 90	90, 82, 74, 66, 58, 50, 42, 34, 26, 18

Stretch zone: 0, 4, 8, 12, 16, 20, 24, 28.

1D Number sequences

Specific learning focus

- Count on in 50s or 100s.

Global skills

- **Creative skills:** exploring
- **Interpersonal skills:** communication

Key vocabulary

- count on, count back

Resources

- number lines marked in 50s to 500 (these can be drawn on the board or use Resource sheet 1.1)

Language support

When students are counting in 10s, 50s, 100s and so on listen and support pronunciation where necessary, encouraging students to listen for patterns in how they recite the numbers.

 **Introductory activity**

Using a number line from 0–500 marked in 50s, count up and back with the class from 0 to 500 in 50s and back. Start at any multiple of 50 and count on in 50s to 500, or count back in 50s to 0.

Ask students to question each other in pairs about multiples of 50, for example, what number is halfway between 100 and 200? 400 and 500? 350 and 450?

Main activity

Write on the board '10'. Ask students to count on 50 from 10. (60) Then count on 50 from 60. (110) Keep adding on 50 to each answer to build a sequence, and write it on the board as it grows: 10, 60, 110, 160, 210, 260, 310, 360… Ask students whether they can see a pattern. *Can you describe the pattern to a partner?*

Now start with 470 and count back in 50s, building the sequence of numbers as you go: 470, 420, 370, 320, 270, 220 and so on. Again, ask students to describe the pattern they notice to a partner.

Start a sequence on the board: 30, 80, 130, 180… *Can you describe the pattern? Can you tell me what it increases by each time? Will 430 be in the sequence? Why? What about 370? Why?*

Ask students to complete the activity on page 23 of the Student Book individually. When describing the patterns

in the digits, check that students have understood that counting on in 100s means the final two digits always stay the same, and that counting on in 50s means the final digit stays the same while the tens digit alternates. Ask questions to encourage students to look at the digits in each number of the sequence: *What changes and what stays the same on each count?*

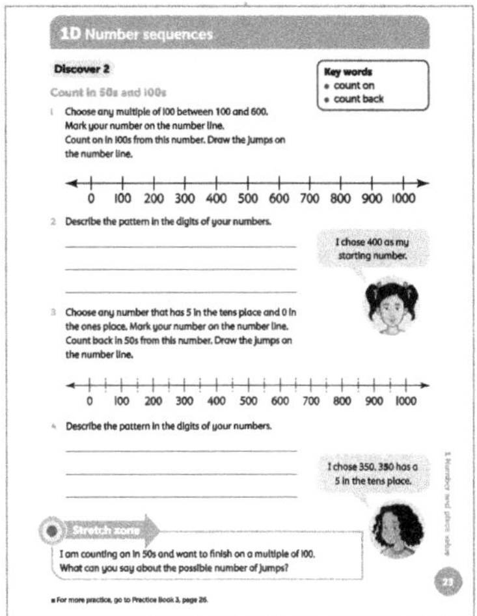

Differentiation

Supporting: Point to the numbers on the number line as students count on or back in 50s.

Consolidating: Encourage students to count on or back in 50s from any multiple of 10.

Extending: Students should build their own 'fifties' sequences starting from any number.

Stretch zone: *I am counting on in 50s and want to finish on a multiple of 100. What can you say about the possible number of jumps?*

Check students' understanding that every two jumps of 50 will make a multiple of 100, as long as you start from a multiple of 100, so the number of jumps from one multiple of 100 to another will always be an even number.

 **Reflection time**

Ask some students to share their sequences and describe anything interesting that they saw. *What did you notice about every other number in your 50s sequences? Can anyone share with the class a really hard sequence counting on or back in 50s? Why is it hard?*

Practice Book: Students complete Practice Book page 26. They can do this directly after the main activity, as homework, or as the focus of a separate mathematics session to help students consolidate their learning and build fluency.

Students complete missing numbers sequences, counting on or back in 50s and 100s. They then make up their own sequences. Encourage them to choose suitable starting numbers for each of their sequences. Look out for students who choose numbers that will end in a sequence with negative numbers!

<table>
<tr><td colspan="2">Differentiated outcomes</td></tr>
<tr><td>All students</td><td>should count on in 50s using a number line.</td></tr>
<tr><td>Most students</td><td>will count on or back in 50s from any multiple of 10.</td></tr>
<tr><td>Some students</td><td>may count on or back in 50s from any number.</td></tr>
</table>

Answers

Student Book page 23

1 and **2** Students' sequences will vary, depending on the starting number. Students should recognise that when counting on in 100s, the final two digits always stay the same.

3 and **4** Students' sequences will vary, depending on the starting number. Students should recognise that, when counting on in 50s the final digit stays the same while the tens digit alternates.

Stretch zone: To finish on a multiple of 100, if the count starts on a multiple of 50, the number of jumps will need to be odd; if the count starts on a multiple of 100, the number of jumps will need to be even.

Practice Book page 26

1 50, 100, **150**, 200, **250**, 300, **350**

2 55, 105, **155**, 205, **255**, 305, **355**

3 340, 390, **440**, 490, **540,** 590, **640**

4 30, 80, **130, 180,** 230, **280,** 330

5 20, 120, **220, 320, 420,** 520, **620**

6 975, 875, **775**, 675, **575,** 475, **375**

7 10, **110,** 210, **310, 410,** 510, **610**

8 **699,** 599, 499, 399, **299, 199, 99**

9 The final digit is unchanged and the tens digit alternates.

10 No, because 75 ends in 5 and, as the ones digit does not change when counting on in 50s, 300 will not be in the sequence because it does not end in a 5.

Stretch zone: Examples of sequences could be: 20, 70, 120, 170, 220 and 835, 735, 635, 535, 434.

1D Number sequences

Explore 1

Student Book pages 24–25 • Practice Book page 27

Specific learning focus

- Count on in fours and eights to at least 100.

Global skills

- **Creative skills:** investigating
- **Real-world skills:** presenting/interpreting information
- **Interpersonal skills:** communication

Key vocabulary

- count on, count back

Resources

- mini whiteboards and markers

Language support

Ask students to describe to you the patterns they notice. Link these patterns to movements on the 100-square. Ask students to explain what they notice when they move 'down one row' on the 100-square; or move 'one column to the right'; or 'one column to the left'. Encourage the use of 'one more than' and 'one less than'. Point to a number. *Can you show me the number that is one/ten more/less than this one?*

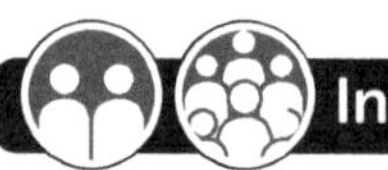 Introductory activity

Students work in pairs to decide whether the answers to the following questions are true or false. When you ask each question, give the pair 30 seconds to discuss their answer. They should then write a 'True (T)' or 'False (F)' on their whiteboards. (Students could answer using thumbs up or thumbs down instead, but mini whiteboards are more effective here because students can also use them to jot down the sequences you're giving them.)

- I start at zero. I count on in fives. 50 is in the sequence.
- I start at 2. I count on in fours. 50 is in the sequence.
- I start at 6. I count on in threes. 30 is in the sequence.
- I start at 5. I count on in threes. 37 is in the sequence.
- I start at 50. I count back in fours. 30 is in the sequence.
- I start at 40. I count back in fours. 0 is in the sequence.

After each statement, ask students whether they found a quick way of finding the answer. Taking the third sequence as an example, students might identify that 6 is in the three times table, so counting up in threes means counting up through the three times table. Thus, 30 will be in the sequence.

 Main activity

Students work individually to shade in the 100-squares in the Student Book on pages 24–25.

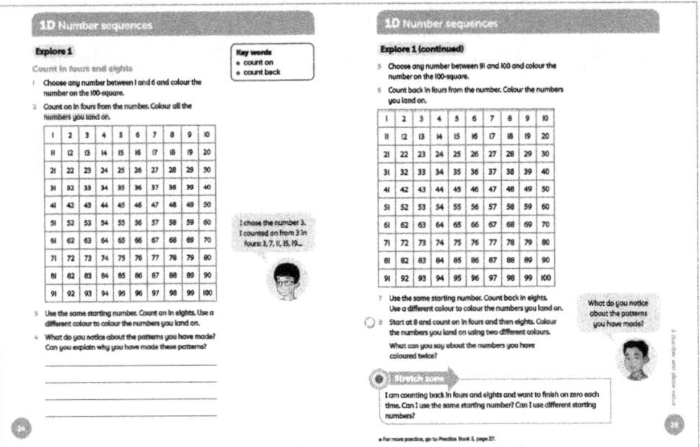

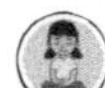

While students are working on the first 100-square, check that they have chosen a starting number that is between 1 and 6. When they have finished questions 1–3, ask them to look a question 4 in pairs, and to discuss why they think some of the numbers are coloured in more than one colour. (The numbers that are in the pattern of counting on in eights from the starting number will all be coloured in twice. This is because they also appear in the pattern of counting on in fours from the starting number.)

While students are working on the second 100-square, check that they have chosen a starting number that is between 91 and 100. When they have finished questions 5–7, ask them to look at question 8 in pairs, and to discuss why they think some of the numbers are coloured in more than one colour. (The numbers that are in the pattern of counting back in eights from the starting number will all be coloured in twice. This is because they also appear in the pattern of counting back in fours from the starting number.)

When they have shaded in the 100-squares, ask them to look at each other's 100-squares to compare the patterns. Encourage the pairs to discuss their patterns, noting the similarities and differences. As students work on the activity, find a pair who will be confident in describing the pattern to the whole class.

Encourage students to spot patterns with other multiples (such as 2 and 12). *Are any of the multiples of 8 also multiples of 2 or 4? Can you think why?*

Differentiation

Supporting: Check the patterns and ask students to check by counting carefully.

Consolidating: Ask students to predict the form the patterns will take.

Extending: Ask students to explore patterns in other multiples, for example: *Would any multiples of 3 be in this pattern of fours and eights? What about multiples of 5? What about multiples of 2? Can you predict what other multiples would be in the pattern?*

Stretch zone: *I am counting back in fours and eights and want to finish on zero each time. Can I use the same starting number? Can I use different starting numbers?*

Check that students are finding the correct numbers to work; that is, multiples of 8.

Reflection time

Use a pair who confidently described the patterns they noticed. Ask them to come to the front and describe the pattern. The rest of the class should check their description against the patterns they have noticed.

Practice Book: Students complete Practice Book page 27. They can do this directly after the main activity, as homework, or as the focus of a separate mathematics session to help students consolidate their learning and build fluency.

Students generate number sequences, counting in fours and eights, that contain specific numbers. This will challenge their knowledge of the patterns in the numbers within these sequences. Ask students to explain how they know they are correct before they write out their number sequences.

Differentiated outcomes	
All students	should complete patterns accurately with support.
Most students	will complete patterns and describe patterns.
Some students	may explore a wider range of number patterns.

Answers

Student Book pages 24–25

1–4 Students' sequences will vary, depending on the starting number. Students should recognise that when counting on in fours and eights, every other number is coloured in twice because all multiples of 8 are also multiples of 4.

5–8 Students' sequences will vary, depending on the starting number. Students should recognise that, as with counting on in fours and eights, when counting back, every other number is coloured in twice because all multiples of 8 are also multiples of 4.

Stretch zone: Yes, you can use the same starting number. It has to be a multiple of 8. The starting numbers can be different, as long as they are both multiples of 8.

Practice Book page 27

Answers will vary because students choose their own starting numbers.

Check that the rules and number sequences that students made up for themselves do match.

Stretch zone: For example, 13, 17, 21, 25, 29, 33, 37, 41, 45, 49, 53, 57, 61 and 13, 21, 29, 37, 45, 53, 61. The second sequence is made from alternate numbers of the first one.

Students' sequences will vary, depending on the starting number. Students should recognise that every other number in the sequence of fours is the same as each number in the sequence of eights.

1D Number sequences

Explore 2 Student Book page 26 • Practice Book page 28

Specific learning focus
- Count on and back in 50s or 100s.

Global skills
- **Creative skills:** investigating

Key vocabulary
- count, more

Resources
- digit cards 0–9

Language support
Remind students that counting in 100s always gives a number that ends in 'hundred', but that counting in 50s gives numbers that end alternately in 'fifty' or 'hundred'.

 **Introductory activity**

Ask a student to come to the front of the class and choose a digit card from a set of digit cards 0–9 (e.g. 7). Then ask the class to count on together in 50s from 7 until they reach a number over 900: 7, 57, 107, 157, 207, 257 and so on. The count will finish at 907. Then count back in 100s from 907: 807, 707, 607 and so on. Repeat with a different digit.

 **Introductory activity**

Ask a student to pick three digit cards at random and make the smallest number they can with the digits. For example, they might pick 4, 9, 2 so they should make 249.

Now build sequences by counting on from 249, firstly in 50s, then in 100s:

249, 299, 349, 399, 449, 499, 549, 599, 649, 699, 749, 799, 849, 899, 949, 999

249, 349, 449, 549, 649, 749, 849, 949

Ask students to look closely at the two sequences. *What is the same and what is different about the numbers in the sequences?*

Repeat using three new digits, making the largest possible number and then building sequences by counting back in 50s and then in 100s. *What is the same and what is different about the numbers in the sequences?*

Now choose a three-digit multiple of 10. Count on and back in 50s or 100s from that number. For example, in 50s from 130:

130, 180, 230, 280, 330, 380, 430, 480, 530

and in hundreds: 130, 230, 330, 430, 530

What is the same and what is different about the two sequences?

Ask students to complete the activities on page 26 in the Student Book individually. They generate new sequences of numbers starting on a multiple of 50 and counting in 50s and 100s. They then use a common starting number to count in 10s, 50s and 100s. *Look at the sequences of 10s and 100s. What is the same and what is different about them? And the 10s and 50s? What about in all three sequences?*

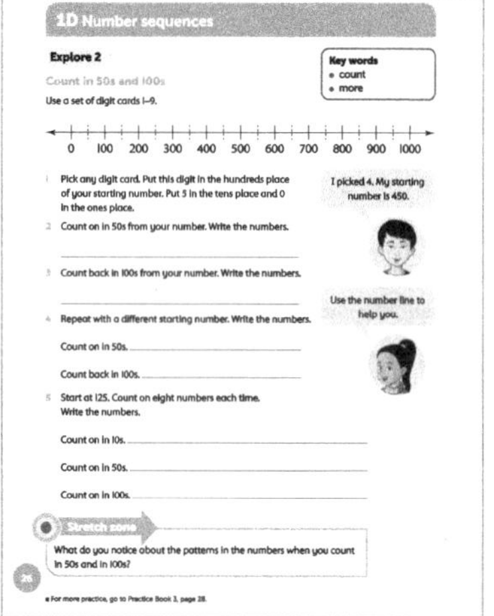

Differentiation
Supporting: Point to the numbers on a number line when counting on in multiples of 50 or 100.

Consolidating: Students could work in pairs, alternating who speaks when they say the sequences.

Extending: Students could create their own sequences that count on or back in 50s or 100s, starting from any number, and then describe the patterns.

Stretch zone: What do you notice about the patterns in the numbers when you count in 50s and in 100s? *Does it matter what the starting number is? Is the pattern always the same?* Check that students notice that when counting in 100s, the tens and ones do not change and when counting in 50s the tens digit alternates between two digits. This is always the same pattern; it does not matter what the starting number is.

 **Reflection time**

Ask students to discuss in pairs what patterns they found when making their sequences. They can offer to share anything interesting they noticed with the whole class. *Do you all agree? Why?*

Practice Book: Students complete Practice Book page 28. They can do this directly after the main activity, as homework, or as the focus of a separate mathematics session to help students consolidate their learning and build fluency.

Students create their own number sequences, counting in 50s and 100s from any 2-digit starting number. They then compare the sequences and write about any patterns that they notice. Check that students notice the consistent patterns in the sequences.

Differentiated outcomes	
All students	should count on in 50s or 100s accurately using a number line.
Most students	will count on or back in 50s or 100s from any multiple of 10.
Some students	may count on or back in 50s or 100s from any number.

Answers

Student Book page 26

1–4 Answers will vary depending on which digits are chosen as the starting number. Check that students have counted on or back in the correct multiples.

5 125   135   145   155   165   175   185   195   205

125   175   225   275   325   375   425   475   525

125   225   325   425   525   625   725   825   925

Stretch zone: Check that students notice that when counting in 100s, the tens and ones units do not change and when counting in 50s the tens digit alternates between two digits. This is always the same pattern; it does not matter what the starting number is. For example, starting from 40:

40, 140, 240, 340, 440

40, 90, 140, 190, 240, 290, 340, 390, 440

Practice Book page 28

Answers will vary depending on which digits are chosen as the starting number. Check that students have counted on in the correct multiples.

Stretch zone: Every other number in the 50s sequence is also in the 100s sequence. The ones digit stays the same in both sequences. In the 50s sequence, the tens digit alternates. In the 100s sequence, the tens digit stays the same each time; only the hundreds digit changes. For example, starting from 70:

70, 120, 170, 220, 270, 320, 370, 420, 470

70, 170, 270, 370, 470

---

1 Number and place value

Connect   Student Book page 27

Big idea

I can use place value to understand the size of a number. I can use rounding and counting in steps to count large numbers of objects or to make estimates.

Global skills:

- **Creative skills:** investigating
- **Interpersonal skills:** communication

Key vocabulary

- estimate, round to the nearest 10

Resources

- mini packets of raisins or similar
- mini whiteboards and markers
- range of jars full of beads or other items for students to estimate

Language support

- Ask open questions that encourage students to develop extended responses, using the vocabulary from the unit, such as 'greater than', 'less than', 'hundreds', 'tens', 'ones', 'estimate', 'groups of ten/five'. You could display the key words for students to see and use throughout the lesson. For example: *Why do you think that is a good estimate?*

 ### Introductory activity

Give each pair a mini packet of raisins (or similar items).

Ask students, in pairs, to estimate how many raisins there are in the packet without opening the packet. They should write their answer on their whiteboards.

Then ask each pair to open the lid, look at the top layer, and close the lid. They can now write down a new estimate if they want to. *Why do you think that is a good estimate?*

Finally, ask pairs to take out one raisin and to use this to work out a new estimate. *Why do you think that is a good estimate?*

List all the final estimates on the board, in order from smallest to largest. Then ask students to count out the raisins (encourage them to group these in fives) from each pack. Record the actual numbers of raisins in each pack on the board, in order. Agree on a sensible approximation, which may be written by the manufacturer on all the packets.

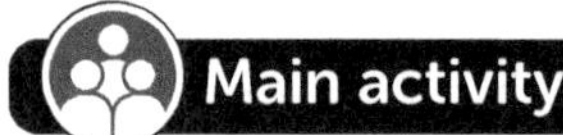

Main activity

Each group should estimate the quantity of objects in a certain jar. It is best if you are able to prepare these for the groups. Alternatively, use the picture in the Student Book on page 27. Look together at this page of the Student Book. If you have access to an IWB you could use this.

Refer students to question 1 in the Student Book. Ask questions to prompt children to share their suggested numbers: *Why are you certain that your number is bigger than the number of beads? How did you decide to choose that number as a smaller number?*

For question 2, remind students to look at the hint in the speech bubble. *Can you use the number of beads in one layer to make a good estimate for the whole jar?*

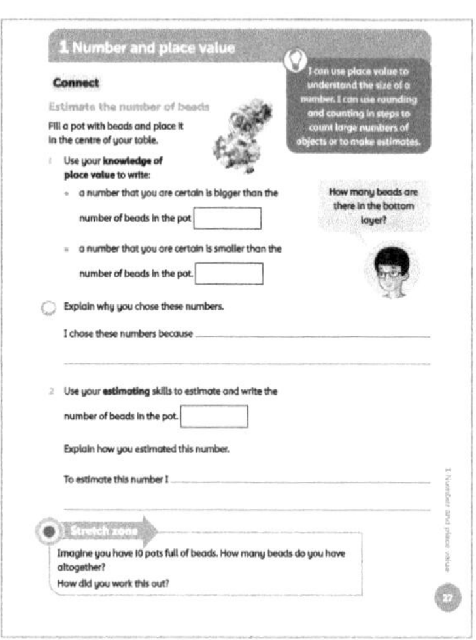

Differentiation

Supporting: Scaffold the activity by asking: *How many do you think are in each layer in the jar? How many layers are there?*

Consolidating: Ask students to explain why they think their estimate is accurate. Ask them whether they used grouping to count the number of beads. *How did you group the beads in one layer, in order to count them?*

Extending: Ask students to calculate totals in different numbers of jars.

Stretch zone: *Imagine you have 10 pots full of beads. How many beads do you have altogether?*

How did you work this out?

Discuss the estimates and ask some students to share their strategy. *Do you all agree? Why?*

 Reflection time

If groups have used jars you have prepared, they should count out the objects and find out whose estimates were closest to the actual number of objects in the jars. For each answer, ask students to round to the nearest 10. If you used the photograph in the Student Book, compare different groups' estimates and compare the strategies they have used. Did each group agree on a similar estimate? Can each group explain their estimate?

Differentiated outcomes	
All students	should estimate objects and count in ones to find the total.
Most students	will estimate objects and count in groups of five (or ten) to find the total. They will estimate the objects accurately.
Some students	may estimate the number of objects to the nearest 10.

1 Number and place value

Global skills

- **Interpersonal skills:** communication
- **Self-development skills:** reflecting on learning

Student Book

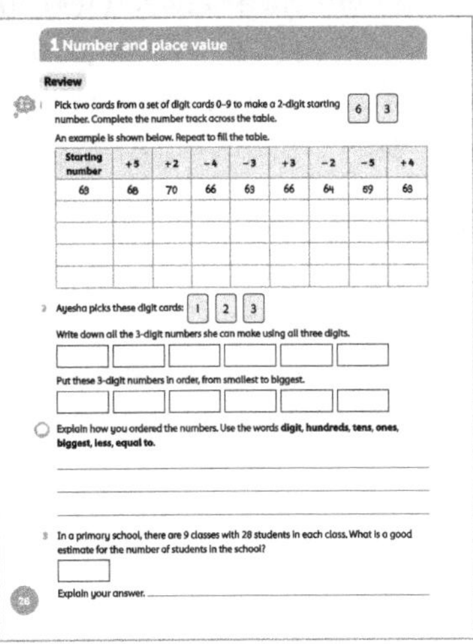

With young children, assessment activities are most effective when carried out as an everyday classroom activity. Students should have digit cards available.

Watch as students choose two cards to make a 2-digit number. Listen to check that they say the number words correctly. As students complete their number tracks, check that they can complete the number sequences for each starting number.

It may help to have 100-squares available to support students as they complete the review.

Encourage children to use a 100-square to help them complete the number sequences in question 1.

Answers

Student Book page 28

1 Students should make number tracks depending on the number they start with, for example: 35, 40, 42, 38, 35, 38, 36, 31, 35.

2 123, 132, 213, 231, 312, 321

Check for an accurate explanation of how students ordered their numbers, for example: First I looked at the hundreds digits. The number with the smallest hundreds digit came first. If some numbers had a hundreds digit equal to other numbers, I looked at the tens digit. If numbers had a hundreds and tens digit equal to other numbers, then I looked at the ones digit. The numbers with the biggest hundreds digit came last.

3 One possible response is 250 – estimate by doing $10 \times 28 - 28$, which is roughly $280 - 30 = 250$. Another possible response is 270 – estimate from 9×30.

Practice Book

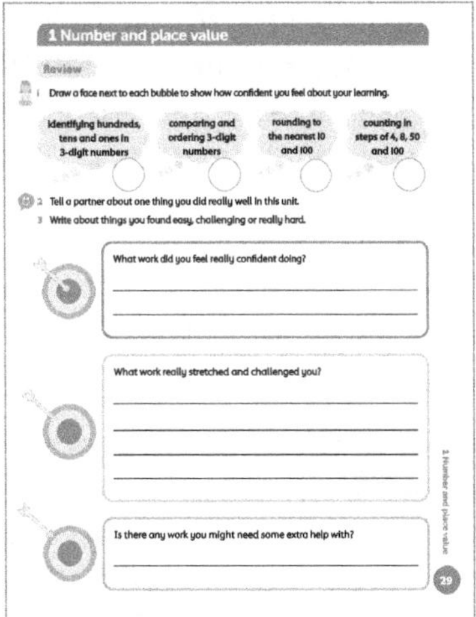

It is appropriate to complete this Practice Book review as a whole-class discussion. You may choose to keep a record of the class discussion or a copy of the review page for your own records. The review provides an opportunity for students to reflect on their learning from the unit, to discuss any areas of mathematics that they feel went particularly well, and any areas that they feel less confident about. Ensure that all students have a copy of the Student Book as a reminder of the areas of mathematics that they have worked on in this unit.

Allow students plenty of time for discussion before asking them to complete the Practice Book page individually, and then, if appropriate, to share their responses with the rest of the class. If students complete this self-assessment at home, encourage them to discuss this with adults. Make a note of areas that students still feel unsure about.

As number and place value are at the heart of many of the other units, you can revisit these areas regularly. You can also build estimating and ordering into everyday practice. For example, estimate how many students are in the class each day; count objects in jars (beads, counters and so on) including estimating larger numbers of things by counting in groups.

Additional material

There are additional end-of-unit assessments available on the *Oxford Owl for School* website.

2 Addition and subtraction

Big idea

This unit begins to develop students' mental methods into written methods. For this to take place, they need to continue to develop good mental images of the number system. This will help students use their understanding of place value to support the more formal written methods that will come later in the programme. In this unit, students use a range of equipment in order to build their mental images of the number system, including base-10 equipment, number lines and 100-squares.

Look out for

- **Students who double count a number on a number line. For example, if they are adding 128 + 6 on a number line, they start on the '128' to begin the addition and might then say '1' while pointing at the 128, and say '2' while pointing to the 129 and so on.** You can overcome this either by modelling the addition with practical objects, such as base-10 equipment and showing students their mistake, or by modelling the counting process on a number line correctly for them. Also look out for students who double count the starting number in a subtraction.

- **Students who may not know the next tens number so are unable to do the first hop on a number line, or may not use bonds to 10 to work out the size of the first hop.** Revise bonds to 10 and then model similar calculations for students, using a number line. Show them that once they have hopped to the next ten, they then hop the rest of the number and add the hops together.

- **Students who may draw the correct hops but then forget to add them to find answer to the addition or subtraction.** Encourage students to write each step of the calculation clearly above the relevant hop (e.g. +4, +40, +1) and then add each hop to find the answer.

Possible misconceptions

- **Students only think of subtraction as taking away or counting back.** Give students experience of counting on (shopkeeper arithmetic) and situations calculating the difference between two numbers.

- **Students draw number line jottings with the largest number at the beginning of the line, because it appears first in the subtraction.** Model how to use a number line to find the difference between two numbers, writing the smaller number at the beginning of the line, and using suitable-sized hops to find the difference.

- **When finding the difference between two numbers, students think they must always take the smaller number from the larger number. For example, in 327 – 129, they start by subtracting 7 ones from 9 ones. This can lead to later difficulties when students are working with negative numbers.** Model the calculation using base-10 equipment and use the correct language when modelling it: *How many more are in this group?* or *How many fewer are in that group? How did you work out the calculation?*

Key vocabulary

- addition, How many?, add, sum, total, altogether, equals
- subtraction, subtract, minus, take away, How many more? What is the difference between?, left over
- addition facts, inverse, complements, calculation, multiples of 10, multiples of 100, 2-digit number, 3-digit number
- 1 less/more, 10 less/more, fewer
- base-10 equipment, number line, row, column
- running total, addition calculation, subtraction calculation, difference
- number story, word problem, estimate, check, strategy, What if?

Coverage in lessons

Learning objective	2A	2B	2C	2D	2E	2F	2G	2H	2I	2J
Add and subtract numbers mentally, including: a 3-digit number and ones a 3-digit number and tens a 3-digit number and hundreds	✓	✓	✓	✓	✓	✓				✓
Add and subtract numbers with up to three digits, using formal written methods of columnar addition and subtraction.							✓	✓		✓
Estimate the answer to a calculation and use inverse operations to check answers.		✓					✓	✓		✓
Solve problems, including missing number problems, using number facts, place value, and more complex addition and subtraction.						✓			✓	✓

2 Addition and subtraction

Engage — Student Book page 29

Big question

- How can I add and subtract 2- and 3-digit numbers?

Global skills

- **Creative skills:** exploring
- **Real-world skills:** interpreting information
- **Interpersonal skills:** teamwork

Key vocabulary

- add, subtract, total, altogether, equals, left over

Resources

- none needed

Language support

As a class, create a poster of words associated with adding and subtracting, including for example: 'addition', 'how many?', 'add', 'sum', 'total', 'how many altogether?', 'subtraction', 'minus', 'take away', 'how many more?' 'what is the difference between?' '1 less/more', '10 less/more', 'fewer', 'row', 'column'. Always support students with using the correct vocabulary for carrying out addition or subtraction. Students may interpret the words wrongly. For example, in 'how many more?' they may think 'more' refers to addition, when it is asking them to find the difference.

 Introductory activity

Ask students to talk to a partner about where, in everyday life, they need to **add** or **subtract** numbers to work something out. You could give them some sentence starters, for example 'I need to add numbers when ____________. I need to subtract numbers when ____________.'

Draw up a list of ideas on the board. These could include money situations, such as how much two or three items cost, or how much change is needed. Suggest ideas relating to cooking and baking: *We often have to measure out ingredients and work out if we have enough of an ingredient, or how much more we need.*

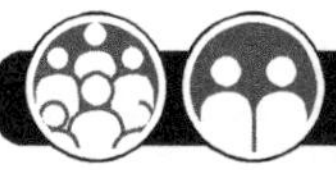 Main activity

Look together at page 29 of the Student Book. If you have access to an IWB you could use this. Ask students to tell you what information they can see about how much of each ingredient can be seen. Make a list of these on the board. Now ask them to focus on the two recipes and see what is needed.

As students work, ask them questions to encourage them to calculate the amounts, for example: *How much flour do we need **altogether**? Do we have that much flour?*

Tell students to work in pairs to decide whether there are enough ingredients to make both the muffins and the cookies. Finally, ask students to use the information in the picture to make up their own question about whether they have enough of each of the ingredients.

Differentiation

Supporting: Use the posters of recipe vocabulary to help students ask and answer questions that use addition and subtraction.

Consolidating: Prompt students by asking: 'How much of ingredient ___ will be **left over**?', for example.

Extending: Students can design their own questions by finding another cake or biscuit recipe and seeing whether the ingredients shown on page 29 would be enough to make it.

 **Reflection time**

Ask some pairs of students to say whether they think there are enough ingredients to make the muffins and the cookies, and why they think so.

Compare calculations for finding out whether there is enough or, where there is too much of an ingredient, how much will be left. Discuss with students the fact that they have used adding and subtracting to help them answer the questions.

2A Mental strategies for addition

Discover Student Book page 30–31 • Practice Book page 30

Specific learning focus

- Know addition and subtraction facts for all numbers to 20 and 200.

Global skills

- **Creative skills:** exploring
- **Real-world skills:** interpreting information
- **Interpersonal skills:** communication

Key vocabulary

- addition fact, total, complements for 20 and 200

Resources

- mini whiteboards and markers

Language support

Use the correct vocabulary of addition. For example, say 'add' rather than 'and'. Model the vocabulary of 'equals' too. For example, if a student says: 'Thirteen and seven is twenty', correct them by saying 'Thirteen add seven equals twenty.'

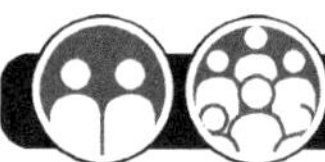 **Introductory activity**

Give each pair two minutes to write down as many ways of adding numbers to make 10 as they can. After two minutes, take feedback from pairs. On the board, write down all the answers they have. At first, write the answers down in the order in which they are given. Then ask: *Can you tell whether we have all the possibilities?* Start to write an ordered list on the board to show how such a list can be used to check that they have all possibilities, for example:

$1 + 1 + 1 + 1 + 1 + 1 + 1 + 1 + 1 + 1$

$2 + 1 + 1 + 1 + 1 + 1 + 1 + 1 + 1$

and so on. There are several hundred possibilities. Students should work to find as many as possible in the time allowed. Suggest that they try to work systematically, for example by finding ways that add two numbers first $(1 + 9, 2 + 8, …)$ then moving on to three numbers $(1 + 1 + 8, 1 + 2 + 7, …)$.

 **Main activity**

Students repeat this Introduction activity but should now write as many ways of adding numbers to make 20 as they can in a given time, say ten minutes. Explain that the key is to work systematically. They will not have time to find all the possibilities so they should just find as many as they can within the given timeframe. They should work in pairs to check each other's answers when they have run out of time.

Students then move on to Student Book page 30, where they find ways of making 200. Refer them to the Think back and remind them to use their facts to 20 to help them with facts to 200. Ask them to talk you through their examples so that they can convince themselves that they are correct.

Ask students to choose one number sentence from their list. *How can this number sentence help you write another one? And another? What is the same and what is different?*

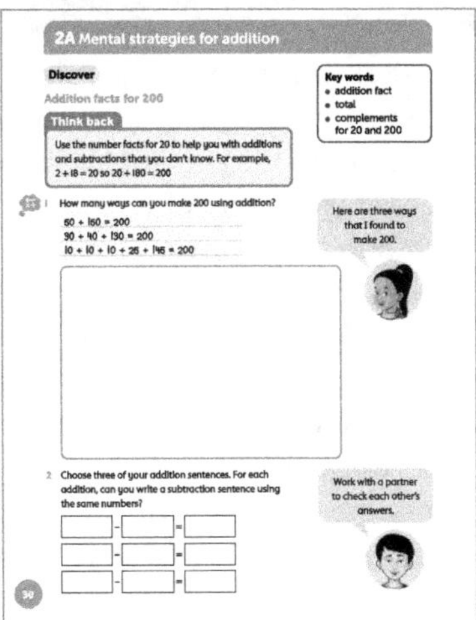

Differentiation

Supporting: Ask students to extend the number of possibilities by offering alternative starting numbers. For example, can they find additions to 10, starting from 7 $(7 + 3, 7 + 2 + 1, 7 + 1 + 2, 7 + 1 + 1 + 1)$.

Consolidating: Ask students whether they can take one addition sentence, and then change two numbers to make another sentence. Repeat, asking students to make another sentence.

Extending: Ask students how they know that they have all the possibilities for a particular starting number.

Stretch zone: *Can you explain to your partner how the number facts for 20 help you to find number facts for 200?*

Students need to be encouraged to clearly describe the method they used to convert facts for 20 into facts for 200.

 **Reflection time**

Tell students that they can use their addition facts for 20 to work out some missing numbers in subtraction calculations. They should work in pairs and write the answers to these missing number questions on their whiteboards. Make sure that you give students thinking time before asking them to show you their answers. Do these examples one at a time. Ask students to describe the methods they used to work out the answer. If students make errors, ask them to talk through their thinking process so that they can see where they have made an error.

$20 - \_\_ = 14$

$20 - \_\_ = 17$

$20 - \_\_ = 8$

$20 - \_\_ = 13, \qquad 200 - \_\_ = 130$

$20 - \_\_ = 5, \qquad 200 - \_\_ = 50$

Practice Book: Students complete Practice Book page 30. They can do this directly after the main activity, as homework, or as the focus of a separate mathematics session to help students consolidate their learning and build fluency.

Encourage students to choose some easy numbers to add and some more challenging ones. For example, they might choose $114 + 3$ as an easy addition and $373 + 50$ as a more challenging one. *How will you work out the answer? What facts can you use to help you?*

Differentiated outcomes	
All students	should find several different ways to make 20 accurately.
Most students	will use a structured approach to find different ways to make 20 accurately.
Some students	may know whether they have all the different possibilities for a particular starting number by using a structured approach.

Answers

Student Book pages 30–31

1 $10 + 190$, $20 + 180$, $30 + 170$ and so on. Also, examples such as $180 + 10 + 10$ and so on.

2 Check the subtraction sentences are completed correctly.

3 Sometimes true: $145 + 55 = 200$, $100 + 100 = 200$

   Sometimes true: $190 + 10 = 200$, $184 + 16 = 200$

   Always true: $185 + 15 = 200$, $175 + 25 = 200$

   Always true: $194 + 6 = 200$ $(4 + 6 = 10)$

   Always true: $174 + 26 = 200$ since $200 - 174 = 26$

Stretch zone: Students should include in their explanation that the facts for 200 are ten times bigger than the facts for 20, but other than that they are exactly the same (e.g. $10 + 10 = 20$ and $100 + 100 = 200$; $3 + 17 = 20$ and $30 + 170 = 200$; $1 + 2 + 5 + 12 = 20$; $10 + 20 + 50 + 120 = 200$).

Practice Book page 30

Answers will vary because students choose their own numbers and operations.

2A Mental strategies for addition

Specific learning focus

- Know addition and subtraction facts for all numbers to 20 and 200.

Global skills

- **Creative skills:** exploring
- **Real-world skills:** interpreting information
- **Interpersonal skills:** communication

Key vocabulary

- addition facts, total, inverse

Resources

- mini whiteboards and markers
- lollipop sticks in a jar – each lollipop stick has a different student's name on it
- empty number lines (these can be drawn on the board or use Resource Sheet 1.1)

Language support

Make sure that you model key language, in particular:

- addition
- subtraction (take away)
- equals
- largest
- smallest
- inverse.

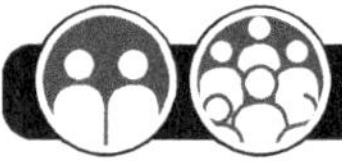 **Introductory activity**

Write on the board the digits 1, 2 and 3. In pairs, students should write as many different addition sentences as they can in two minutes using all the digits once and only once. For example, they might write:

$1 + 2 + 3 = 6$

$12 + 3 = 15$

$31 + 2 = 33$

Take feedback. Each time a pair gives an example, ask them the method they used to work out the answer. Repeat this activity, this time asking for subtraction sentences. Say that you only want number sentences that give a positive answer (or an answer greater than 0). So, $31 - 2 = 29$ is acceptable but $2 - 31$ is not as this would give you a negative answer. When you explain

this, avoid saying that 'you can't do' $2 - 31$. Instead, say: *This would give you a negative answer or an answer less than zero.*

 **Main activity**

Ask students to work in pairs on the activities on Student Book pages 32–33. Working in pairs will support language development. One student in each pair should say an addition calculation that gives the number in the centre of the 'bug' as an answer. The other student checks the calculation using an **inverse** calculation, and if the students agree it is correct then they both write it on their whiteboards. The other student in the pair then suggests a different addition calculation and so on. While the pairs work on the activity, ask students to explain the strategies they are using. *Are you using known facts? Are you using a number line? When using a number line, are you counting on or counting back?* You could model the use of a number line for those students who need it. Provide number lines for these students.

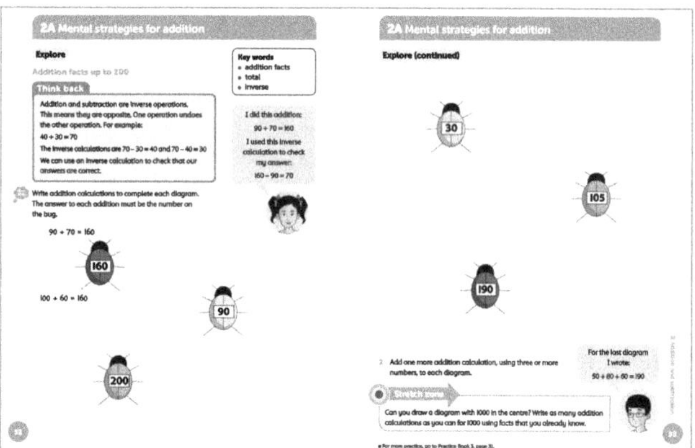

Differentiation

Supporting: Provide practical materials such as base-10 equipment and number lines to support students. Model how to use the equipment to help with calculations.

Consolidating: Ask students to use a wider range of calculations, including some with more than two numbers.

Extending: Ask students whether they can think of calculations using 3-digit numbers, or with three or more separate numbers.

Stretch zone: *Can you draw a diagram with 1000 in the centre? Write as many addition calculations as you can for 1000 using facts you already know.*

Check that students' number facts are accurate and that they have used a range of different calculations, some with three or more numbers added together.

 **Reflection time**

Pick a lollipop stick (with a student's name on it) and ask that student to come to the front and read out one of their questions that gave the answer 160. They should write it on the board and explain how they carried out the calculation. Model the calculation on an empty number line for the rest of the class. Repeat this for each of the numbers.

Practice Book: Students complete Practice Book page 31. They can do this directly after the main activity, as homework, or as the focus of a separate mathematics session to help students consolidate their learning and build fluency.

Encourage students to be creative when creating their diagrams and use more than two numbers added together for some of their facts.

Differentiated outcomes	
All students	should complete the diagrams using two numbers added together.
Most students	will use a range of more complex additions, including some with more than two numbers added together.
Some students	may use imaginative responses using 3-digit numbers and three or more numbers added together.

2B Mental strategies for subtraction

Discover Student Book page 34 • Practice Book page 32

Specific learning focus

- Know addition and subtraction facts for all numbers to 20.

Global skills

- **Creative skills:** investigating
- **Interpersonal skills:** communication
- **Self-development skills:** reflecting on learning

Key vocabulary

- subtraction, calculation

Resources

- mini whiteboards and markers

Language support

Use the vocabulary of subtraction so say 'subtract' rather than 'less'. Model the vocabulary of 'equals' too. For example, if a student says: 'Twenty less five is fifteen', correct them by saying 'Twenty subtract (or take away) five is fifteen'.

Answers

Student Book pages 32–33

Students write their own addition calculations so their answers will vary. They should check their answers, in pairs, using a subtraction calculation.

Practice Book page 31

Students write their own addition calculations so their answers will vary. Check that the calculations correctly give the answers in the rectangles. Encourage students to write calculations involving several numbers.

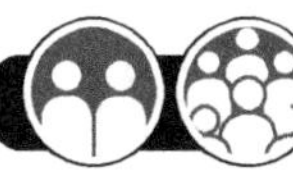 **Introductory activity**

Give each pair two minutes to write down as many ways of subtracting numbers to make 5 as they can, using any two numbers up to 20. Then take feedback from pairs. On the board, write down all the answers they give. At first, write the answers down in the order in which they are given. Then ask: *Can you tell whether we have all the possibilities?* Write an ordered list on the board to show how a systematic way of listing the possibilities can be used to check that they have them all:

$20 - 15 = 5$	$16 - 11 = 5$	$12 - 7 = 5$	$8 - 3 = 5$
$19 - 14 = 5$	$15 - 10 = 5$	$11 - 6 = 5$	$7 - 2 = 5$
$18 - 13 = 5$	$14 - 9 = 5$	$10 - 5 = 5$	$6 - 1 = 5$
$17 - 12 = 5$	$13 - 8 = 5$	$9 - 4 = 5$	$5 - 0 = 5$

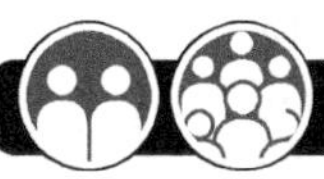 **Main activity**

Students repeat the introductory activity but should now write as many ways of subtracting any numbers to make 8 as they can, using any two numbers up to 20. They should work in pairs to check each other's answers when they think they have all possibilities. *What patterns can you see in the subtractions that you have written?*

Students then move on to page 34 of the Student Book, writing **subtractions** with the answer 10. This activity does not restrict students to only using numbers to 20. As there are an infinite number of ways to do this, ask: *How many ways of making 10 do you think there are? What are the biggest numbers you can think of to use for your subtraction?*

Ask students to talk you through their examples so that they can convince themselves that they are correct. *What patterns in the digits convince you that the **calculation** is correct?*

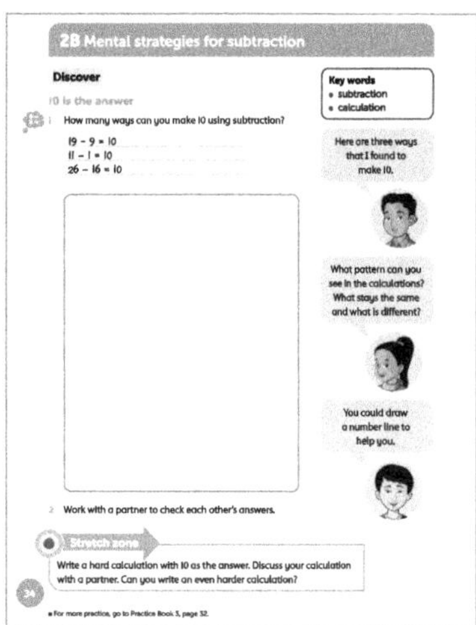

Differentiation

Supporting: Ask students to extend the number of possibilities by offering starting numbers. For example, ask them to find subtractions to 10 starting from 13 ($13 - 3$, $14 - 4 = 10$ and so on).

Consolidating: Ask students whether they can take one subtraction sentence, and then change two digits to make another sentence. Repeat, asking students to make another sentence.

Extending: Ask students whether they think they can find all the possible subtraction sentences in the time they have in the lesson. Ask them to explain their answer.

Stretch zone: *Write a hard calculation with 10 as the answer. Discuss your calculation with a partner. Can you write an even harder calculation?*

Students need to be encouraged to describe their method clearly. They may use large numbers, as large as they feel confident with, perhaps $200 - 190 = 10$, or they may suggest using three numbers such as $100 - 85 - 5 = 10$).

 Reflection time

Ask whether anyone thinks they found all the subtraction sentences for 10. *Why has no one found all the possible subtraction sentences?* (There is an infinite number of possibilities.) Tell the class that they can use their subtraction facts to work out some missing numbers in addition calculations. They should work in pairs and write the answers to these missing number additions on their whiteboards. Make sure to give thinking time before asking students to show you their answers. Do these examples one at a time. Ask students to describe the methods they used to work out the answer. If students make errors, ask them to talk through their thinking process so that they can see where they have made an error.

$10 + \_\_ = 19$

$10 + \_\_ = 18$

$10 + \_\_ = 17$

$10 + \_\_ = 16$

$10 + \_\_ = 15$

$10 + \_\_ = 14$

Practice Book: Students complete Practice Book page 32. They can do this directly after the main activity, as homework, or as the focus of a separate mathematics session to help students consolidate their learning and build fluency.

Students are given the answers to questions and have to write their own questions. The questions should involve subtractions with two or more numbers.

Differentiated outcomes	
All students	should find several different ways to make a given number accurately.
Most students	will use a structured approach to find different ways to make given numbers accurately.
Some students	may know whether they have all the different possibilities for a given number by using a structured approach.

Answers

Student Book page 34

Answers may vary depending on how large the numbers are that students feel confident to work with. Answers may include, for example:

$12 - 2 = 10$, $21 - 11 = 10$, $30 - 20 = 10$, $57 - 47 = 10$ and so on.

Practice Book page 32

Answers may vary depending on how large the numbers are that students feel confident to work with. Answers may include subtractions using two numbers (e.g. $33 - 23 = 10$, $17 - 2 = 15$, $160 - 20 = 140$) or three numbers (e.g. $80 - 30 - 5 = 45$).

Stretch zone: Examples could include $1000 - 1 = 999$, $1001 - 2 = 999$, $2000 - 1001 = 999$.

2B Mental strategies for subtraction

Explore Student Book page 35 • Practice Book page 33

Specific learning focus

- Know addition and subtraction facts for all numbers to 20.

Global skills

- **Creative skills:** problem solving
- **Interpersonal skills:** communication

Key vocabulary

- subtraction, calculation

Resources

- lollipop sticks in a jar – each lollipop stick has a different student's name on it

Language support

Make sure that you model key language, in particular:

- subtraction (take away)
- equals
- largest
- smallest.

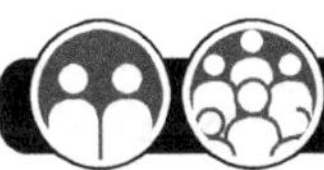 **Introductory activity**

Write on the board the digits 4, 5 and 6. In pairs, students should write as many different addition sentences as they can in two minutes using all the digits once and only once. For example, they might write:

4 + 5 + 6 = 15

45 + 6 = 51

64 + 5 = 69

Take feedback. Each time a pair gives an example, ask them what method they used to work out the answer. Repeat this activity, this time asking for subtraction sentences. Say that you only want number sentences that give a positive answer (or an answer greater than 0). So, 64 − 5 = 59 is acceptable; but 5 − 64 is not as this would give you a negative answer. When you explain this, avoid saying that 'you can't do' 5 − 64. Instead, say: *This would give you a negative answer* or *an answer less than zero*.

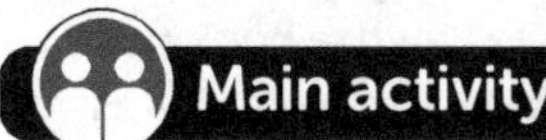 **Main activity**

Ask students to work in pairs on the activity in the Student Book on page 35. Working in pairs will support language development. One student in each pair should say a subtraction calculation that gives the number in the centre of the 'bug' as an answer. The other student checks the calculation using an inverse calculation and, if the students agree that it is correct, they both write it down. The other student in the pair then suggests a different subtraction calculation and so on. While the pairs work on the activity, ask students to explain the strategies they are using. Are they using known facts? Are they using a number line? When using a number line, are they counting on or counting back? You could model the use of a number line for those students who need it. Provide number lines for these students.

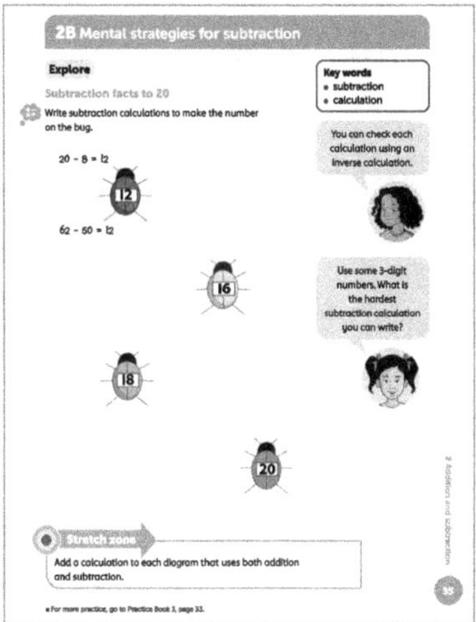

Differentiation

Supporting: Provide practical materials such as base-10 equipment and number lines to support students. Model the use of the equipment to help them with calculations.

Consolidating: Ask students to use calculations with more than two numbers.

Extending: Ask students whether they can think of calculations that involve 3-digit numbers and calculations with three or more numbers.

Stretch zone: *Add a calculation to each diagram that uses both addition and subtraction.*

Students write their own addition and subtraction calculations so answers will vary (e.g. 50 − 40 + 2 = 12). Check that the calculations are correct. Encourage students to use 2-digit numbers where appropriate.

 **Reflection time**

Pick a lollipop stick (with a student's name on it) and ask that student to come to the front and read out one of their questions that gave the answer 16. They should write it on the board and explain how they carried out the calculation. Model the calculation on an empty number line for the rest of the class. Repeat this for 18 and 20.

Practice Book: Students complete Practice Book page 33. They can do this directly after the main activity, as homework, or as the focus of a separate mathematics session to help students consolidate their learning and build fluency.

Students complete three different diagrams, writing subtractions that give the answer in the centre of the diagram. Encourage them to be creative and use more than two numbers for some of their calculations.

Differentiated outcomes	
All students	should complete the diagrams using single- and 2-digit numbers.
Most students	will use a range of 2-digit numbers to make each answer.
Some students	may use imaginative responses possibly using 3-digit numbers and more than two numbers.

Student Book page 35

Students write their own calculations so their answers will vary. Check that students' calculations are correct and encourage them to use 2- and 3-digit numbers where appropriate.

Practice Book page 33

Students write their own calculations so their answers will vary. Check that the calculations correctly give the answers in the rectangles. Encourage students to write calculations involving larger numbers and more than two numbers.

2C Adding and subtracting multiples of 10 and 100

Discover 1 Student Book pages 36–37

Specific learning focus

- Add and subtract 10 (and multiples of 10) from 2-digit and 3-digit numbers.
- Add 100 (and multiples of 100) to 3-digit numbers.
- Find complements to 100.

Global skills

- **Creative skills:** problem solving
- **Interpersonal skills:** communication

Key vocabulary

- multiples of 10, multiples of 100

Resources

- sticky notes with numbers 1–6 written on them placed in a small bag
- 100-squares

Language support

As students work on the activities in the Student Book, move around the class and ask them to say the calculation aloud. This reinforces the vocabulary of addition and subtraction. It also provides an opportunity to check that students are pronouncing the tens numbers correctly. Listen for students saying '-ty' at the end of these numbers and not '-teen'.

Introductory activity

Pick a number from the bag. Write the multiple of 10 that you make with this number on the board. For example, if you pick a 3 you write 30. Each pair then has five minutes to write as many addition calculations with the answer 30 as they can. Model the language of addition as students give you feedback, as 'X add Y equals Z, so the sum of X and Y is Z'. For example, you might say: *14 add 16 equals 30, so the sum of 14 and 16 is 30.*

Repeat this activity. Pick a number from the bag again, and this time students should find subtraction calculations with this answer. Model language carefully, such as 'X minus Y equals Z, so the difference between X and Y is Z'. For example, you might say: *35 minus 15 equals 20 so the difference between 35 and 15 is 20.*

Main activity

Refer back to the first example in the introductory activity. *We know that 14 + 16 = 30. Can we use this calculation to write some other calculations with the answer 30? What if we changed the 14 to 15, what would we need to change the 16 to? How could we use the original addition calculation to come up with a new calculation that uses three numbers? What other addition calculations could we write?*

Can you tell me a subtraction calculation with the answer 30? Does that help us to write other subtractions with the answer 30? Can anyone tell me a subtraction calculation with an answer of 30 that uses three numbers?

As each calculation is suggested, ask students whether they agree with the suggestion and whether they can say a related calculation. As students work on the activity in the Student Book on pages 36–37, ask them how they are carrying out the calculation or how they know they are correct. Refer students to the examples in the Student Book. They may notice that:

50 − 10 − 10 − 10 = 20 because 10 + 10 + 10 = 30 and 50 − 30 = 20, for example.

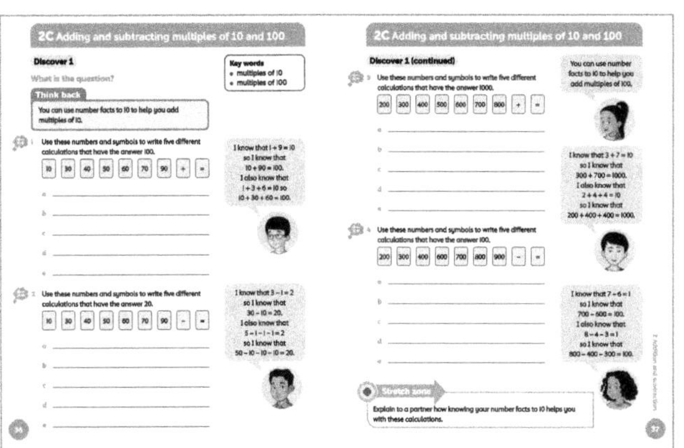

Differentiation

Supporting: Ask students to focus on pairs of numbers and use a 100-square for support.

Consolidating: Encourage students to use three numbers using a 100-square for support.

Extending: Encourage students to use three or four numbers to create their calculations.

Stretch zone: *Explain to a partner how knowing your number facts to 10 helps you with these calculations.*

Students should be able to explain how these number facts are ten times bigger than number facts to 10 but, other than that, they are the same.

 **Reflection time**

Pick three numbers from the bag to create a 3-digit number. Write this number on the board. Then pick a number from the bag again and make a **multiple of 10**. Write this on the board. For example, you may pick 4, 3, 4 to make 434, and then 6 to make 60. So, you would write:

434   60

In pairs, students should find 60 more than and 60 less than 434. When all students show their answers, select a pair who got the answer correct to explain the method to the whole class. Repeat this activity five times.

Practice Book: There is no Practice Book page for this lesson.

Differentiated outcomes	
All students	should use pairs of numbers to create their calculations.
Most students	will use pairs and sets of three numbers to create their calculations.
Some students	may use more than three numbers to create their calculations.

Answers

Student Book pages 36–37

1   Answers will vary because students choose their own numbers from a given set to make five addition calculations. Check that students' calculations are additions with the answer 100 and that they are correct.

2   Answers will vary because students choose their own numbers from a given set to make five subtraction calculations. Check that students' calculations are subtractions with the answer 20 and that they are correct.

3   Answers will vary because students choose their own numbers from a given set to make five addition calculations. Check that students' calculations are additions with the answer 1000 and that they are correct.

4   Answers will vary because students choose their own numbers from a given set to make five subtraction calculations. Check that students' calculations are subtractions with the answer 100 and that they are correct.

2C Adding and subtracting multiples of 10 and 100

Discover 2 — Student Book page 38

Specific learning focus

- Add and subtract 10 or 100 using the 100-square and number lines.

Global skills

- **Creative skills:** problem solving

Key vocabulary

- multiples of 10, multiples of 100, 10 more, 10 less

Resources

- large 100-square, small 100-squares for students

Language support

As students work on the activities in the Student Book, move around among the class and ask them to say the calculation aloud. This reinforces the vocabulary of addition and subtraction and encourages students to use the language of place value such as 'ones', 'tens' and 'digit'.

Introductory activity

Highlight a number on the large 100-square and ask students to tell you the number that is 10 more, then highlight their answer. Repeat several times, then ask students to describe the pattern they see. Check that they have noticed that 10 more is always the number below the one you start from. Ask them to look the patterns of the digits of all the numbers. Check that they notice that the tens number changes each time but that the ones digit stays the same.

Main activity

Repeat the activity on the 100-square for 10 less, then for 100 more and 100 less. Make sure that students recognise that the number that is 10 more is below the chosen number, on the next row but in the same column, and the number that is 10 less is on the previous row but in the same column.

Highlight a number on the 100-square (e.g. 74). Ask students to say how many tens and how many ones make the number 74. Then, ask how many hundreds it has. (0) *What would happen to the number if 1 hundred was added?* Students should say that it becomes 174 (which is not on the 100-square).

Repeat with some other numbers (e.g. 38, 22, 6, 91 and 50). Students should be able to see that adding 100 means placing a 1 in front of the numbers, which is placing a 1 in the hundreds column.

Ask students to complete the activities in the Student Book on page 38, which extend the main activity by asking students to add 100 to a number that is not on the 100-square. Refer to the second speech bubble on the Student Book page. *Can anyone describe the pattern in the digits when we keep adding 100 to a number?*

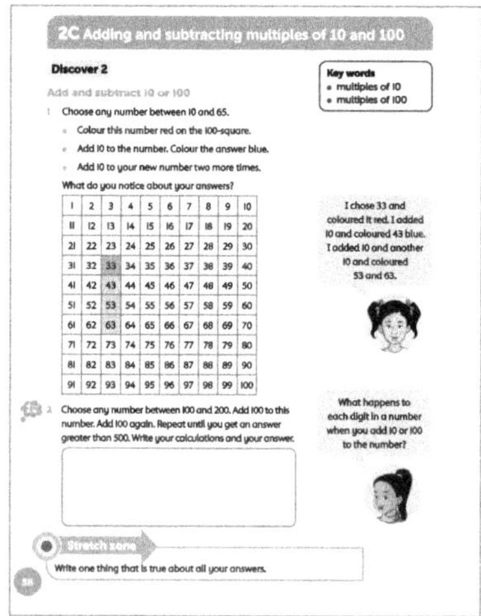

Differentiation

Supporting: Model the use of the 100-square for finding 10 more and 10 less than a given number.

Consolidating: Help students to move over tens boundaries when adding 10 and 100, for example when adding 10 to 43 to get to 53, or adding 100 to 43 to get 143.

Extending: Encourage students to use the patterns they notice in numbers when adding 10 and 100, to add and subtract 9, 11, 90 and 110 from numbers on the 100-square and beyond.

Stretch zone: *Write one thing that is true about all your answers.*

Answers may vary but check them for accuracy. Students may say that, when adding or subtracting tens, the ones digit does not change, and when adding or subtracting 100, the tens and the ones digits do not change.

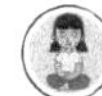

Reflection time

Ask pairs of students to explain how they added or subtracted 10 or 100 to numbers. Ask students how they could find 20 more/less, then 30 more/less.

Differentiated outcomes	
All students	should use a 100-square to find 10 more or less than any number, and understand the pattern in digits when adding 100 to a number.
Most students	will be able to carry out the calculation mentally, without the visual aid of the 100-square.
Some students	may be able to use similar methods to add 9, 90, 11 or 110 and subtract 9, 90 or 110.

2C Adding and subtracting multiples of 10 and 100

Explore 1 Student Book pages 39–40 • Practice Book page 34

Specific learning focus

- Add and subtract 10 (and multiples of 10) from 2-digit and 3-digit numbers.
- Add 100 (and multiples of 100) to 3-digit numbers.
- Find complements to 100.

Global skills

- **Creative skills:** problem solving

Key vocabulary

- multiples of 10, multiples of 100

Resources

- sticky notes with numbers 1–6 written on them
- a small bag
- 100-squares

Language support

As students work on the activities in this lesson, ask individuals to say the calculations aloud. This reinforces the vocabulary of addition and subtraction.

Answers

Student Book page 38

Students' answers will vary depending on which number they picked. Check they have added tens to the number in sequence and coloured the numbers appropriately.

 **Introductory activity**

Pick two numbers from the bag and write down the 2-digit number you make. If you picked 3 and 5, you would write 35. From this number, count on in 10s by going around the class and asking different students to say the number that is 10 more than the previous one. The first student would say 35, the second would say 45, the third would say 55 and so on. Repeat all the way around the class. Support students in saying the number names.

- Repeat by making a 3-digit number and counting back in tens.
- Repeat by making a 3-digit number and counting on in hundreds to 1000.
- Repeat by making a 3-digit number that has 9 in the hundreds column, and counting back in hundreds until you get to a 2-digit number.

 **Main activity**

Look together at page 39 of the Student Book. If you have access to an IWB you could use this. Demonstrate the first activity in the Student Book. Read the example given in the Student Book speech bubble and work through the following route as a class. Start in the centre at 30, then move up to 50, writing the first part of the number sentence as you move: $30 + 50$. Then, move right to 40 and record this move: $30 + 50 + 40$. Finally, move diagonally up to 20 and record the final part of the number sentence with the answer: $30 + 50 + 40 - 20 = 100$.

Students should work individually on the rest of the activities in the Student Book on pages 39–40. Encourage them to make rough notes as they work out a route to reach a particular number. For question 3, encourage students to fill in the grid in question with numbers that would challenge them, rather than choosing the simplest examples.

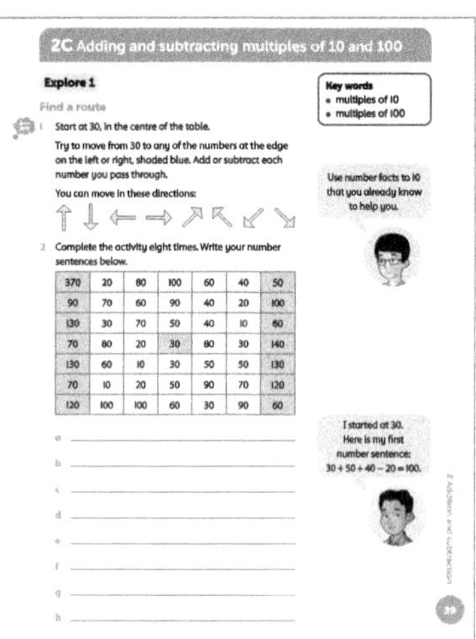

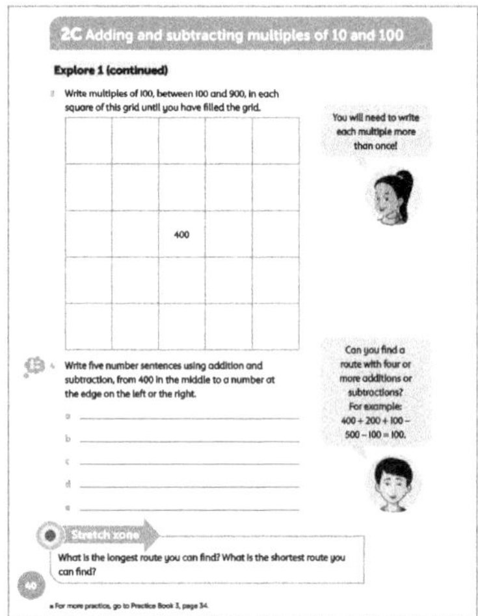

Practice Book: Students complete Practice Book page 34. They can do this directly after the main activity, as homework, or as the focus of a separate mathematics session to help students consolidate their learning and build fluency.

Students continue to write number chains using addition or subtraction of multiples of 10 and 100 to reach a target number.

Differentiated outcomes	
All students	should create an addition or subtraction calculation using two multiples of 10 or 100.
Most students	will create addition and subtraction calculations to reach the target number.
Some students	may create complex calculations to reach the target number.

Answers

Student Book pages 39-40

1 and **2** Answers will vary because students choose their own routes and calculations to start at 30 and reach one of the blue numbers at the edges of the grid.

3 and **4** Answers will vary because students choose their own numbers for the grid, and then their own routes and calculations to start at 400 and reach an outside number.

Practice Book page 34

Answers will vary because students choose their own calculations to take them from the starting number to the finishing number for each number chain.

Differentiation

Supporting: Allow students to use just the centre number and one other number to create the calculation.

Consolidating: Ask students to explain their thinking as they select the numbers to use in their route.

Extending: Challenge students to create a complex route through the target board in question 3.

Stretch zone: *What is the longest route you can find? What is the shortest route you can find?*

Students should be able to explain how they found the longest and shortest routes.

 **Reflection time**

Use the bag of numbers from the introductory activity. Choose three numbers from the bag and make a 3-digit number. Students should work in pairs to write down an addition calculation using three numbers that, when summed, give this 3-digit number. Encourage them to use multiples of 10 and 100 to help them. For example, if you make the number 526, they could write: '500 + 20 + 6' or '490 + 30 + 6'.

Repeat, but this time ask for subtraction calculations. Model the language of addition and subtraction as you take feedback from students.

2C Adding and subtracting multiples of 10 and 100

Explore 2 Student Book page 41

Specific learning focus

- Add and subtract 10 (and multiples of 10) from 2-digit and 3-digit numbers.
- Add 100 (and multiples of 100) to 3-digit numbers.

Global skills

- **Creative skills:** exploring
- **Interpersonal skills:** communication

Key vocabulary

- multiples of 10, multiples of 100

Resources

- 100-squares
- sticky notes with numbers 1–6 written on them
- small bag

Language support

As students work on the activities in the lesson, ask individuals to say the calculations aloud. This reinforces the vocabulary of addition and subtraction and supports students in describing numbers using place value vocabulary such as 'ten', and 'hundred'.

 **Introductory activity**

Ask a student to choose three digits from the bag to make a 3-digit number (e.g. 416).

Replace all numbers in the bag and then pick out three different numbers, one at a time. Each number tells you how many tens to add, in turn. For example, if the numbers are 2, 5 and 3, then the calculations will be:

416 + 20 = 436   436 + 50 = 486   486 + 30 = 516

Repeat with three new digits to determine how many hundreds to add, for example for the numbers 1, 3 and 2:

516 + 100 = 616   616 + 300 = 916   916 + 200 = 1116

Repeat, starting with a number made from 1000 and three number cards (e.g. 1325) then use three new numbers to indicate how many hundreds and then tens to subtract.

 **Main activity**

Students should work individually on the activities in the Student Book on page 41. They need to complete question 2 in their notebook. Look together at this page of the Student Book. If you have access to an IWB you could use this. Talk through the worked example so that students can see how to complete the place-value grid in their notebook.

When students have completed the grid in question 1, refer them to the first speech bubble in the Student Book. *Look at the first row of the table. What patterns can you see in the numbers? Compare the first row and the second row. What patterns can you see in these two rows? Can you see any other patterns in the numbers?*

Encourage them to complete the grids in question 2 with numbers that would challenge them, rather than choosing the simplest examples.

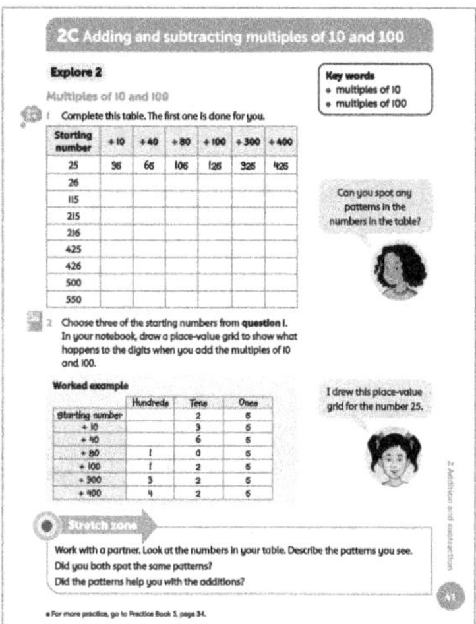

2C Adding and subtracting multiples of 10 and 100

Explore 2

Multiples of 10 and 100

1 Complete this table. The first one is done for you.

Starting number	+10	+40	+80	+100	+300	+400
25	35	65	105	125	325	425
26						
115						
215						
216						
425						
426						
500						
550						

2 Choose three of the starting numbers from **question 1**. In your notebook, draw a place-value grid to show what happens to the digits when you add the multiples of 10 and 100.

Worked example

	Hundreds	Tens	Ones
Starting number		2	5
+ 10		3	5
+ 40		6	5
+ 80	1	0	5
+ 100	1	2	5
+ 300	3	2	5
+ 400	4	2	5

Stretch zone

Work with a partner. Look at the numbers in your table. Describe the patterns you see.
Did you both spot the same patterns?
Did the patterns help you with the additions?

For more practice, go to Practice Book 3, page 34.

Differentiation

Supporting: Allow students to use base-10 equipment or a number line to help where needed.

Consolidating: Ask students to explain their thinking as they show the effect on the digits of their numbers.

Extending: Challenge students to show what happens to the digits of numbers above 1000 when adding multiples of 10 or 100.

Stretch zone: *Work with a partner. Look at the numbers in your table. Describe the patterns you see.*

Did you both spot the same patterns?
Did the patterns help you with the additions?

Students should be able to reason about and describe the patterns they noticed. They may spot patterns in rows where the tens and ones digits are the same (e.g. rows 3 and 4) or where the hundreds and tens digits are the same (e.g. rows 4 and 5).

 Reflection time

Ask students to share the number grids they drew for question 2. Ask whether they noticed anything about the numbers they made and how adding tens or hundreds changes the digits of the numbers.

Practice Book: There is no Practice Book page for this lesson.

Differentiated outcomes	
All students	should add multiples of 10 to 2- and 3-digit numbers.
Most students	will add multiples of 10 and 100 to 2- and 3-digit numbers.
Some students	may add multiples of 10 or 100 to numbers up to and beyond 1000.

2D Adding several small numbers

Discover Student Book page 42

Specific learning focus

- Add several small numbers.
- Reorder an addition to help with calculation.

Global skills

- **Creative skills:** investigating
- **Interpersonal skills:** communication

Key vocabulary

- addition fact, total

Resources

- sticky notes with numbers 1–6 written on them (enough for all pairs to have their own)
- a small bag
- mini whiteboards and markers

Language support

Throughout this activity, use the word 'sum' when talking about the answer. Remind students that 'sum' always means the total of an addition; it is not another word for 'question' or 'calculation'.

Answers

Student Book page 41

1						
25	35	65	105	125	325	425
26	36	66	106	126	326	426
115	125	155	195	215	415	515
215	225	255	295	315	515	615
216	226	256	296	316	516	616
425	435	465	505	525	725	825
426	436	466	506	526	726	826
500	510	540	580	600	800	900
550	560	590	630	650	850	950

2 Answers will vary because students choose their own numbers to write in place value grids.

 Introductory activity

Each student should have a mini whiteboard. Pull one number out of the bag and record that number on the board. Then put it back in the bag. Repeat this 10 times. Students should try to keep a running total in their heads and only write down the sum of all 10 numbers at the end. When all students have written down their totals, ask them to give you their answers. List all the different answers on the board (if there are different answers). Check the correct answer by adding the numbers you have written. Show students that they can use their knowledge of number bonds to 10 to help with the addition.

Repeat the activity, but this time write all 10 numbers on the board before you ask students to add them up. *How did you rearrange the numbers in the calculation to make it easier?*

Main activity

Students work in pairs using their own set of numbers 1–6 on sticky notes, and a bag. They place the numbers back in the bag after each selection, so a number can be chosen more than once. They discuss and choose the quickest strategies to add the ten numbers and find the total each time. *What is the largest possible total? What is the smallest?* (If they choose 6 ten times, the total will be 60, but if they choose 1 ten times, the total will be 10.)

Students then work on the activity in the Student Book on page 42. They first choose a route through a maze of numbers and then rearrange the numbers to make them easier to add. While students are working, ask them to explain why they chose particular 'quick ways' to add some numbers together.

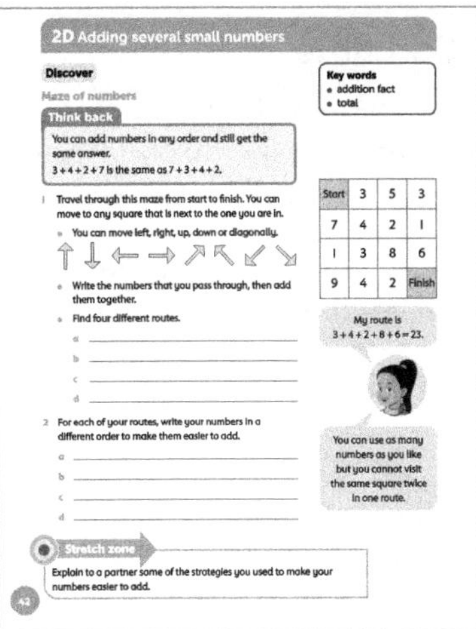

Differentiation

Supporting: Some students should start by finding totals to 10 and underlining these. You can support them in the calculation from this point.

Consolidating: Encourage students to rewrite the calculation with the addition pairs to 10 at the beginning.

Extending: Encourage students to 'hold the numbers in their head' after they have found the totals to 10.

Stretch zone: *Explain to a partner some of the strategies you used to make your numbers easier to add.*

Students should be able to explain how they added their numbers and found an efficient method, such as using pairs to 10 first and then adding the other numbers in order from largest to smallest.

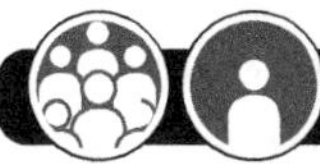 **Reflection time**

Take feedback on the answers that pairs have written for the smallest and largest possible totals. Link the largest possible total to the 6 times table. (The largest possible total comes from 6 being picked 10 times, which is the same as $6 \times 10 = 60$.)

Ask pairs for examples of ways in which they rearranged the numbers from the maze in the Student Book to make the calculation simpler. Highlight methods, including finding pairs of numbers that total 10, doubling and using near doubles.

Practice Book: There is no Practice Book page for this lesson.

Differentiated outcomes	
All students	should find pairs of numbers that total 10.
Most students	will rearrange calculations to successfully add several numbers.
Some students	may find pairs to 10 and calculate totals mentally.

Answers

Student Book page 42

Answers will vary because students choose their own route through the maze.

Check that students' calculations are correct.

2D Adding several small numbers

Explore Student Book page 43 · Practice Book page 35

Specific learning focus

- Add several small numbers.
- Reorder an addition to help with calculation.

Global skills

- **Creative skills:** problem solving

Key vocabulary

- addition, calculation

Resources

- sticky notes with numbers 1–6 written on them
- a small bag

Language support

Focus particularly on the vocabulary of 'sum' and 'difference' in this lesson. *What is the sum of these numbers? What is their difference?* Encourage students to use a sentence in their answer. For example, they might say: 'The sum of 25 and 32 is 57. The difference between 25 and 32 is 7.'

You could write this sentence on the board to offer support to all students.

Introductory activity

Pick two numbers from the bag and create a 2-digit number. Replace the numbers in the bag, then pick another four numbers. These will be four single-digit numbers. Ask students, using their whiteboards, to add the four single-digit numbers to the 2-digit number. They should keep a running total so that each calculation is adding a single-digit number to a 2-digit number. They can check the total with a partner after each addition of a single-digit number.

Demonstrate to students that they will get the same answer if they add all the single-digit numbers together and then add this to the 2-digit number. Model this calculation on an empty number line. For example, if your 2-digit number is 27 and the numbers picked are 6, 6, 6, 1, you would get the same answer by calculating '27 + 6 + 6 + 6 + 1' as you would by calculating '27 + 19'. Repeat, but this time start off with a 3-digit number.

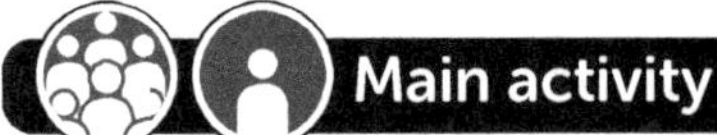

Main activity

Look together at page 43 of the Student Book. If you have access to an IWB you could use this. Ask students to look at the artwork of fruit. *How many oranges are there? Does anyone know what this fruit is called? What is this one?* Ensure that students know what each fruit is: oranges, apricots, strawberries, cherries, red/purple grapes, pineapple, bananas, apples, kiwi fruit and green grapes. Look at question 1. *What calculation do we need to write to answer this question?* (7 + 5 + 10 + 3 =) Refer students to the first speech bubble. *How can we make this calculation easier?* Draw out that looking for pairs to 10 (7 + 3) will make this calculation easier. Students should complete the rest of the activity in the Student Book individually.

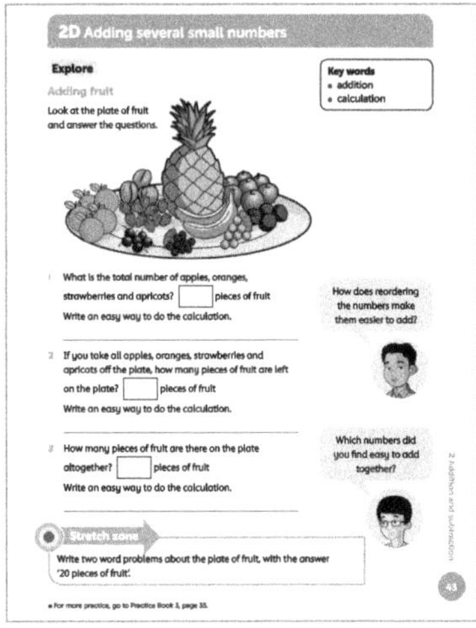

Differentiation

Supporting: Model the calculations for students, using counters or cubes, helping them count by touching or moving each counter or cube as it is counted.

Consolidating: Ask students to talk through their thinking as they find ways to add the numbers.

Extending: Challenge students to form their own calculation question about the fruit for a partner to solve.

Stretch zone: *Write two word problems about the plate of fruit, with the answer '20 pieces of fruit'.*

Check that students' problems are accurate and have the required answer. Students could swap their problems with a partner. Their partner checks that their problem does have the answer 20. *How did you make the calculation easier to do to check the answer?*

 **Reflection time**

Ask students who have written problems for the Stretch zone to share some of their problems. The rest of the class can suggest 'quick ways' of finding the answer.

Practice Book: Students complete Practice Book page 35. They can do this directly after the main activity, as homework, or as the focus of a separate mathematics session to help students consolidate their learning and build fluency.

Students write their own additions and solve them. They need to reason about which numbers to use to find the biggest and smallest totals.

Differentiated outcomes	
All students	should successfully add two single-digit numbers with support.
Most students	will add two or more single-digit numbers.
Some students	may add several single-digit numbers efficiently.

Answers

Student Book page 43

1 25

2 30

3 55

Check that students have written easy ways to do the calculations.

Practice Book page 35

Answers will vary because students make their own numbers from the digits 1, 2, 3, 4, 5 and 6 and then use these to write addition calculations.

Check that students' calculations are correct.

Stretch zone: The largest possible answer is 156 (61 + 52 + 43 or 63 + 51 + 42, for example).

The smallest possible answer is 3 (2 + 1).

2E Adding pairs of 2- and 3-digit numbers

Discover 1
Student Book page 44 • Practice Book page 36

Specific learning focus

- Add and subtract pairs of 2-digit numbers.
- Add 3-digit and 2-digit numbers using notes to support.

Global skills

- **Creative skills:** problem solving

Key vocabulary

- 2-digit number, base-10 equipment

Resources

- base 10 equipment
- digit cards 0–9

Language support

Encourage students to read the questions aloud. This reinforces the vocabulary of addition and supports students in describing numbers using place-value vocabulary. Encourage students to describe how they partition numbers into hundreds, tens and ones to do addition calculations.

 **Introductory activity**

Ask a student to choose three digit cards at random, for example 1, 3, 6. Write the digits on the board. Ask students to make a number with two of the digits and then partition it. For example, they may make 36 and tell you that it is 3 tens and 6 ones.

Repeat until all six possible **2-digit numbers** using those digits have been made. (13, 16, 31, 36, 61, 63) Repeat with a new set of digits.

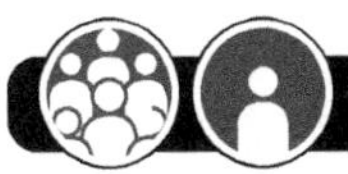 **Main activity**

Write the following calculation on the board: 42 + 37. Ask students to use **base-10 equipment** to represent each number. Check that they have used 4 tens-rods and 2 ones-cubes for 42 and 3 tens-rods and 7 ones-cubes for 37.

How many tens-rods are there altogether? (4 + 3 = 7) *How many ones-cubes are there altogether?* (2 + 7 = 9) *How many rods and cubes do you have in total?* (7 rods and 9 cubes) *So what is the total of 42 + 37?* (79).

Now use the calculation 38 + 26. Repeat the process of making each number with tens-rods and ones-cubes.

How many rods and cubes do you have in total? (5 rods and 14 cubes) *Can you exchange some of the 14 cubes to make a tens-rod and some cubes?* Students should be able to make 1 tens-rod and 4 ones-cubes. *Now how many rods and cubes do you have altogether?* (5 + 1 = 6 tens-rods, and 4 ones-cubes) *So what is the total of 38 + 26?* (64)

Ask students to complete the calculations in the Student Book on page 44. As they work, check that they are carrying out the process of exchanging cubes for rods when the total of cubes is 10 or more.

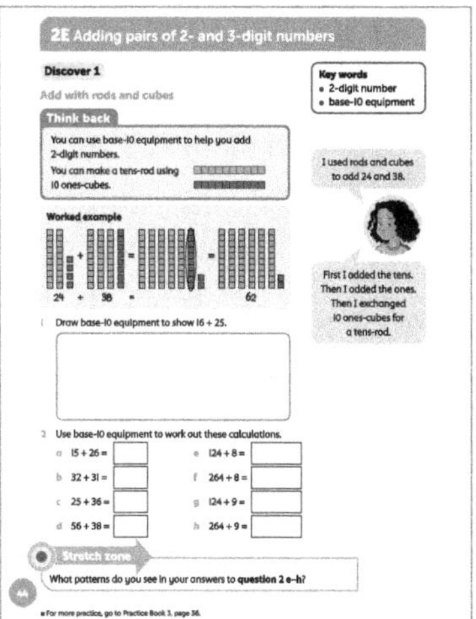

Differentiation

Supporting: Model the use of the rods and cubes to represent 2-digit numbers to help students add the numbers together.

Consolidating: Help students represent numbers and exchange 10 ones-cubes for a tens-rod where necessary.

Extending: Ask students to continue the process for 3-digit numbers.

Stretch zone: *What patterns do you see in your answers to question 2 e–h?*

Students should be able to reason that adding numbers that end in 4 and 8 will always give an answer ending in 2, for example, because 4 + 8 = 12, and that adding numbers that end in 4 and 9 will always give an answer ending in 3. They may also notice the relationship between the answers to e and g, and f and h.

 **Reflection time**

Ask students whether, when adding the ones digits in their calculations, they ever needed to take two tens-rods when they exchanged the ones-cubes for tens-rods. *Can you explain why not? What is the biggest total of cubes you would need to exchange and what does it become?* (9 + 9 = 18, which is 1 ten rod and 8 cubes)

Practice Book: Students complete Practice Book page 36. They can do this directly after the main activity, as homework, or as the focus of a separate mathematics session to help students consolidate their learning and build fluency.

Students write their own additions and solve them using partitioning. Encourage them to write some challenging calculations that involve exchanging ones for tens and tens for hundreds.

Differentiated outcomes	
All students	should use base-10 equipment to represent 2-digit numbers for adding.
Most students	will be able to carry out the calculation by totalling the rods and cubes and exchanging where necessary.
Some students	may be able to use the equipment to help them add 2-digit numbers that give a 3-digit total, or to add 3-digit numbers.

Student Book page 44

1 Check that students have drawn a total of 4 rods and 1 cube.

2 a 41 **b** 63 **c** 61 **d** 94
e 132 **f** 272 **g** 133 **h** 273

Practice Book page 36

Answers will vary depending on which digits are chosen. Check that in each case the numbers have been partitioned correctly and that the totals are correct.

Stretch zone: Check students' strategies for solving the additions.

2E Adding pairs of 2- and 3-digit numbers

Discover 2
Student Book pages 45–46 • Practice Book page 37

Specific learning focus

- Add two or more 2-digit numbers.
- Add 3-digit and 2-digit numbers using notes to support.

Global skills

- **Creative skills:** problem solving/exploring
- **Interpersonal skills:** communication

Key vocabulary

- addition fact, total, addition, calculation, 2-digit number, 3-digit number

Resources

- small pieces of scrap paper (enough for one for each student)
- base-10 equipment

Language support

Encourage students to read the questions aloud to reinforce the language of addition. Model language of place value by saying the calculation aloud, for example: *We are adding two 2-digit numbers. Each number has some tens and some ones.*

 Introductory activity

This activity is best done in a school hall or an outside space. Ask each student to write down their birth date (for example 3 if their birthday is 3 January) on their piece of scrap paper. Students should then move about the space randomly. When you say 'Stop!', they should pair up with the person nearest and calculate the total of their two numbers.

Find out which pair has:

- the largest total
- the smallest total

Check the answers and then repeat twice more, ensuring that each student finds a different pair each time. Then ask students to arrange themselves in groups of three so that the sum of their numbers is less than 45. Repeat, and ask for groups of four students with sums less than 50.

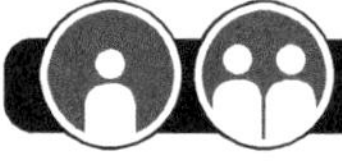 Main activity

Students should work on the activity in the Student Book page 45 individually but encourage them to talk to each other when they have finished to explain what they notice about the totals.

For question 1, encourage students to use an empty number line to support their mental calculations. Alternatively, remind students that they could use partitioning to add numbers. For example:

$$15 + 22 + 29 = 10 + 5 + 20 + 2 + 20 + 9$$
$$= 10 + 20 + 20 + 5 + 2 + 9 = 50 + 16 = 66$$

Will the total of all the rows or the total of all the columns be greatest? Convince me.

For question 2, look together at page 46 of the Student Book. If you have access to an IWB you could use this. Show students the worked example, using base-10 equipment to add the numbers.

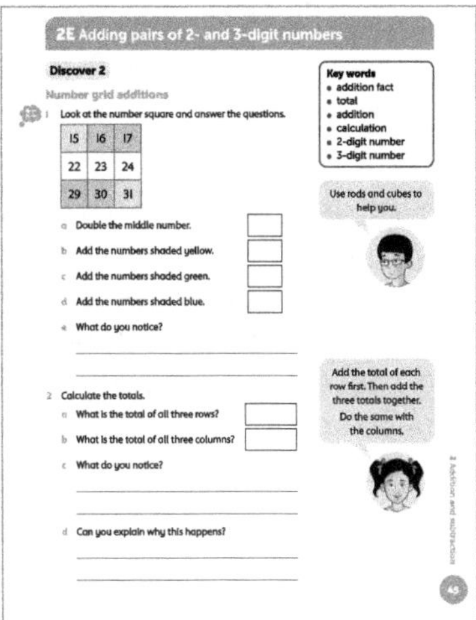

Differentiation

Supporting: Model the use of the number line, base-10 equipment and partitioning to find the totals. Students should use all methods and tell you which they think is the most efficient.

Consolidating: Ask students to describe to their partner their strategies for calculating.

Extending: Ask students to explain why they think the patterns they notice occur when adding the totals in the grid.

Stretch zone: *Which was your biggest answer? Is it possible to make a bigger answer using the same numbers?*

Which was your smallest number? Is it possible to make a smaller answer using the same numbers?

Students should reason about why they are convinced they have made the largest and smallest totals.

 **Reflection time**

Select someone who is confident in explaining their strategy for adding 2- and **3-digit numbers** to share this with the rest of the class.

Ask students to look at the total of the numbers in a row or a column in the grid that passes through the centre (middle vertical, middle horizontal and the two diagonals). *Can you explain what you notice? Can you explain why this happens?* The reason the totals of the middle row and the middle column are the same can be seen by looking at the grid in this way:

$n - 8$	$n - 7$	$n - 6$
$n - 1$	n	$n + 1$
$n + 6$	$n + 7$	$n + 8$

so we see that:

Middle row = 3n

Middle column = 3n

Both diagonals = 3n

Students may notice that the total of the middle column is three times the middle number, the total of the left-hand column is three less than this and the total of the right-hand column is three more than this.

Practice Book: Students complete Practice Book page 37. They can do this directly after the main activity, as homework, or as the focus of a separate mathematics session to help students consolidate their learning and build fluency.

Encourage students to use their knowledge of place value to calculate the totals. *Why do you think it is important to estimate your answer first?*

Differentiated outcomes	
All students	should carry out the calculations with support.
Most students	will carry out the calculations independently.
Some students	may carry out the calculations accurately and explain the patterns they notice.

Answers

Student Book pages 45–46

1 a 46

  b 46

  c 46

  d 46

  e The middle number, the sum of the yellow numbers, the sum of the green numbers and the sum of the blue numbers are all the same, 46.

2 a top row = 48, middle row = 69, lower row = 90, total = 207

  b left-hand column = 66, middle column = 69, right-hand column 72, total = 207

  c The total of the 3 columns and the total of the 3 rows are the same.

  d Students may say they think this happens because they are adding all the numbers in the grid together and so it doesn't matter what order they add them in – row by row, or column by column – the total of all the numbers will be the same.

3 Answers will vary because different numbers will be chosen by students. Check that they have added the numbers correctly.

1 97 + 86 (or 96 + 87) = 183

2 46 + 57 (or 47 + 56) = 103

3 467 + 589 = 1056

4 849 + 657 = 1506

5 985 + 76 = 1061

6 458 + 69 = 527

Stretch zone: Check students' strategies for solving the subtractions.

2E Adding pairs of 2- and 3-digit numbers

Explore 1 Student Book page 47 • Practice Book page 38

Specific learning focus

- Add pairs of 2-digit numbers.
- Add 2- and 3-digit numbers using notes to support.

Global skills

- **Creative skills:** exploring
- **Interpersonal skills:** communication

Key vocabulary

- 2-digit number, 3-digit number, running total

Resources

- set of digit cards 0–9 for each student
- set of loop cards made by photocopying and cutting out the cards from Resource sheet 2.1

Language support

Work with individual students and read out their calculations. Model the correct vocabulary carefully and ask them to say the calculations aloud.

 **Introductory activity**

Give students one loop card each, but keep the cards for 'Who has 67 + 32?' and 'I have 18'. You start the process by saying: *Who has 67 + 32?* All students work out the answers, and the student who is holding the card with the answer '99' then reads the next question. Continue until the loop is completed.

If students find this activity challenging, ask them to work in pairs and give each pair two cards. You can also make sure that you differentiate the activity by selecting particular cards for individual students.

Main activity

Students complete the activity on page 47 of the Student Book individually. Point out the importance

of writing down the **running total** each time, to keep track of the totals. Encourage students to check each other's calculations. As you move around the class, ask individuals: *How did you work that out? How do you know that is the correct answer?* Focus your attention on the students who need support when doing the calculations. You may need to model using an empty number line, using base-10 equipment or partitioning to calculate. If students are using base-10 equipment, refer them to the second speech bubble and remind them to exchange 10 ones-cubes for one tens-rods when necessary.

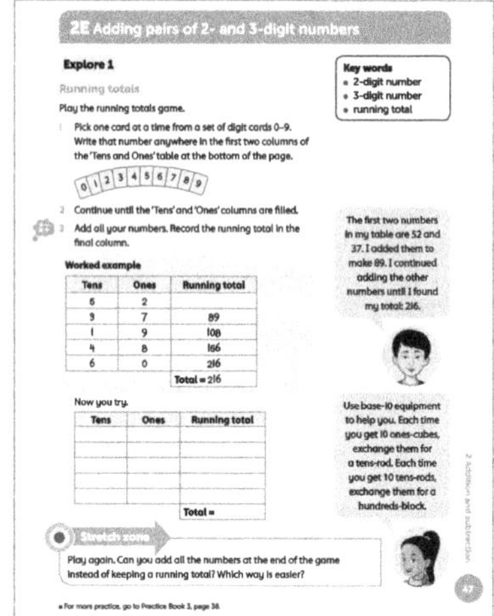

Differentiation

Supporting: Model the use of base-10 equipment, the number line and partitioning to find the totals. Students should use all three methods. *Which do you think is the most efficient?*

Consolidating: Ask students to describe their strategies for calculating to their partners.

Extending: Ask students to explain how they calculate their running total and how they can check it, perhaps by adding the numbers in a different order.

Stretch zone: *Play again. Can you add all the numbers at the end of the game instead of keeping a running total? Which way is easier?*

Students should be able to explain which method they find easier and why.

 **Reflection time**

Ask the class to check each addition, one at a time. Ask individual students to explain the methods used as clearly as they can. Support them by asking questions such as: *Which digits did you add first? How did you know that was correct? How did [the number line, base-10 equipment/partitioning] help you?*

Practice Book: Students complete Practice Book page 38. They can do this directly after the main activity, as homework, or as the focus of a separate mathematics session to help students consolidate their learning and build fluency.

You may need to explain the rules of addition pyramids before students complete the activity. *How will you make sure that the top number is close to 100 when you write your own addition pyramid?*

Differentiated outcomes	
All students	should carry out the calculations with support.
Most students	will calculate independently.
Some students	may successfully keep their running total and check by adding in a different order.

2E Adding pairs of 2- and 3-digit numbers

Explore 2 Student Book page 48 · Practice Book page 39

Specific learning focus
- Add pairs of 2-digit numbers.
- Add 2- and 3-digit numbers using notes to support.

Global skills
- **Creative skills:** investigating

Key vocabulary
- 2-digit number, 3-digit number, running total

Resources
- digit cards 0–9 for each student
- mini whiteboards and markers

Language support

Work with individual students and read out their calculations. Model the correct vocabulary carefully and ask them to say the calculations aloud. Help them reach the target number by asking: *How many more tens do we need?* and so on.

Introductory activity

Tell students that they are going to make pairs of 2-digit numbers using four different digits from a set of 0–9 cards. They should try to make two numbers that total 50 or get as close to 50 as they can (e.g. 38 + 12).

Student Book page 47

Answers will vary because students make their own numbers to add. Check that the running totals are correct. Ask students to explain the methods they used to add the numbers.

Practice Book page 38

1 Middle: 52, 72   Top: 124

2 Middle: 117, 102   Top: 219

3 Middle: 40, 51   Top: 91

4 Middle: 30, 60   Top: 90

5 Middle: 198, 198   Top: 396

Can they make 50 in other ways using four digits in two 2-digit numbers? (36 + 14, 34 + 16, 32 + 18) Ask whether they can explain why they can't make 50 in any other ways using four different digits.

Main activity

Work in pairs on the activities in the Student Book on page 48. As they build up the running total, encourage students to think about how many tens they need and what digits they have left after each line. Encourage them to use their knowledge of bonds to 10 and 100 when thinking about which digits to choose. For question 1, see how close they can get to 200 as the total.

Students could use their whiteboards to make jottings. They can then explore changing digits to see what effect it has on the total.

For question 2, ask students to think about which digits should be in the tens column to make the smallest total.

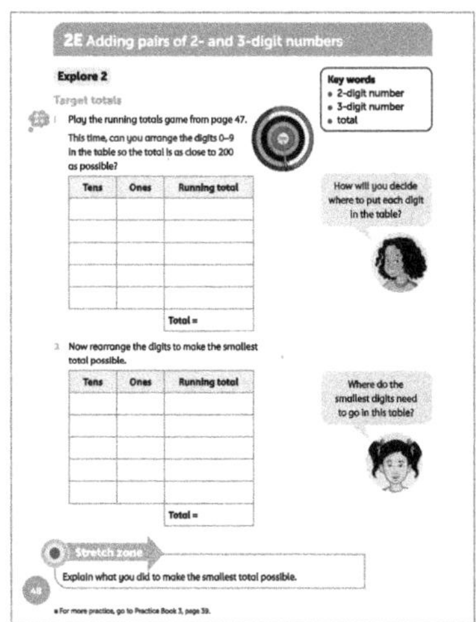

Differentiation

Supporting: Model the use of base-10 equipment to help students build their total. Suggest they start with 200 (that is, 20 tens-rods) and see whether they can divide it up into five 2-digit numbers with a different digit in each 'place'.

Consolidating: Encourage students to write down the calculation they do at each stage, along with the running total.

Extending: Ask students to explain how they know they are as near 200 as possible for the first question.

Stretch zone: *Explain what you did to make the smallest total possible.*

Students should be able to explain how they made the smallest possible total, giving reasons. *Did everyone do it the same way? Who has a different strategy?*

 **Reflection time**

Discuss students' strategies for placing the digits in the table. Did they find it easier to think about the tens or the ones when aiming for 200? Ask pairs of students to talk through their calculations and explain their reasoning at each stage.

Practice Book: Students complete Practice Book page 39. They can do this directly after the main activity, as homework, or as the focus of a separate mathematics session to help students consolidate their learning and build fluency.

Students must use what they know about subtracting tens and ones to make the targets given. *How will you decide where to place each digit? Why is it important to estimate the answer first?*

Differentiated outcomes	
All students	should choose 2-digit numbers using digit cards and add them, keeping a running total.
Most students	will be able to get a total close to the target number.
Some students	may be able to calculate mentally and reach the target number.

Answers

Student Book page 48

1 Answers may vary. Check that students have made a calculation with a total close to 200 or exactly 200. (For example, $89 + 24 + 63 + 15 + 07 = 198$)

2 Smallest total is 135, for example: $09 + 18 + 27 + 36 + 45 = 135$

Practice Book page 39

1 $98 - 45 = 53$

2 $74 - 69 = 5$, for example, but can be done other ways.

3 $958 - 467 = 491$, for example.

4 $964 - 758 = 206$

5 $987 - 45 = 942$

6 $456 - 98 = 354$

Stretch zone: Check students' strategies for solving the subtractions.

2F Adding and subtracting numbers

Discover 1
Student Book page 49 · Practice Book page 40

Specific learning focus

- Add single-digit numbers to 2- and 3-digit numbers.

Global skills

- **Creative skills:** investigating
- **Interpersonal skills:** communication

Key vocabulary

- number line, add, multiple of 10

Resources

- sets of 1–9 digit cards, one per pair of students, one large set for the teacher
- lollipop sticks in a jar – each lollipop stick has a different student's name on it

Language support

Ask students to read the calculations aloud. If they struggle, read the calculation out to them first, and ask them to repeat it.

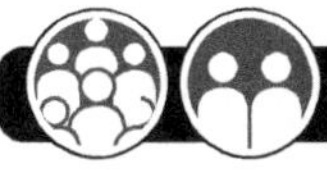

Introductory activity

Ask three students to pick a digit card from your pack. Make a 3-digit number using these digits and replace the cards in the pack. Write this number on the board. For example, if the first three students chose 5, 2, 8, you would write 528.

Then ask individual students to select another digit card. Ask pairs to add this to the 3-digit number on the board. If the student chooses the card showing 6, ask the pairs to calculate 528 + 6. Ask students the method they used. Here, they might say, 'I know that 8 + 6 = 14, so I know that 528 + 6 = 534.' Write this sentence on the board for support. Repeat several times, with new digit cards each time.

Main activity

Select five more students to pick numbers. Model adding these to the larger number. If there are examples that bridge the tens boundary, show students how they can do these additions by partitioning. For example, if the second student picks a 7, the calculation would be:

528 + 2 = 530 (to get to the next multiple of 10)

530 + 5 = 535 (to add on the remainder of the single-digit number)

Here, we partition the 7 into 5 + 2.

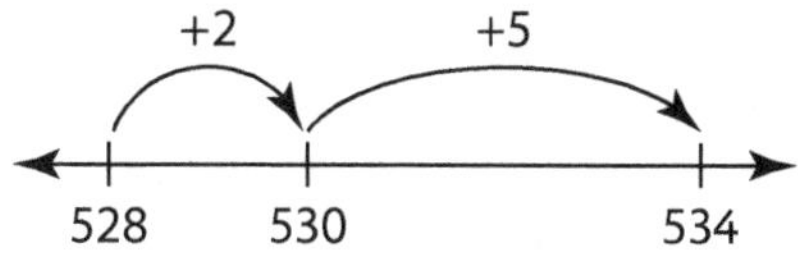

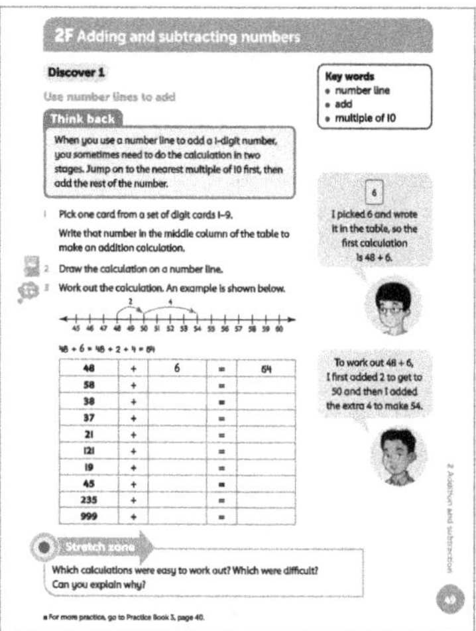

Students complete the activity on page 49 of the Student Book individually. As you move around the class, focus on those calculations which bridge a tens or a hundreds boundary. Make sure that students understand how they can use partitioning and an empty **number line** to help with the calculation. While students are working, ask them to explain how they answered particular questions. *How did you partition the numbers? For adding 9, did you add 10 and count back 1?*

Differentiation

Supporting: Model the use of the number line and partitioning to carry out the additions as many times as is needed. Talk through the steps as students draw an example for themselves.

Consolidating: Ask students to describe to their partners their strategies for adding.

Extending: Ask students to predict the answer before they carry out the calculation.

Stretch zone: *Which calculations were easy to work out? Which were difficult? Can you explain why?*

Students should be able to give you reasons for how they calculated.

 Reflection time

Select an individual by picking a lollipop stick out of the jar. Ask them to pick one of the calculations they found difficult. They should come to the front and talk through the method they used and say why they found it difficult. Support them by modelling the correct use of language or by asking questions such as: *What did you do first? What was the next step?*

Practice Book: Students complete Practice Book page 40. They can do this directly after the main activity, as homework, or as the focus of a separate mathematics session to help students consolidate their learning and build fluency.

Ensure that students notice that the calculations include additions and subtractions. *Make sure you look carefully at the calculation before you solve it. Are you adding or subtracting? Will you need to bridge a multiple of 10 or 100?*

Differentiated outcomes	
All students	should carry out the calculations with support.
Most students	will carry out the calculations independently.
Some students	may use mental methods to carry out the calculations drawing on known facts.

2F Adding and subtracting numbers

Discover 2   Student Book page 50 · Practice Book page 41

Specific learning focus

- Subtract single-digit numbers from 3-digit numbers.

Global skills

- **Creative skills:** investigating
- **Interpersonal skills:** communication

Key vocabulary

- number line, subtraction, multiple of 10

Resources

- sets of 1–9 digit cards, one per pair of students, one large set for the teacher
- lollipop sticks in a jar – each lollipop stick has a different student's name on it

Language support

Ask students to read the calculations aloud. If they struggle, read the calculation out to them first, and ask them to repeat it.

Answers

Student Book page 49

Answers will vary because students pick cards to make their own calculations.

Check that students' calculations are correct.

Practice Book page 40

1 305

2 483

3 986

4 115

5 297

6 504

 **Introductory activity**

Ask three students to pick a digit card from your pack. Make a 3-digit number using these digits and replace the cards in the pack. Write this number on the board. For example, if the first three students chose 5, 2, 8, you would write 528.

Then ask individual students to select a card. Ask pairs to subtract this from the 3-digit number on the board. If the student chooses the card showing 6, ask the pairs to calculate 528 − 6. Ask students the method they used. Here, they might say, 'I know that 8 − 6 = 2, so I know that 528 − 6 = 522.' Write this sentence on the board for support.

 **Main activity**

Select five more students to pick numbers and subtract each one of these from the larger number. If there are examples that bridge across the tens boundary, show students how they can do these subtractions by partitioning. For example, if the second student picked a 9, the calculation would be:

528 − 8 = 520 (to get to the nearest multiple of 10)

520 − 1 = 519 (to subtract the remainder of the single-digit number)

Here, we partition the '9' into '8 + 1'. See below:

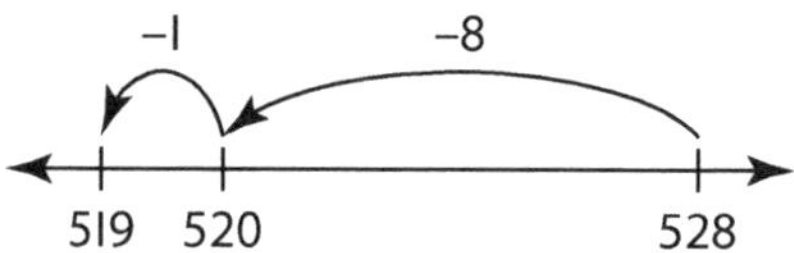

Students complete the activity on page 50 of the Student Book individually. As you move around the class, focus on those calculations that bridge across a tens or a hundreds boundary. Make sure that students understand how they can use partitioning and an empty number line to help with the calculation. While students are working, ask them to explain how they answered particular questions. How did they partition the numbers? For subtracting 9, did they subtract 10 and add 1?

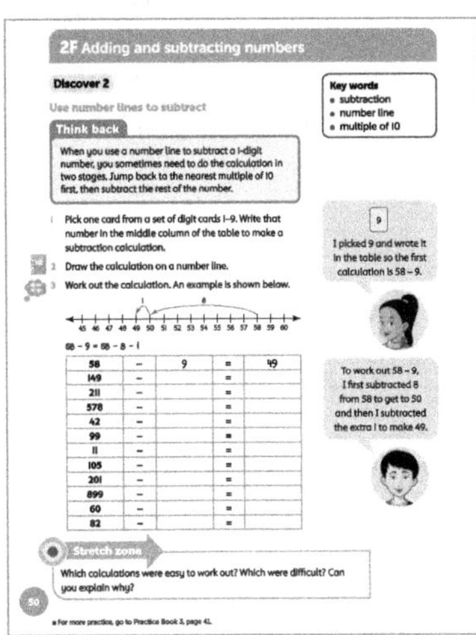

Differentiation

Supporting: Model the use of the number line and partitioning to carry out the subtractions as many times as is needed. Talk through the steps as students draw an example for themselves.

Consolidating: Ask students to describe their strategies for subtracting to their partners.

Extending: Ask students to predict the answer before they carry out the calculation.

Stretch zone: *Which calculations were easy to work out? Which were difficult? Can you explain why?*

Students should be able to give you reasons for how they calculated.

 Reflection time

Select an individual by picking a lollipop stick out of the jar. Ask them to pick one of the calculations that they found difficult. They should come to the front and talk through the method they used and say why they found it difficult. Support them by modelling the correct use of language or by asking questions such as: *What did you do first? What was the next step?*

Practice Book: Ask students to look at Rosa's calculations and use a number line to check whether her answers are correct or not. Can students explain where Rosa made her mistakes?

Students complete Practice Book page 41. They can do this directly after the main activity, as homework, or as the focus of a separate mathematics session to help students consolidate their learning and build fluency. Note: it may be more suitable to do the second part of this activity once students have done Unit 2G and are familiar with the column method.

Differentiated outcomes	
All students	should carry out the calculations with support.
Most students	will carry out the calculations independently.
Some students	may use mental methods to carry out the calculations drawing on known facts.

Answers

Student Book page 50

Answers will vary because students will choose different digits to subtract. Check that each line is correct for the digits used.

Practice Book page 41

1 **a** Marked with a cross   **b** Marked with a cross
   c Ticked

2 **a** Rosa wrote the carried-over hundreds from the tens column directly into the hundreds column (instead of carrying them and then adding them on to the other hundreds). She then wrote the hundreds in the thousands column.

   b Rosa carried over the hundreds to the hundreds column but she also wrote them in the main answer of the hundreds column, so she got the same answer as in a.

3 **d** Marked with a cross   **e** Marked with a cross
   f Marked with a cross

4 **d** Rosa subtracted the smallest digit from the largest digit in each column each time, regardless of which number it was part of.

   e Rosa did not subtract the hundreds correctly.

   f Rosa did not exchange a ten for 10 ones but exchanged a hundred for 10 ones.

2F Adding and subtracting numbers

Explore 1
Student Book page 51 · Practice Book page 42

Specific learning focus
- Add and subtract pairs of 2-digit numbers.
- Add and subtract single-digit numbers from 3-digit numbers.

Global skills
- **Creative skills:** problem solving
- **Real-world skills:** interpreting information/financial literacy
- **Interpersonal skills:** communication

Key vocabulary
- addition calculation, subtraction calculation, altogether, difference

Resources
- base-10 equipment

Language support
Ask students to read aloud their calculations as they work. Reinforce the language of addition and subtraction, including 'altogether' and 'difference'.

 **Introductory activity**

Show students the following information on the board, based on a traffic survey:

Cars – 58   Lorries – 35   Motor bikes – 29
Buses – 17   Tractors – 9

Ask students how they would calculate how many cars and buses there were altogether. What calculation would they use? (58 + 17) Work this out using base-10 equipment or a number line (75) and describe the method.

Now ask how many more lorries there were than motor bikes. *How can you work this this out? What calculation would you use?* (35 – 29). Work this out using base-10 equipment or a number line (6) and describe the method.

Ask students to make up one **addition calculation** and one **subtraction calculation** for their partner, based on the traffic survey data, and give them time to do the calculations using a strategy of their choice.

 **Main activity**

Look together at page 51 of the Student Book. If you have access to an IWB you could use this. Show students the worked example, which demonstrates the different strategies that students could use to work out the calculations. Students should work in pairs to answer the questions on the Student Book page. They should also check each other's work. While students work on this activity, move around the room and pick two different pairs who are able to explain their thinking. Make base-10 equipment and number lines available for students to support their calculations.

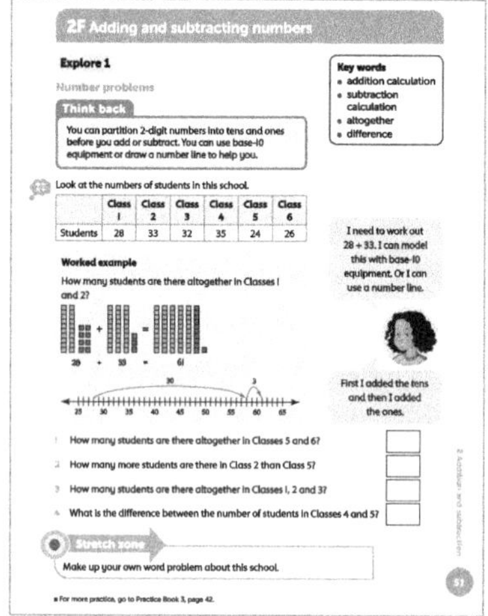

Differentiation
Supporting: You can help write the calculation for students who tell you their answers. Check the answers using practical materials.

Consolidating: Ask students to explain their strategies for checking the calculation.

Extending: Encourage students to create two-step problems. For example: how many of the cars and lorries can fit on a ferry if 11 of them have to wait for the next ferry?

Stretch zone: *Make up your own word problem about this school.*

Students should be able to make some suitable problems that they or a partner can answer using a range of addition or subtraction calculations.

 **Reflection time**

Ask pairs of students to explain how they answered each of the questions in the Student Book on page 51. *How did you decide which type of calculation to use? Which words in the questions helped you decide which calculation to use?*

Practice Book: Students complete Practice Book page 42. They can do this directly after the main activity, as homework, or as the focus of a separate mathematics session to help students consolidate their learning and build fluency.

Differentiated outcomes	
All students	should be able to partition 2-digit numbers to help them solve addition or subtraction calculations.
Most students	will be able to carry out the calculations using number lines or base-10 equipment.
Some students	may be able to write their own problems involving addition or subtraction.

Answers

Student Book page 51

1	50	**3**	93
2	9	**4**	11

Practice Book page 42

1	290	**5**	574
2	478	**6**	260
3	381	**7**	378
4	655	**8**	808

2F Adding and subtracting numbers

Explore 2
Student Book pages 52–53 · Practice Book page 43

Specific learning focus

- Add and subtract pairs of 2-digit numbers.
- Add and subtract single-digit numbers from 3-digit numbers.

Global skills

- **Creative skills:** problem solving
- **Real-world skills:** interpreting information/ financial literacy
- **Interpersonal skills:** communication

Key vocabulary

- number story, word problem

Resources

- none needed

Language support

Encourage students to tell you the number story before they try to write it down. You can help write the answers for these students. Write down the number story as a number sentence and encourage students to read the number sentence aloud.

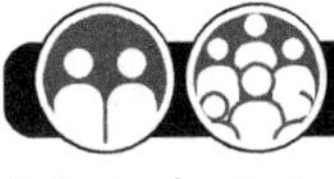 Introductory activity

Ask students to look at question 1 in the Student Book on pages 52–53. Each pair should make up a question based on the image. Explain that they do not have to use the same numbers that are in the worked example. Take feedback – make sure that you find at least three

different stories to match the image. Each pair should show how they calculated the answer and should model the calculation on a number line. Make sure that each pair checks the calculation by carrying out the inverse operation (if they have carried out an addition they can check by subtracting, and they can check a subtraction by adding). On the board, model how to check the answer using a number line.

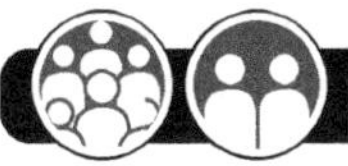 Main activity

Look together at page 52 of the Student Book. If you have access to an IWB you could use this. Show students the worked example so students are familiar with what they need to do. Students should work in pairs to discuss possible **number stories** for each image, select the one they think is best and write it in the Student Book. They can also check each other's work. While students work on this activity, move around the room and pick different pairs who are able to explain their thinking. Encourage students to come up with a range of different stories using both addition and subtraction. *Can you think of a different story for that image? And another? If that is your addition story, can you tell me a subtraction story using the same numbers?*

When pairs have written their number stories, they should swap books with another pair and solve each other's number problems.

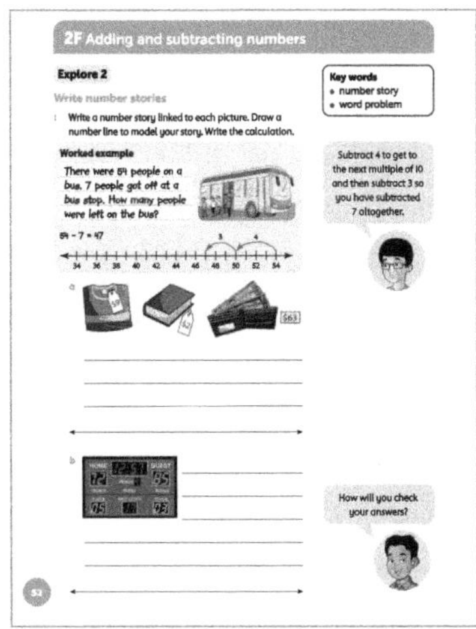

Differentiation

Supporting: You can help write the calculation for students who tell you their number stories. Check the answer using concrete materials.

Consolidating: Ask students to tell you their strategies for checking the calculation.

Extending: Encourage students to create two-step problems.

Stretch zone: *Make up your own number story. Draw a picture, write the story and then draw the story on a number line.*

Encourage students to draw their own picture on which to base their number story. Check students' stories for accuracy and ensure that their number line shows the calculation correctly.

 **Reflection time**

Choose two pairs of students to share one of their number stories with the rest of the class. As they do this, model the correct use of vocabulary. Write on the board the calculation needed to solve the number story and model the calculation using a number line. Ask a student to check your calculation using an inverse calculation.

Practice Book: Students complete Practice Book page 43. They can do this directly after the main activity, as homework, or as the focus of a separate mathematics session to help students consolidate their learning and build fluency. Encourage students to write a mixture of single-step problems using addition or subtraction and two-step problems using one or both operations, using the items from the shop. Notice that this task introduces the phrase '**word problem**'.

Differentiated outcomes	
All students	should write simple number problems using either addition or subtraction.
Most students	will write both addition and subtraction number problems and use the inverse calculation to check their answers.
Some students	may be able to write two-step problems.

Answers

Student Book pages 52–53

Answers will vary because students make up their own number stories for each image, choosing the numbers for their calculations and whether to use addition or subtraction.

Check that students' number stories fit the images and that their calculations are correct.

Practice Book page 43

Answers will vary because students make up their own number stories for the items and the available dollar bills. Check that students' calculations are correct.

Stretch zone: Check students' strategies for solving the calculations.

2G Adding pairs of 3-digit numbers

Specific learning focus

- Add pairs of 3-digit numbers.

Global skills

- **Creative skills:** problem solving
- **Real-world skills:** interpreting information/ financial literacy

Key vocabulary

- word problem, estimate

Resources

- place-value counters or base-10 equipment

Language support

Remind students of the language associated with adding and subtracting such as base-10 equipment: ones-cubes, tens-rods, hundreds-blocks and thousands-blocks. Say:

- You can exchange 10 ones-cubes for 1 tens-rod.
- You can exchange 10 tens-rods for 1 hundreds-block.
- You can exchange 10 hundreds-blocks for 1 thousands-block.

Students will use this knowledge when learning the compact method of addition. While they will not physically use the base-10 equipment to add and subtract, they will still be 'exchanging' in the same way.

 **Introductory activity**

Explain that, so far, students have added and subtracted using mental methods, supported by concrete resources, such as base-10 equipment, or by jottings, such as number lines. Explain that they can also use a written method to solve additions and subtractions.

Write this calculation on the board:

```
   57
+  28
```

Model how to solve the calculation in two different ways:

```
50  +   7            57
20  +   8        +   28
70  +  15            85
                     1
```

Answer: 85

The first method uses partitioning. The second method is called the compact method. Explain that the small 1 represents the one 10 that is exchanged from the ones (10 ones) to the tens. *Adding the ones gives 15 ones. You can exchange 10 of the ones cubes to give 1 tens-rod and 5 ones-cubes. This gives the answer 85 as before.* Model the calculation using base-10 equipment. Using practical equipment, or visual images, helps students to explain the method.

 **Main activity**

Repeat the introductory activity with more 2-digit and 3-digit numbers. Monitor students' confidence with the method. Ask a more-confident student to demonstrate the compact column method, talking through what happens. Then ask another student to demonstrate each step with the base-10 equipment. When you think that they have the confidence to work on their own, ask students to complete the activities in the Student Book on page 54. You may choose to work with a small group of the less-confident students to continue modelling the use of base-10 equipment. Look together at page 54 of the Student Book. If you have access to an IWB you could use this. Talk through the worked example before students answer the questions.

Encourage students to **estimate** the answer to each question before they solve it. This helps them to see whether their answer is reasonable. For example, in the worked example both values are less than $200, so we know that our total must be less than $400.

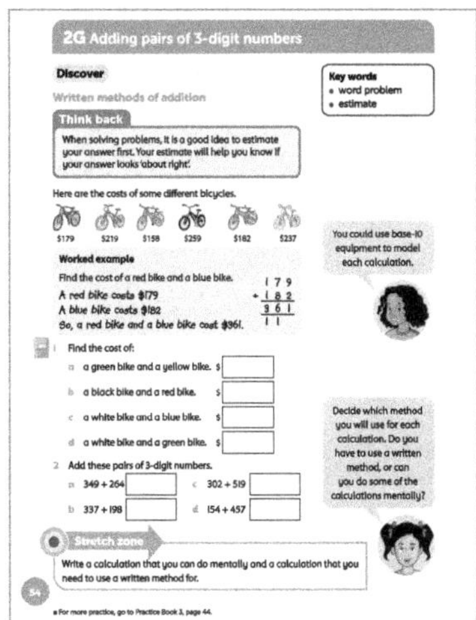

Differentiation

Supporting: Students should continue to use base-10 equipment to support their written calculations.

Consolidating: Students should choose to use the partitioning written method or the compact method to solve calculations.

Extending: Students should use the compact method and explain the process.

Stretch zone: *Write a calculation that you can do mentally and a calculation that you need to use a written method for.*

Check that students have calculated correctly and that they have made sensible decisions about working mentally and using a written method, depending on the numbers involved.

 Reflection time

Revisit the bikes question from the Student Book. Here are two ways of recording the cost of the two most expensive bikes, by partitioning and using the compact method:

```
$                                    $
259 ──▶ 200  +  50  +  9            2 5 9
+237 ──▶ 200  +  30  +  7          +2 3 7
        ─────────────────          ──────
        400  +  80  +  16           4 9 6
        400  +  90  +  6  ──▶496       1
```

Ask students to visualise adding the ones by exchanging the 16 ones-cubes for a tens-rod and six ones-cubes. *Which method do you prefer? Why?*

Practice Book: Students complete Practice Book page 44. They can do this directly after the main activity, as homework, or as the focus of a separate mathematics session to help students consolidate their learning and build fluency.

Remind students about the importance of estimating first. They are asked to use a partitioning written method to solve the calculations, but they may also use base-10 equipment if needed.

Differentiated outcomes	
All students	should use base-10 equipment to help them solve addition calculations by written methods.
Most students	will solve addition problems using a written method.
Some students	may solve addition calculations using a compact written method.

Answers

Student Book page 54

1 a $377

 b $438

 c $419

 d $456

2 a 613

 b 535

 c 821

 d 611

Practice Book page 44

1 Estimate: 350 + 430 = 780
Answer: 773

2 Estimate: 410 + 230 = 640
Answer: 640

3 Estimate: 510 + 180 = 690
Answer: 698

4 Estimate: 460 + 310 = 770
Answer: 770

5 Estimate: 310 + 220 = 530
Answer: 528

6 Estimate: 190 + 210 = 400
Answer: 401

2G Adding pairs of 3-digit numbers

Explore Student Book page 55 • Practice Book page 45

Specific learning focus

- Add pairs of 3-digit numbers.
- Recognise odd and even numbers.

Global skills

- **Creative skills:** problem solving
- **Interpersonal skills:** communication

Key vocabulary

- estimate, total

Resources

- place-value counters or base-10 equipment

Language support

Remind students of the language associated with adding and subtracting such as base-10 equipment: ones-cubes, tens-rods, hundreds-blocks and thousands-blocks.

- You can exchange 10 ones for 1 ten.
- You can exchange 10 tens for 1 hundred.
- You can exchange 10 hundreds for 1 thousand.

They will use this knowledge when learning the compact method of addition. While they will not physically use the base-10 equipment to add and subtract, they will still be 'exchanging' in the same way.

 Introductory activity

Ask students to work in pairs to explain to each other what an even number is and what an odd number is. After one minute, take feedback from the group and make sure that all students understand the difference between an even and an odd number.

Ask each student to write their birth date (e.g. 28 if their birthday is 28 March) on a small piece of paper. They should move around the class and when you ask them to stop, they should find a partner. Ask them to mentally calculate the sum of and the difference between their two birth dates. Repeat three times. During the next round, ask students to select a partner to form a pair with, so that the sum of their birth dates is an even number. Ask pairs what they notice about their two birth dates.

- Repeat to form an odd total.
- Repeat to form an even difference.
- Repeat to form an odd difference.

 Main activity

Ask students what they notice about the total when adding two even numbers, two odd numbers, or an odd and an even number. Students should notice that adding two even numbers or adding two odd numbers gives an even total. Adding an odd and an even number gives an odd total. Similarly, the difference between two odds or two evens is even whereas the difference between an odd and an even is odd.

Ask students whether the rules for adding odd and even numbers work for 3-digit numbers. Give them 124 + 253 to test, then ask them to make up examples of their own.

Look together at page 55 of the Student Book. If you have access to an IWB you could use this. Ask students to work in pairs to complete the activities. Talk through the worked example before students answer the questions. Encourage the use of the compact written method but allow students to use the written partitioning method if they prefer. Again, you may choose to work with a small group of the less-confident students to continue modelling the use of base-10 equipment and to support them in understanding how to use the compact method for recording. Refer students to the first speech bubble on the Student Book page, which reminds them that they can use an inverse calculation to check answers.

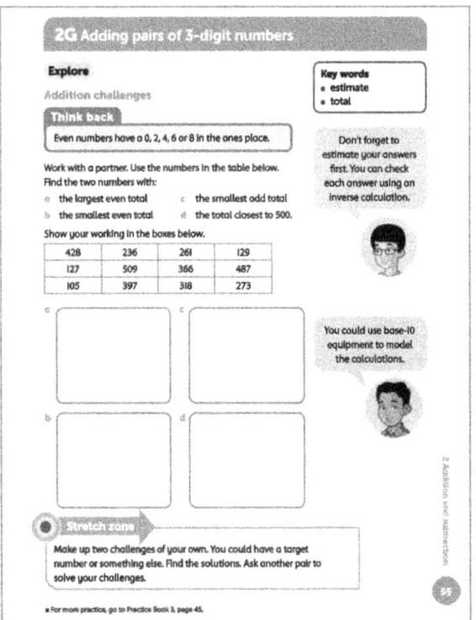

Differentiation

Supporting: Student should use base-10 equipment to support them with the exchanging of ones for tens and tens for hundreds.

Consolidating: Encourage students to use the compact method, or written partitioning, to model calculations.

Extending: Students should use the compact method and explain the process.

Stretch zone: *Make up two challenges of your own. You could have a target number or something else. Find the solutions. Ask another pair to solve your challenges.*

Check that students made suitable challenges and solved them correctly. An example could be: Use any two numbers from the box to get a total as close to 800/1000 as possible.

 **Reflection time**

Share some students' answers to the Student Book questions. *How did you decide which numbers to add to reach your target totals? How did you know those two numbers would make an even number?* Ask confident students to model the use of the compact method for the rest of the class.

Practice Book: Students complete Practice Book page 45. They can do this directly after the main activity, as homework, or as the focus of a separate mathematics session to help students consolidate their learning and build fluency.

Did your work in the Student Book mean that you got quicker at finding the target numbers? Can you explain how you choose which numbers to add?

Differentiated outcomes	
All students	should carry out written additions of odd and even numbers using base-10 equipment for support.
Most students	will begin to understand the use of the compact method to add and subtract odd and even numbers and know whether the answer will be odd or even.
Some students	may use and explain the compact method to find largest and smallest odd and even totals of odd and even numbers.

Student Book page 55

1 a $509 + 487 = 996$

   b $127 + 105 = 232$

   c $105 + 236 = 341$

   d $397 + 105 = 502$

Practice Book page 45

1 $611 + 527 = 1138$

2 $129 + 107 = 236$

3 $622 + 611 = 1233$

4 $218 + 107 = 325$

Stretch zone: $527 + 468 = 995$

2H Subtracting 2- and 3-digit numbers

Discover
Student Book page 56 • Practice Book page 46

Specific learning focus
- Subtract pairs of 3-digit numbers.

Global skills
- **Creative skills:** problem solving
- **Real-world skills:** financial literacy
- **Interpersonal skills:** communication

Key vocabulary
- estimate, check, subtraction, strategy, word problem

Resources
- base-10 equipment

Language support

Remind students of the language associated with adding and subtracting such as base-10 equipment: ones-cubes, tens-rods, hundreds-blocks and thousands-blocks. Focus on the language relevant to 'exchanging' that is used when modelling the subtraction using the base-10 equipment and also when using a written method.

- You can exchange 10 ones for 1 ten.
- You can exchange 10 tens for 1 hundred.
- You can exchange 10 hundreds for 1 thousand.

Introductory activity

You can build effectively on student's knowledge of place value and the concept of 'exchange' between 1 ten and 10 ones, using base-10 equipment. Reverse the process they learned for addition. Start with subtracting a 2-digit number from a 3-digit number. In pairs, ask students to use the base-10 equipment to calculate $372 - 38$.

Both students make the numbers using the base-10 equipment. To begin the calculation, they must exchange one of the tens-rods that make the 70 so they have $300 + 60 + 12$. They can now carry out this calculation, using either a written partitioning method or a compact written method:

$$372 \longrightarrow 300 + 60 + 12$$
$$-38 \longrightarrow 30 + 8$$
$$\overline{ 300 + 30 + 4 = 334}$$

$$3 \; {}^{6}\!\!\not{7} \; {}^{1}2$$
$$- \; 3 \; 8$$
$$\overline{3 \; 3 \; 4}$$

Repeat for $257 - 28$ and then for $372 - 148$.

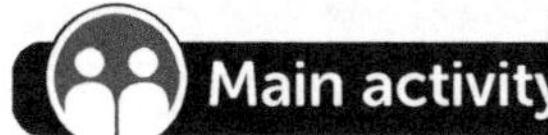

Main activity

Ask students to work in pairs to complete the activities in the Student Book. Again, you may choose to work with a small group of the less-confident students to continue modelling the use of base-10 equipment and to support them in understanding how use one of the written methods for recording. Encourage students to use the compact method if they are confident.

Encourage students to **check** their calculations using an addition calculation. You could refer them to the example given in the first speech bubble in the Student Book. *What calculation would you do to check your answer to 850 – 245? Can you explain why?*

Encourage pairs to discuss their strategies with a partner, whether they are using base-10 equipment or a written method. Their partner should do an alternative calculation to check their answer.

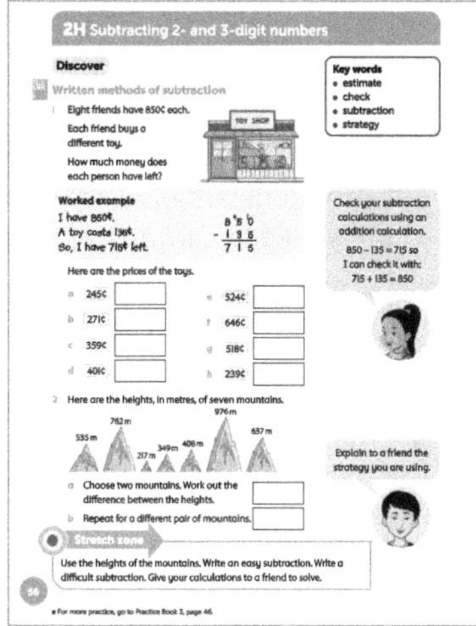

Differentiation

Supporting: Model the calculations using base-10 equipment and encourage students to use base-10 equipment until they are confident with exchanging tens for ones or hundreds for tens.

Consolidating: Encourage students to use base-10 equipment to partition numbers when doing calculations. Model how to move on to a written subtraction method.

Extending: Students should use the compact method and explain the process, and check their calculations using an inverse calculation.

Stretch zone: *Use the heights of the mountains. Write an easy subtraction. Write a difficult subtraction. Give your calculations to a friend to solve.*

Students should be able to write suitable subtraction calculations and solve each other's calculations using a **strategy** they are confident with.

 **Reflection time**

Ask pairs of students to work on this question: What is the difference in height between the lowest and highest mountain in question 2? Model the answer using base-10 equipment and the compact method.

Ask pairs of students to create a **word problem** that involves subtraction and has the answer 150. *How did you decide on the numbers to use? How did you know what your ones digits needed to be?*

Practice Book: Students complete Practice Book page 46. They can do this directly after the main activity, as homework, or as the focus of a separate mathematics session to help students consolidate their learning and build fluency.

Explain that estimating the answers before working them out is vital, especially when students come to work with larger numbers and decimals. It allows them to see whether their answer looks 'about right'. Students can also use estimates when checking with inverse calculations.

Differentiated outcomes	
All students	should be able to carry out the calculations using base-10 equipment.
Most students	will begin to understand the use of the compact method.
Some students	may use and explain the compact method and check using an inverse calculation.

Student Book page 56

1 a 605   **e** 326

   b 579   **f** 204

   c 491   **g** 332

   d 449   **h** 611

3 Answers will vary because students choose which two heights to compare.

Practice Book page 46

1 Estimate: $460 - 220 = 240$
Answer: 241

2 Estimate: $960 - 430 = 530$
Answer: 535

3 Estimate: $910 - 310 = 600$
Answer: 603

4 Estimate: $580 - 250 = 330$
Answer: 334

5 Estimate: $650 - 100 = 550$
Answer: 548

6 Estimate: $150 - 50 = 100$
Answer: 101

2H Subtracting 2- and 3-digit numbers

Specific learning focus

- Subtract a 2-digit number from a 3-digit number.
- Subtract pairs of 3-digit numbers.

Global skills

- **Creative skills:** problem solving
- **Real-world skills:** interpreting information
- **Interpersonal skills:** communication

Key vocabulary

- estimate, check, subtraction, strategy

Resources

- base-10 equipment

Language support

Remind students of the language associated with adding and subtracting, such as base-10 equipment: ones-cubes, tens-rods, hundreds-blocks and thousands-blocks. Focus on the language relevant to 'exchanging' that is used when modelling the subtraction using the base-10 equipment and also when using a written method.

- You can exchange 10 ones for 1 ten.
- You can exchange 10 tens for 1 hundred.
- You can exchange 10 hundreds for 1 thousand.

 **Introductory activity**

Model an example of the compact method of subtraction involving 3-digit and 2-digit numbers. At each stage of recording, use the base-10 equipment to explain what is happening. Practise putting the calculations into context to help students with the context used in the Student Book. For example:

$225 - 176 = 49$

A plane had 225 passengers, 176 were adults. How many were children? (49 were children)

 **Main activity**

Ask students to work in pairs to complete the activities in the Student Book. You may choose to work with a small group of the less-confident students to continue

modelling the use of base-10 equipment and to support them in understanding how to use the compact method for recording.

Direct students to the speech bubble, encouraging them to estimate their answers first to check that they look about right. Go through an example, such as: 269 is approximately 270 and 94 is approximately 100 so 269 − 94 is going to be around 170.

Discuss with students whether looking at numbers in a context such as this makes the calculations easier to do, or to understand. *How they are different from straightforward calculations?* Ask students to think of other contexts for subtraction calculations.

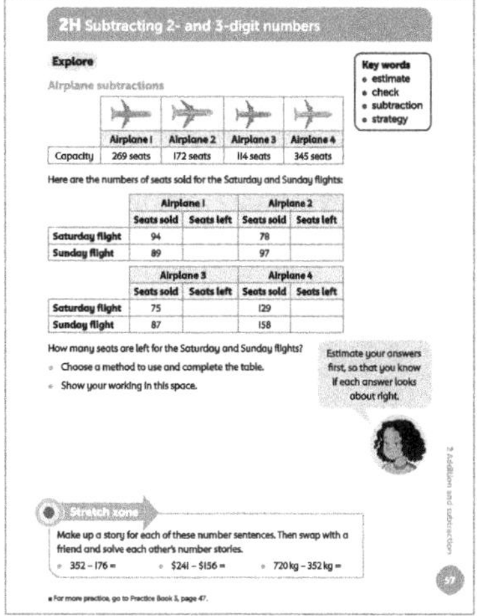

Differentiation

Supporting: Model the calculations using base-10 equipment and encourage students to use base-10 equipment until they are confident with exchanging tens for ones or hundreds for tens.

Consolidating: Encourage students to continue to use base-10 equipment to partition numbers when doing calculations. Model the written subtraction method.

Extending: Students should continue to use the compact method and check their subtraction calculations using an inverse calculation.

Stretch zone: *Make up a story for each of these number sentences. Then swap with a friend and solve each other's number stories.*

$352 - 176 =$

$\$241 - \$156 =$

$720\,\text{kg} - 352\,\text{kg} =$

Check that students have made up appropriate stories and can solve the ones their partner provides.

 **Reflection time**

Select two or three of the word problems from the main activity to go through. Students should explain their methods and the rest of the class can check them in pairs.

Practice Book: Students complete Practice Book page 47. They can do this directly after the main activity, as homework, or as the focus of a separate mathematics session to help students consolidate their learning and build fluency.

Ensure that students know to look at the ones digits when thinking about odd and even target numbers. *If I choose 622 and 102, will my answer be odd or even? How do you know?*

Differentiated outcomes	
All students	should be able to carry out calculations using base-10 equipment.
Most students	will begin to understand the use of the compact method.
Some students	may use and explain the compact method and begin to estimate their answers before calculating.

21 Addition and subtraction puzzles

Discover Student Book page 58 • Practice Book page 48

Specific learning focus

- Use addition and subtraction of numbers up to 3-digits in solving puzzles and problems.

Global skills

- **Creative skills:** problem solving
- **Interpersonal skills:** communication

Key vocabulary

- addition fact, total

Resources

- none needed

Language support

Look out for opportunities to practise and reinforce the language of addition and subtraction in problem-solving contexts. Help students by asking them questions such as *Which calculation can help you here?* Talk through with students the vocabulary associated with 'magic squares' such as 'row', 'column', 'diagonal', 'total'.

Answers

Student Book page 57

1	Plane 1		Plane 2		Plane 3		Plane 4	
	Seats sold	Seats left	Seats sold	Seats left	Seats sold	Seats left	Seats sold	Seats left
Saturday flight	94	175	78	94	75	39	129	216
Sunday flight	89	180	97	75	87	27	158	187

Practice Book page 47

1 611 − 107 = 504

2 320 − 318 = 2

3 622 − 107 = 515

4 622 − 611 = 11

5 611 − 412 = 199

6 468 − 412 = 56

 **Introductory activity**

Play the game of 21. Explain that students will take turns adding 1, 2 or 3 to a running total, with the aim of being the person to 'land on 21'. Each time a student gets to 21 on their turn, they score a point. See who has the most points after several rounds.

For example:

 **Main activity**

Show students a square grid made up of 3 × 3 squares, and use digits 1–9. Ask students to see whether they can write the digits in the grid so that each row, column and diagonal add up to 15.

Encourage them to use a trial-and-error approach. If a row does not work, they can change the digits and try again.

Students complete the activities on Student Book page 58. The first magic square is the same as the one used in the main activity above. How quickly do students realise that using bonds to 10, with 5 in the centre square, will allow them to solve this puzzle? Encourage them to use a similar strategy of pairing lower numbers with higher numbers when they try to solve the larger puzzle, and looking for rows, columns or diagonals with two or three numbers already filled.

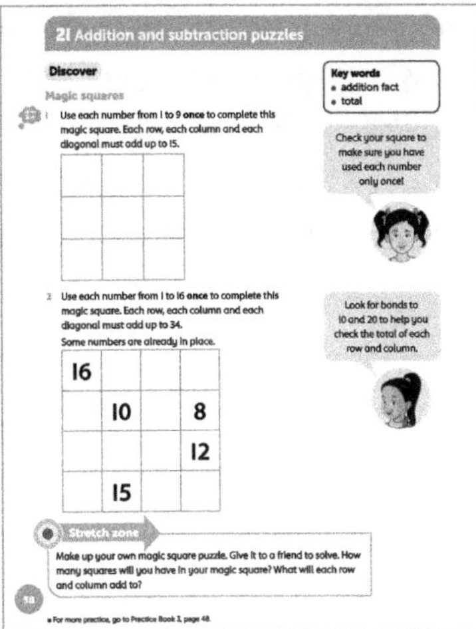

Differentiation

Supporting: Help students with the different row and column totals to check that they are all 15.

Consolidating: Ask students to explore the magic squares for patterns and connections.

Extending: Encourage students to complete magic squares on a 3 × 3 and 4 × 4 grid for different 'magic' totals.

Stretch zone: *Make up your own magic square puzzle. Give it to a friend to solve. How many squares will you have in your magic square? What will each row and column add to?*

Students can explore the patterns in magic squares to find the connections between the digits used, the centre number in odd-sized squares, and the row and column totals. Complex formula exist for solving 4 × 4 magic squares, but students should use the numbers already filled and then 'trial and error' to complete the square.

 **Reflection time**

Ask students to explain how they decided where to put each digit, and how they checked to see that all rows and columns made the correct total. If the totals were not correct, what strategies did they try to correct them? For example, did they try to see the effect of swapping two digits in the same column or row?

Practice Book: Students complete Practice Book page 48. They can do this directly after the main activity, as homework, or as the focus of a separate mathematics session to help students consolidate their learning and build fluency.

Students apply what they have learned in the Student Book to solve and create magic squares of their own. Ask them to explain how they created their magic squares.

Differentiated outcomes	
All students	can place digits to make some of the rows or columns total 15.
Most students	will complete the entire magic square for 15.
Some students	may be able to use similar methods to complete larger magic squares with different totals and explain how they solved them.

Answers

For a 3 × 3 magic square using digits 1–9, each row, column and diagonal add to 15. Since the centre square is 5, then each pair around the centre must add to 10.

Student Book page 58

1 There are many correct versions for this magic square. Check that every row, column and diagonal total 15.

2 There are many correct versions for this magic square. Check that every row, column and diagonal total 34.

Practice Book page 48

There are many correct versions for this magic square. Check that every row, column and diagonal total 34.

Stretch zone: Check that students making up their own magic square have got the totals correct.

21 Addition and subtraction puzzles

Specific learning focus

- Use addition and subtraction of numbers up to 3-digits in solving puzzles and problems

Global skills

- **Creative skills:** problem solving
- **Interpersonal skills:** communication

Key vocabulary

- addition, total, what if?

Resources

- digit cards 1–9, one for each pair

Language support

To help students with filling the magic square, refresh with them the vocabulary of the 3 × 3 grid, for example 'row', 'column', 'diagonal' and 'total'. Encourage students to use the language associated with reasoning. So, for example, when you are playing the game with a partner, and thinking about where to place the numbers, use some 'What if' questions such as *If I do this, what can they do?*, and *What if I put my number there?* Model questions such as these for students.

Introductory activity

Place some counters in three separate piles. The number in each pile does not matter. Explain to students that you are going to teach them how to play a version of the game of Nim. Ask a student to come out to play against you. The rules are that you each take turns to choose a pile of counters, then remove 1, 2 or 3 counters from that pile. Continue to choose a pile and remove counters each time it is your turn until only one counter remains. In this version, the person left with the last counter loses the game.

Main activity

Direct students to the Student Book page 59. Each pair of students needs a set of digit cards 1–9 between them. The rules are given on page 59 and the game is similar to the magic square activity seen previously. However, in this version as soon as a player makes a line of 15, that player is the winner. Encourage students to think ahead before they place their digit each time, trying to make 15 but also prevent their partner being able to make 15.

What if you write 5 in that square, will that help your partner? What if your partner writes 8 in that square, how does that affect the total for that row?

After playing the 15 game twice, students work in pairs to complete the calculations in question 3, using their digit cards.

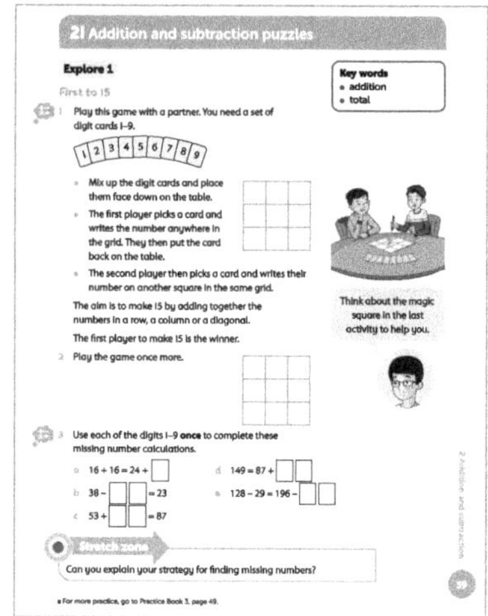

Differentiation

Supporting: Model the use of 'What if?' questions to help students decide where to place their numbers.

Consolidating: Ask students to describe their strategies to win the 15 game. For example, are they looking for pairs that make 10?

Extending: Encourage students to use good reasoning skills to win the games and place the numbers correctly.

Stretch zone: *Can you explain your strategy for finding missing numbers?*

Students should be able to reason about how they arranged the digit cards to fit all the calculations. For example, can they calculate the missing ones digits first? Can they use inverse calculations to help them?

 Reflection time

Ask pairs of students to explain their strategies when playing the version of Nim in the introductory activity. *Did you play randomly until there were a few counters left? Did it make a difference once only two piles remained?* Ask students to say whether the 'What if?' questions helped them.

Practice Book: Students complete Practice Book page 49. They can do this directly after the main activity, as homework, or as the focus of a separate mathematics session to help students consolidate their learning and build fluency.

Students play another version of Nim. Can they apply the strategies they learned in the main activity and the Student Book activities to guarantee they win the game each time? *What strategies did you use? If your partner used the same strategy as you, who would win?*

<table>
<tr><td colspan="2">Differentiated outcomes</td></tr>
<tr><td>All students</td><td>should be able to play the games and solve the puzzles with support.</td></tr>
<tr><td>Most students</td><td>will complete the game and the calculations independently.</td></tr>
<tr><td>Some students</td><td>may be able to develop strategies for winning.</td></tr>
</table>

21 Addition and subtraction puzzles

Explore 2   Student Book page 60 · Practice Book page 50

Specific learning focus

- Use addition and subtraction of numbers up to 3-digits to make loop cards with addition and subtraction calculations.

Global skills

- **Creative skills:** problem solving
- **Interpersonal skills:** communication

Key vocabulary

- addition, subtraction, total, difference

Resources

- A4 paper

Language support

In making loop card sets, model for students the idea of giving a statement about a number and then asking a question to generate a new number. Reinforce that the question style: 'Who is___' does not refer directly to a person, but to the person who has that answer.

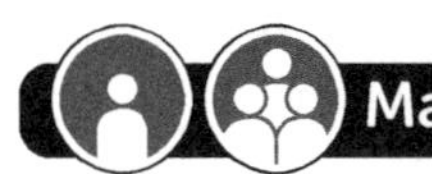

Introductory activity

Allocate each student in the class a number from 1 to however many students there are, including yourself. For example if there are 36, including yourself, write the numbers from 1–36 on the board, with someone's name next to each number. You start by saying *I am* (your number, e.g. 36), then look at the list on the board and choose another number from the list. Ask, for example, *Who is 20 – 5?* The student with that number comes to the front of the class, crosses off their number from the list and makes up a calculation with one of the remaining numbers as the answer.

Student Book page 59

3 a 8   **b** 15   **c** 34   **d** 62   **e** 97

Practice Book page 49

Observe students playing this version of Nim and talk to them about their strategy and tactics.

Stretch zone: Check students' strategies for winning the game. Share any good strategies with the class.

The game continues with each student taking a turn as their number is called. They come to the front of the class to choose a new number from those remaining and ask a question. Eventually, each student will have been to the front of the class, the final student being left with 36, which in this example is the teacher's number.

Main activity

Students will work individually on the task in the Student Book on page 60. They will each create a set of loop cards, using the structure of the game played in the introductory activity. Remind them to use a mixture of addition and subtraction problems, and to include single-, 2- and 3-digit numbers. Make sure that the questions and answers flow from one card to the next.

When everyone is finished, form small groups and use one set of loop cards to play the game. Each student will have a small number of the cards and so will have to check their cards each time a question is asked.

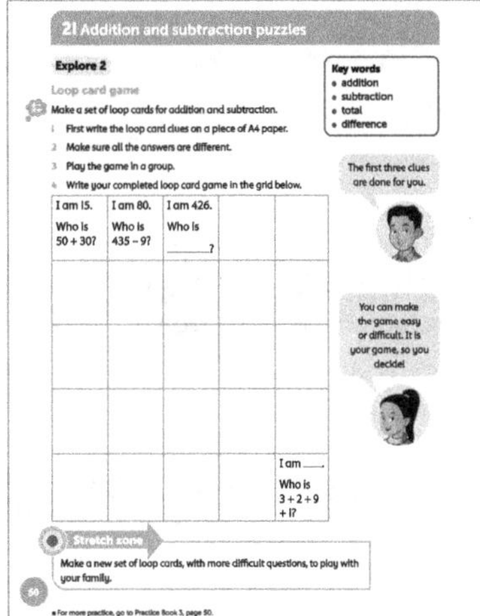

Differentiation

Supporting: Model the questions and responses.

Consolidating: Suggest that students use 2- and 3-digit numbers.

Extending: Encourage students to explore a range of question types, for example: 'Who is 10 more than 436 – 27?'

Stretch zone: *Make a new set of loop cards, with more difficult questions, to play with your family.*

Later, ask students to give feedback on how their cards worked at home.

 ## Reflection time

Ask students to recall their reasoning as they built their sets of loop cards and describe how they chose some of their calculations. *What was your hardest calculation? What was the easiest?*

Choose one or two students and use their sets of cards with the whole class.

Practice Book: Students complete Practice Book page 50. They can do this directly after the main activity, as homework, or as the focus of a separate mathematics session to help students consolidate their learning and build fluency.

This gives students another opportunity to create a set of loop cards. Encourage them to write more difficult questions, or a greater variety of questions, on these cards so that they really challenge themselves.

Differentiated outcomes	
All students	should complete a set of loop cards, perhaps with support.
Most students	will make a set of loop cards without support.
Some students	may be able to make a set of more complex calculations for their loop cards.

Answers

Student Book page 60

Each set of loop cards will vary. Students should check as they play that each calculation links to the next card until returning to the first card.

Practice Book page 50

Answers will vary because students make their own sets of loop cards. They should check as they play that the questions and answers link up.

2J Word problems

Discover Student Book page 61 • Practice Book page 51

Specific learning focus

- Solve word problems using addition and subtraction calculations.

Global skills

- **Creative skills:** problem solving
- **Real-world skills:** interpreting information
- **Interpersonal skills:** communication

Key vocabulary

- word problem, addition, subtraction, total, difference

Resources

- base-10 equipment, number lines, 100-squares

Language support

As students work on word problems, help them to understand what is being asked by using the RUCSAC acronym. Help them to interpret the words in the problems that imply what calculations are needed and to make sense of the answer by referring back to the wording in the problem. For example, problems asking 'how many more?' refer to finding a difference, 'what is the total?' means doing an addition and so on.

 ## Introductory activity

Ask students to think about the following problem: 'I collected 23 stickers for my book in October and another 18 stickers in November. How many stickers did I collect in total over those 2 months?'

Ask students to find a word that suggests which calculation is needed. ('Total', which suggests addition.) Then ask which numbers need to be added, as there are three different numbers written in the problem. Students should identify that 23 and 18 need to be added. *Why is 2 is not part of the calculation, even though it is in the wording of the problem?*

 ## Main activity

Remind students of the steps in the RUCSAC method:

- **R**ead the question – what is the important information?
- **U**nderstand the question – what do you need to find out?
- **C**hoose the correct method of calculation and operation(s).
- **S**olve the problem – make sure that you follow the steps.
- **A**nswer the question – what were you meant to find out?
- **C**heck your answer – use the inverse to check your working out.

Now ask students to work in pairs to solve the word problems on page 61 of the Student Book. As they work through each problem, remind them to follow the stages in the RUCSAC method. For the calculations, students can use base-10 equipment, number lines or a 100- square. Look together at page 61 of the Student Book. If you have access to an IWB you could use this. Talk through how a number line can help with the calculations needed to solve the problems.

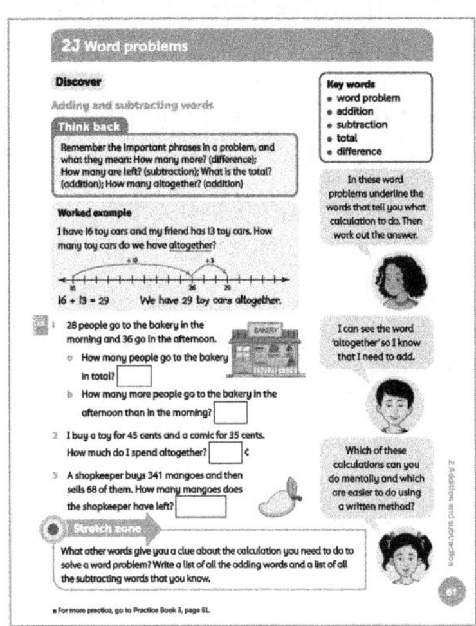

Differentiation

Supporting: Model the use of RUCSAC and help students follow the steps as you work through a problem with them.

Consolidating: Encourage students to work through the problem, in pairs, using RUCSAC. They should discuss each step in their pairs.

Extending: Challenge students to solve problems in ways that make sense to them. For example, if there is more than one way to do a calculation can they choose the most efficient way?

Stretch zone: *What other words give you a clue about the calculation you need to do to solve a word problem? Write a list of all the adding words and a list of all the subtracting words you know.*

Students should be able to explain that they understand that different words imply using a particular calculation, for example:

- addition: total, altogether, sum
- subtraction: difference, how many more, how many are left?, less than.

Note: When students are finding a difference, they may use addition (counting up to find a difference) or subtraction. Ensure that they are able to provide a suitable explanation to go with their list of words.

 **Reflection time**

Talk through with students their methods for solving the word problems. Ask a student to come to the front and to describe the process they used. Can they write or draw their calculation on the board for other students to see?

Practice Book: Students complete Practice Book page 51. They can do this directly after the main activity, as homework, or as the focus of a separate mathematics session to help students consolidate their learning and build fluency.

Students need to recognise the key word in the word problem, which tells them whether to add or subtract. *What is the key word here? So what calculation will you do?* In the second part of the activity, students put this into practice to solve addition and subtraction problems, choosing the correct calculation to do each time.

Note that some students might write 'subtract' if a problem involves finding a difference. But some might write 'add' because they would count up from, for example, 152 to 175 to find the answer. Either could be correct, so always ask students to explain their answer to confirm that they have chosen a correct method.

Differentiated outcomes	
All students	should be able to follow RUCSAC, perhaps with support.
Most students	will solve the problems using RUCSAC.
Some students	may be able to solve word problems in different ways.

Answers

Student Book page 61

1 a 64 **b** 8

2 80 cents

3 273

Practice Book page 51

1 Add

2 Add

3 Subtract

4 Subtract

5 Add

6 255

7 150

8 120

2J Word problems

Specific learning focus

- Solve word problems using addition and subtraction calculations.

Global skills

- **Creative skills:** problem solving
- **Real-world skills:** interpreting information

Key vocabulary

- word problem, addition, subtraction, total, difference

Resources

- base-10 equipment, number lines, 100-squares

Language support

The introductory activity focuses on the key words of addition and subtraction so you can emphasise these to help students with the main activity. As students work on word problems, help them to understand what is being asked by using the RUCSAC acronym. Help them interpret the words in the problems that imply what calculations are needed and to make sense of the answer by referring back to the wording in the problem.

Introductory activity

Ask students to suggest words and phrases that they know tell them to do a particular calculation. Make a class list. They might suggest, for example, 'total', 'altogether', 'how many more than'. Once the list is created, sort the words or phrases into two columns, one for addition and one for subtraction and display this for all students to see. Students can refer to this as they work on word problems in the Student Book.

Main activity

Ask students to look at the first problem on page 62 of the Student Book. Work on this as a class, using RUCSAC, and identify what type of calculation is needed from the word list created in the introductory activity. *What is the problem asking you to find out? What word in the problem is written in one of our lists? So what calculation should we do for this problem? Which of the numbers do we use in the calculation? How do you know?*

Once students have worked through the calculation, remind them of the final steps of RUCSAC. *What did you need to find out? So does your answer look about right? What calculation can we do to check that it is correct?* Remind students that they can check their answer using an inverse calculation.

Students then work on the other word problems on the page. They can draw number lines or use base-10 equipment to help them solve the calculations.

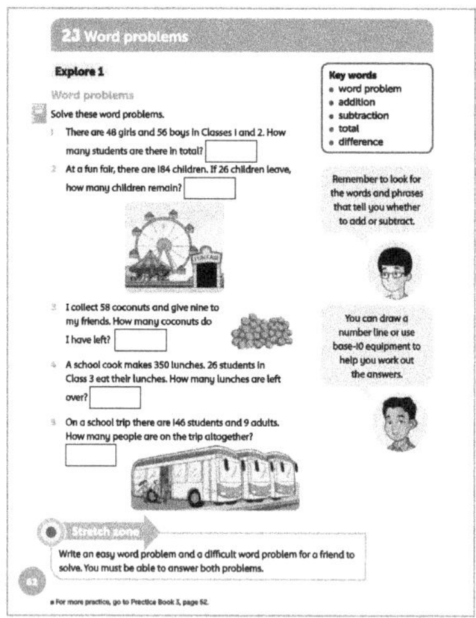

Differentiation

Supporting: Model the use of RUCSAC and help students how to find which calculation to use.

Consolidating: Encourage students to work through the problem, using RUCSAC. They could discuss each step in pairs.

Extending: Ask students to solve the problems using any method they feel confident with and check their answers using inverse calculations.

Stretch zone: *Write an easy word problem and a difficult word problem for a friend to solve. You must be able to answer both problems.*

Students should be able to write and solve two word problems. Ask them to explain what makes a problem easy or difficult. For example, do they think subtraction is harder than addition? Do their problems have calculations that require exchanging ones and tens or tens and hundreds?

Reflection time

Talk through with students their methods for solving the word problems. Ask a student to come to the front and to describe the process they used. Can they write or draw their calculation on the board for other students to see?

Practice Book: Students complete Practice Book page 52. They can do this directly after the main activity, as homework, or as the focus of a separate mathematics session to help students consolidate their learning and build fluency.

Students solve addition and subtraction problems, choosing the correct calculation to do each time. Remind them to work through the steps of RUCSAC, and to look for the words that give them clues about which calculation they need to do.

Differentiated outcomes	
All students	should identify the calculations needed in word problems.
Most students	will be able to carry out the calculations.
Some students	may be able to solve word problems and check them using inverse calculations.

Student Book page 62

1 104   **4** 324

2 158   **5** 155

3 49

Practice Book page 52

1 423

2 67

3 115

4 187

2J Word problems

Explore 2   Student Book page 63 · Practice Book page 53

Specific learning focus

- Write and solve word problems using addition and subtraction calculations.

Global skills

- **Creative skills:** problem solving
- **Real-world skills:** research/financial literacy
- **Interpersonal skills:** communication

Key vocabulary

- word problem, addition, subtraction, total, difference

Resources

- base-10 equipment, number lines, 100-squares

Language support

Before students start to write word problems, use the RUCSAC acronym to identify the key parts of a word problem and the words that are needed. Support students by using writing frames to help them structure their word problems with a context, information and a question.

Introductory activity

Write on the board the following four amounts: 85¢, 32¢, 94¢ and 68¢. Ask students, in pairs, to choose two of the amounts. Ask them to write a word problem about the amounts that includes a calculation. You could use

a writing frame to help. For example, the writing frame could be:

A __________ costs ____¢. A __________ costs ____¢.

How much more does the __________ cost than the __________?

Then they might say: 'A pencil costs 94¢ and an eraser costs 68¢. How much more does the pencil cost than the eraser?' Ask some students to share their word problem and see whether the class can work out the answer.

Main activity

Students should work on the activity on page 63 of the Student Book, making word problems from the set of items as shown. They should make sure that their word problem includes the words or phrases from earlier lessons that indicate whether the calculation is an addition or subtraction. You could include writing frames for addition and subtraction or you could leave it completely up to students to write their own.

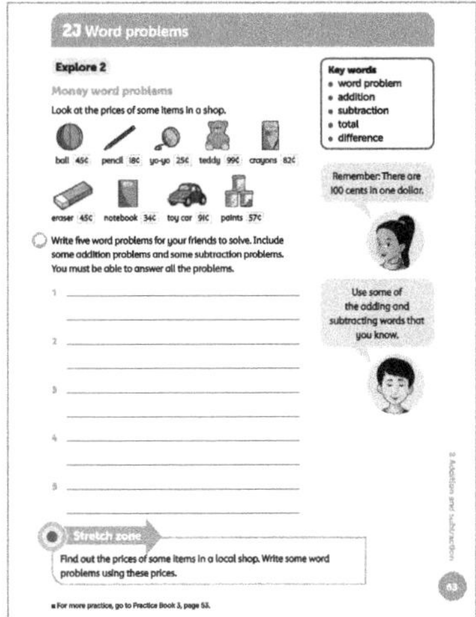

Differentiation

Supporting: Help students use the writing frames to write a word problem involving addition or subtraction. Prompt them to use the vocabulary that indicates which calculation to use.

Consolidating: Guide students to complete writing frames for compiling word problems.

Extending: Challenge students to write one- and two-step word problems, solve them and check the answers using inverse calculations.

Stretch zone: *Find out the prices of some items in a local shop. Write some word problems using these prices.*

Students should be able to write and solve word problems using prices from a local shop. Ask them to think about the vocabulary in their word problems so that anyone solving the problem can decide which calculation to use.

 ### Reflection time

Talk through with students their methods for writing the word problems. Ask them to describe the process they used. They can also come to the front and share their word problem on the board for other students to see, discuss and solve.

Practice Book: Students complete Practice Book page 53. They can do this directly after the main activity, as homework, or as the focus of a separate mathematics session to help students consolidate their learning and build fluency.

Students solve addition and subtraction problems, choosing the correct calculation to do each time. Remind them to work through the steps of RUCSAC, and to look for the words that give them clues about which calculation they need to do.

Differentiated outcomes	
All students	should write word problems using writing frames for support.
Most students	will write word problems, referring to writing frames if necessary.
Some students	may be able to write word problems and solve them using a range of methods.

Answers

Student Book page 63

Answers will vary because students devise their own word problems. Check that for the problems they write, they can work out the correct answer.

Practice Book page 53

1 $254

2 $114

3 $284

4 No

Stretch zone: Check students' answers. A family will get three cars if they decide to buy the Hornet, the Superstar and the Speedstar so that might be the best for a family to buy.

2 Addition and subtraction

Connect Student Book page 64

Big idea

I can add and subtract 2- and 3-digit numbers using number facts to 10, base-10 equipment, empty number lines or a written column method.

Global skills

- **Creative skills:** problem solving/exploring/ investigating
- **Real-world skills:** research/presenting/ interpreting information/financial literacy
- **Interpersonal skills:** communication/teamwork

Key vocabulary

- word problem, addition, total

Resources

- base-10 equipment, number lines, 100 squares

Introductory activity

Look together at page 64 of the Student Book. If you have access to an IWB you could use this. Equip students with some knowledge about the school. For example, they can count how many students are in their class, and you can tell them how many classes there are in the school. Ask them to suggest ways of estimating and then finding out the actual number of students in the school. Share the strategies and guide them if you see any possible problems with the methods they suggest. *If there are 30 students in our class and 7 classes in the school, approximately how many students are there in our school altogether? Do you think all the students are at school today? Why might some students not be here? Do you want to revise your estimate?*

Main activity

Arrange students into small groups. Students in each group work together to decide how to estimate and then find out the number of students in the school on that day. Students may suggest walking around school and asking for the number of students in each class, or they may suggest checking the register for each class. Guide them as appropriate for your school.

How will you record the numbers for each class? How will you find the total of all the classes?

If it works better for your school, you could share some of the calculations out between the groups.

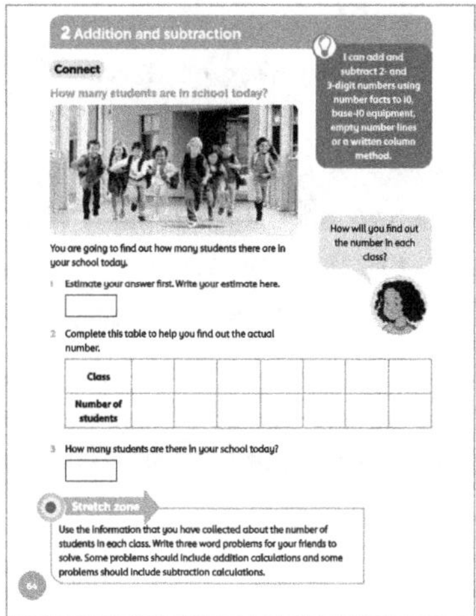

Differentiation

Students can work together in mixed-attainment groups, with the more-able students supporting the others and everyone in the group being involved in making the estimates and doing an accurate count.

Stretch zone: *Use the information that you have collected about the number of students in each class. Write three word problems for your friends to solve. Some problems should include addition calculations and some problems should include subtraction calculations.*

Students should be able to describe their problem clearly. They should demonstrate how they have used vocabulary that leads to particular calculations, and be able to give a solution to the problem. You can refer them back to the writing frames from the previous lesson, if necessary.

 ### Reflection time

Allow each group time to present their findings. Ask them to explain how they carried out the task and what information they used to calculate their estimate and final number of students in the school. Students from other groups can ask questions to find out more about how each group calculated the total. Compare the totals from each group to see whether they agree. If they do not, ask students to suggest why they might have different answers.

Differentiated outcomes	
All students	should be engaged in group work to estimate and count the number of students in the school.
Most students	will be able to suggest methods for getting estimates and counts.
Some students	may be able to use all the skills they have learned in the unit.

2 Addition and subtraction

Global skills

- **Creative skills:** problem solving
- **Self-development skills:** reflecting on learning

Student Book

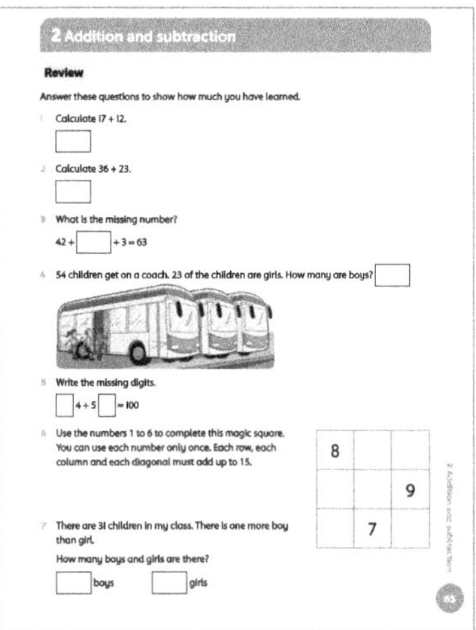

With young children, assessment activities are most effective when carried out as an everyday classroom activity. Watch as students add and subtract 2- and 3-digit numbers. They should have base-10 equipment, number lines and 100-squares available for support. Provide daily practice in adding and subtracting mentally, to continually enhance mental capacity to add and subtract with confidence.

Look out for students applying the skills of calculation to solving word problems in different contexts, and solving mathematical puzzles involving addition and subtraction. Refer students to RUCSAC to support them if they have difficulties in interpreting the word problems.

Answers

Student Book page 65

1 29

2 59

3 18

4 31

5 $44 + 56 = 100$

6

8	3	4
1	5	9
6	7	2

7 16 boys and 15 girls

Practice Book

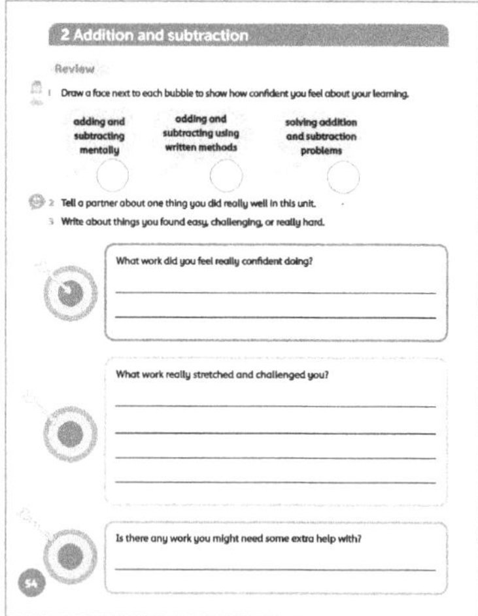

It is appropriate to complete this Practice Book review as a whole-class discussion. You may choose to keep a record of the class discussion or a copy of the review page for your own records. The review provides an opportunity for students to reflect on their learning from the unit, to discuss any areas of mathematics that they feel went particularly well, and any areas that they feel less confident about. Ensure that all students have a copy of the Student Book as a reminder of the areas of mathematics that they have worked on in this unit.

Allow students plenty of time for discussion before asking them to complete the Practice Book page individually, and then, if appropriate, to share their responses with the rest of the class. If students complete this self-assessment at home, encourage them to discuss this with adults. Make a note of areas that students still feel unsure about. For example: how do students feel about mental methods compared to written methods? Does it depend on what method they use? How will students practise adding and subtracting in other units? What equipment should they use if they need support?

Additional material

There are additional end-of-unit assessments available on the *Oxford Owl for School* website.

3 Multiplication and division

Big idea

There are several big ideas contained in this unit. Students develop their knowledge of multiplication tables, looking at patterns in times tables and associated division facts. They develop their mental calculation strategies, using known facts to multiply and divide larger numbers, including partitioning numbers into tens and ones, then progress to using more formal written methods for multiplication and division. They look at division calculations that include remainders.

They are also reminded of the commutative rule, the rule which states that whichever order you multiply in, you get the same answer. For example, $5 \times 3 \times 2 \times 4 = 3 \times 4 \times 5 \times 2$.

Students continue to represent multiplication and division as arrays and use these to look at associated facts. An egg box is an example of an array:

This egg box array is made up of six rows and two columns, and represents the following multiplication and division 'fact family':

$2 \times 6 = 12$

$6 \times 2 = 12$

$12 \div 6 = 2$

$12 \div 2 = 6$

This also illustrates the fact that multiplication and division are inverse operations.

Look out for

- **Students who do not recognise multiples easily and rely on guessing rather than being secure with times tables.** This can be supported by taking opportunities through the school day to practise the times tables, not just in mathematics lessons, but perhaps lining up in twos, threes, fours and so on, or counting in multiples during PE or other lessons.

- **Students who do not understand the idea that some divisions leave remainders. For example, some students might say, '6 into 13 doesn't go'.** Because previous examples on division have all involved whole number answers, this will need explaining. You can overcome this by using practical examples where there are 'left overs'. For example: 'I have 13 sweets to share between 6 children. How many does each child get and how many are left over?'

Possible misconceptions

- **Students may think that division is commutative like multiplication. For example, they may think that $35 \div 5$ has the same answer as $5 \div 35$.** In this case, use cubes or counters to model the calculation that each would represent.

- **Students may only understand division as sharing rather than grouping. For example, they may try to share 40 between 10, rather than thinking how many tens are in 40.** Once again, this can be modelled using cubes, counters or base-10 equipment to represent the calculations. Ensure that you model the language for each situation.

Key vocabulary

- ×, times, multiply, multiplication, multiplied by, multiple of, product, multiplication fact
- once, twice, three times (and so on)
- repeated addition, product, partition, fact family
- array, row, column
- double, halve, even, pattern
- ÷, divide, division, divided by, divided into, dividend, divisible by, division fact, share, share equally, equal groups of, multiple
- left, left over, remainder
- dividend, quotient, divisibility rules
- grid method, equal groups, regroup
- inverse, inverse operation, commutativity, commutative

Learning objective	E	3A	3B	3C	3D	3E	3F	3G	C	R
Recall and use multiplication and division facts for the 3, 4 and 8 multiplication tables.	✓	✓	✓						✓	✓
Write and calculate mathematical statements for multiplication and division using the multiplication tables that they know, including for 2-digit numbers times 1-digit numbers, using mental and progressing to formal written methods		✓	✓	✓	✓	✓	✓	✓		✓
Solve problems, including missing number problems, involving multiplication and division, including positive integer scaling problems and correspondence problems in which n objects are connected to m objects.					✓	✓	✓	✓	✓	✓

3 Multiplication and division

Engage Student Book page 66

Big question

- What facts can I use to help me multiply and divide?

Global skills

- **Creative skills:** problem solving
- **Interpersonal skills:** communication

Key vocabulary

- array, rows, columns, multiplication, division, inverse operation, commutativity, commutative

Resources

- magazines and newspapers

Language support

Organise students into groups so that you have stronger English speakers with less-confident English speakers. This allows those who are less confident to hear others using the vocabulary.

As you work with the groups on producing their posters, encourage students to read the words aloud. You should model the correct pronunciation for specific words such as 'array', 'row', 'column', 'multiplication', 'division', 'fact' and so on.

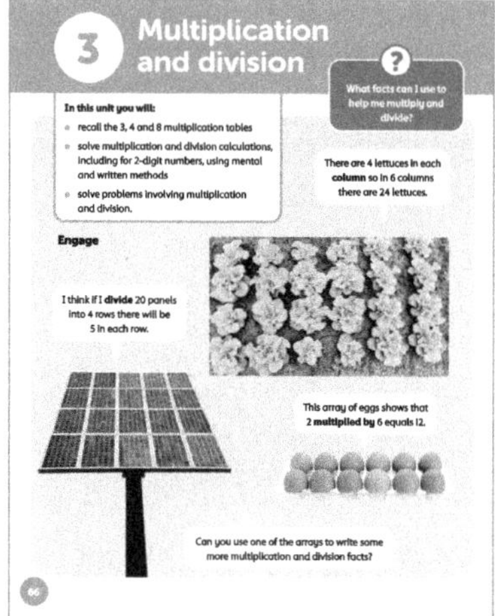

### Introductory activity

Look together at page 66 of the Student Book. If you have access to an IWB you could use this. Use the example of the egg box at the bottom of the page to highlight the following words:

- **array**
- **row**
- **column.**

Ask students to discuss in pairs all the different ways they can describe what the egg box array shows. Here are some examples.

- Six lots of two are twelve.
- Six multiplied by two equals twelve.
- Twelve divided by two is six.
- Twelve shared into two equal groups is six.
- Two rows of six makes twelve.
- Six columns of two is twelve.
- Twelve divided into six equal columns is two.

- Twelve shared into two equal rows is six.
- An array of six columns of two shows twelve in total.

Write the vocabulary on the board and then sort it into two lists so that you have a list of 'multiplication words' and a list of 'division words'.

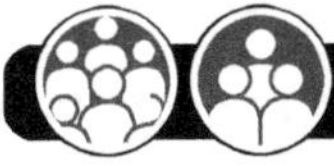

Main activity

Extend the discussion from the introductory activity to introduce the idea of **multiplication** and **division** as **'inverse operations'**. This means that they are 'opposites', one 'undoes' the other, so:

$2 \times 6 = 12 \qquad 12 \div 6 = 2$

Also discuss **commutativity**. This means that multiplication (like addition) can be carried out in any order, so $2 \times 6 = 12$ and $6 \times 2 = 12$. However, the order *does* matter for division. Therefore, division (like subtraction) is not **commutative**.

In groups of three or four, students should look through magazines and newspapers to find examples of arrays. They cut these out and make posters that use all the key vocabulary from the introductory activity. Display these posters for the rest of the unit.

Differentiation

Supporting: You can help write the key vocabulary, or let other students write it, for those students who are less confident with writing and more confident *saying* the multiplication and division facts. Students should find arrays and write down related multiplication and division facts.

Consolidating: Encourage students to find the full fact family for each array and to use language of multiplication and division confidently.

Extending: Ask students to describe concepts of inverse operations and commutativity to others.

 ### Reflection time

Students should swap their posters with each other and check the vocabulary. Ask students to bring some of the posters to the front and read out the vocabulary to model the correct pronunciation to the class. Ask more-confident learners to explain the ideas of 'inverse operations' and 'commutativity', making connections to addition and subtraction.

3A Multiplication and division facts

Discover 1   Student Book page 67 · Practice Book page 55

Specific learning focus

- Know multiplication/division facts for the 2, 3, 5 and 10 times tables.
- Begin to know the 4 times table.

Global skills

- **Creative skills:** investigating

Key vocabulary

- divisible by, division fact, product, multiplication fact

Resources

- mini whiteboards and markers
- digit cards 0–9

Language support

Reinforce the vocabulary of **multiplication facts** and help students to say the facts aloud as they write them. Remind them of the use of '**product**' as the answer to a multiplication calculation.

 ### Introductory activity

Recap the 2, 5 and 10 times tables with students, by reciting each table forward and backwards, and then by asking random questions from each table.

Display the 4 times table on the board and begin to help students to learn it by first reciting it, in order, and then asking random questions. Prompt students to look at the pattern of the answers and to compare it with the 2 times table.

 ### Main activity

Write the number 12 on the board. Ask students to think of any multiplication facts that have an answer of 12. Remind them that they can have the same facts written in reverse order so, for example, 3×4 is the same answer as 4×3.

Check students' answers by asking them to show you their whiteboards. Then repeat for 16 as the answer.

Refer students to the Think back box on Student Book page 67. Remind them that multiplication is commutative, so if they know, for example, that $6 \times 2 = 12$, they also know that $2 \times 6 = 12$. Before students complete the activity, ask them to refer to the example given in the speech bubble but to use a number other than 20 for their numbers. Students should then work on completing the activities on page 67 of the Student Book.

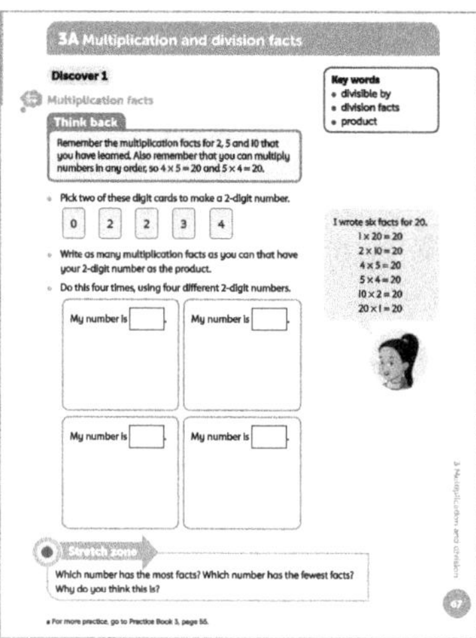

Differentiation

Supporting: Model how to use counters or cubes arranged in arrays to support the activity.

Consolidating: Encourage students to describe the patterns in the multiplication tables, for example the 4 times table is double the 2 times table.

Extending: Ask students to describe their strategies for finding multiplication facts.

Stretch zone: *Which number has the most facts? Which number has the fewest facts? Why do you think this is?*

See whether students can work systematically in finding the multiplication facts. *Can you explain why (e.g. 23) only has two facts? Do even numbers always have more facts than odd numbers?*

 ## Reflection time

Ask students to share the multiplication facts for the numbers they made using the digit cards. Ask them to explain how they found them, and how they know that they have found all the facts for their numbers.

Ask whether any students found multiplication facts for numbers ending in 3. What did they notice about the multiplication facts for these numbers?

Practice Book: Students complete Practice Book page 55. They can do this directly after the main activity, as

homework or as the focus of a separate mathematics session to help them consolidate their learning and build fluency.

Students write number chains using multiplication and **division facts** combined with addition and subtraction facts.

Differentiated outcomes	
All students	should find a multiplication fact for each number.
Most students	will find all the multiplication facts for their numbers.
Some students	may notice connections between different multiplication facts.

Answers

Student Book page 67

Answers will vary because students will choose different numbers but the possible numbers and multiplication facts are:

20: 1×20, 2×10, 4×5, 5×4, 10×2, 20×1

22: 1×22, 2×11, 11×2, 22×1

23: 1×23, 23×1

24: 1×24, 2×12, 3×8, 4×6, 6×4, 8×3, 12×2, 24×1

30: 1×30, 2×15, 3×10, 5×6, 6×5, 10×3, 15×2, 30×1

32: 1×32, 2×16, 4×8, 8×4, 16×2, 32×1

34: 1×34, 2×17, 17×2, 34×1

40: 1×40, 2×20, 4×10, 5×8, 8×5, 10×4, 20×2, 40×1

42: 1×42, 2×21, 3×14, 6×7, 7×6, 14×3, 21×2, 42×1

43: 1×43, 43×1

Practice Book page 55

Answers will vary because students make their own chains.

3A Multiplication and division facts

Specific learning focus

- Know multiplication/division facts for the 2, 3, 5 and 10 times tables.
- Begin to know the 4 times table.

Global skills

- **Creative skills:** investigating
- **Interpersonal skills:** communication

Key vocabulary

- factor, multiple, product, quotient, divisible by, division fact

Resources

- mini whiteboards and markers
- digit cards 0–9

Language support

Display the words 'divisible' and '**multiple**'. Ask questions that model their use, saying: *Is that divisible by 3 – how do you know? How can you tell if a number is a multiple of 5? How can you tell if a number is a multiple of another number?*

Introductory activity

Say to students that you are going to read out a series of instructions. They should follow the instructions to come up with a final number that they will write on their whiteboards. Say:

I am starting with 15

I double it

I divide it by 3

I multiply it by 4

I halve it

I divide it by 10

I add 13

What number do I have? (15)

When you have checked all students' answers by asking them to show you their whiteboards, go through the calculations step by step. Ask students to identify where they made an error if their final answer was incorrect. Encourage a positive attitude to mistakes by emphasising how we can all learn from other people's mistakes.

Main activity

Look together at page 68 of the Student Book. If you have access to an IWB you could use this. Show students the worked example and discuss some of the multiplication and division facts for the number 32. Students should work on the Student Book activity in pairs. They take it in turns to complete each statement, with their partner checking that they are correct. As they work through the statements, go around the class asking questions: *Tell me how you know that is correct. Tell me another number that is divisible by that number. And another?* Ask students to explain their thinking, for example as follows.

- If a number has a 2, 4, 6, 8 or 0 in the ones column, it is **divisible by** 2.
- If a number has a 0 in the ones column, it is **divisible by** 10.

If students need support, refer them to the first speech bubble's suggestion of drawing an array. *Can you draw an array with equal columns and rows? If not, what does that tell you?*

For their explanations, do not accept sentences such as 'If a number ends in 0, it is divisible by 10.' This does not work for decimals. Here are the divisibility rules.

- Any even number divides by 2.
- Any number with a digital sum that is divisible by 3 is itself divisible by 3.
  So 15 327 becomes $1 + 5 + 3 + 2 + 7 = 18$.
  18 is divisible by 3, so 15 327 is too.
- If the last two digits are divisible by 4, then the whole number is too.
- Any whole number with a 0 or 5 in the ones column is divisible by 5.
- If a number has a 0 in the ones column, it is divisible by 10.

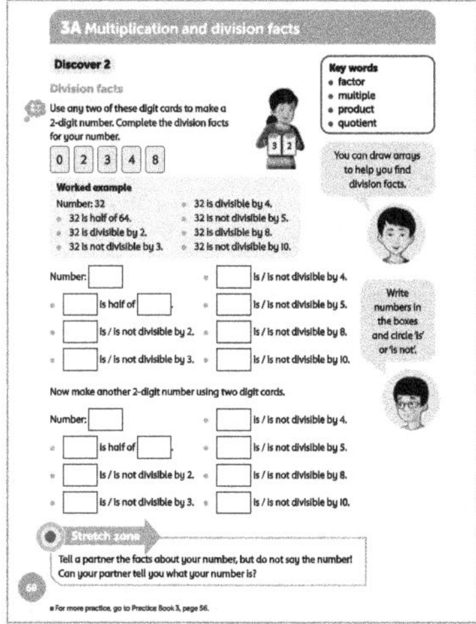

Differentiation

Supporting: Draw arrays to support students with the activity.

Consolidating: Ask students to explain their strategies and to provide other examples of numbers that will also divide by a given number.

Extending: Ask students to share and explain the divisibility rules with the rest of the class.

Stretch zone: *Tell a partner the facts about your number, but do not say the number!*

Can your partner tell you what your number is?

Encourage students to give facts that include factors and multiples of their number.

 **Reflection time**

Ask students to name some numbers they found that were divisible by 2, 3, 4, 5, 8 and 10. Each time, they should say the number, what it is divisible by and how they know.

Practice Book: Students complete Practice Book page 56. They can do this directly after the main activity, as homework or as the focus of a separate mathematics session to help them consolidate their learning and build fluency.

Encourage students to use their knowledge of **factors** and multiples to write appropriate calculations.

Differentiated outcomes	
All students	should complete a range of facts for each number.
Most students	will complete all the facts for their numbers.
Some students	may complete all the facts and be able to explain the divisibility rules.

Answers

Student Book page 68

Answers will vary because students make their own 2-digit numbers to investigate. Check that the facts they have written for their numbers are correct.

Practice Book page 56

Answers may vary as they could involve different calculations. For example, 12 could be 6×2 or 3×4 or $24 \div 2$.

Stretch zone: Answers will vary but could include multiplication: $5 \times 2 \times 5 = 50$, or a combination of other operations such as: $(2 + 3) \times 10 = 50$.

3A Multiplication and division facts

Explore 1   Student Book page 69 • Practice Book page 57

Specific learning focus

- Know multiplication/division facts for the 2, 3, 5 and 10 times tables.
- Begin to know the 4 times table.

Global skills

- **Creative skills:** exploring
- **Interpersonal skills:** communication

Key vocabulary

- multiple, multiplication fact

Resources

- 0–9 digit cards

Language support

Continue to ask questions that model the use of the word 'multiple' saying, for example: *How can you tell whether a number is a multiple of 5?, How can you tell whether a number is a multiple of another number?, What do the multiples of 5 have in common?*

 **Introductory activity**

Ask a student to choose a digit card at random from a set containing 2, 3, 4, 5, 8 and 10. Whichever digit is chosen (4, for example), ask students to recite that times table and as they do, you write the multiples on the board in sequence.

When they have finished as far as $12 \times 4 = 48$, ask them to look at the numbers you have written and tell you whether they can see any patterns in the numbers. They might notice that the numbers are all even, or that the ones digits repeat in a pattern – 4, 8, 2, 6, 0. Explain that the numbers they are looking at are called multiples of 4, because they are made from multiplying 4 by different numbers.

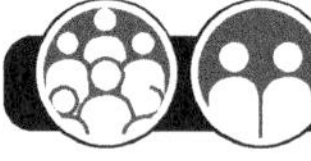 **Main activity**

Ask students to tell you some multiples of 5. As they offer their responses, ask them to justify why their number is a multiple of 5 by giving the **multiplication fact** that makes it. For example, they might say '35 is a multiple of 5 because 7×5 is 35'.

Repeat for multiples of 3 and for multiples of 8.

Students should then complete the activity on page 69 of the Student Book in pairs. Move around the class and check that students are locating correctly the different multiples of each number. Refer them to the second speech bubble on the Student Book page and make the link between the divisibility rules from the previous lesson and whether a number is a multiple of another number.

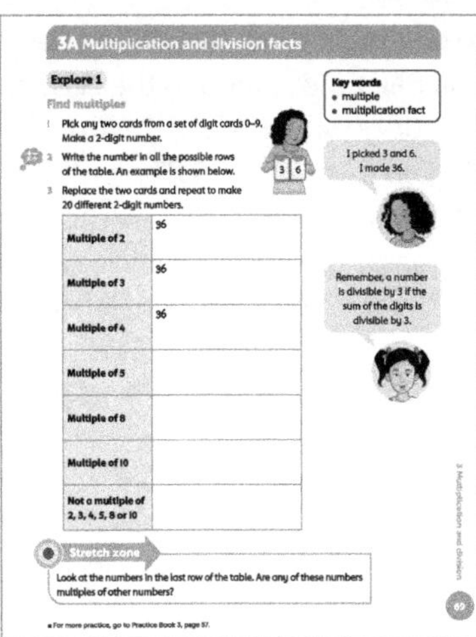

Differentiation

Supporting: Check that students know the times tables and help them by providing a written version if needed.

Consolidating: Ask students to explain what a multiple is and give examples.

Extending: Ask students to explore numbers that are multiples of more than one number, for example 15 is a multiple of 5 and 3 (because $5 \times 3 = 15$).

Stretch zone: *Look at the numbers in the last row of the table. Are any of these numbers multiples of other numbers?*

Look to see whether students can reason why the numbers are in other multiplication tables.

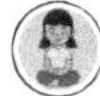 **Reflection time**

Ask stdents to explain how they allocated the numbers they made to the different rows of the table. *How did you know your number is a multiple of X? Did you 'just know' or did you use the divisibility rules that you know?*

Ask students to look at their rows and see whether they can think of any more numbers that can go in each row. *And another? How do you know which numbers will fit?*

Practice Book: Students complete Practice Book page 57. They can do this directly after the main activity, as homework, or as the focus of a separate mathematics session to help students consolidate their learning and build fluency.

Students complete a multiplication grid, using patterns in multiples where they don't know the multiplication facts by heart.

Differentiated outcomes	
All students	should be able to give some multiples of 2, 3, 5 or 10.
Most students	will find multiples of 2, 3, 4, 5, 8 and 10.
Some students	may explore connections between multiplication facts.

Answers

Student Book page 69

Answers may vary but can include:

multiples of 2: 10, 12, 14, 16, 18, 20, …

multiples of 3: 12, 15, 18, 21, 24, 27, 30, 33, 36, …

multiples of 4: 12, 16, 20, 24, 28, 32, 36, 40, 44, 48, …

multiples of 5: 10, 15, 20, 25, 30, 35, 40, 45, 50, …

multiples of 8: 16, 24, 32, 40, 48, 56, 64, 72, 80, …

multiples of 10: 10, 20 30, 40, 50, 60, 70, 80, 90, 100.…

numbers that are not multiples of 2, 3, 4, 5, 8, 10: 11, 13, 17, 19, 23, 29, …

Practice Book page 57

Check that the correct numbers are in the multiplication grid:

x	2	3	4	8
4	8	12	16	32
6	12	18	24	48
8	16	24	32	64
10	20	30	40	80
12	24	36	48	96

Stretch zone: The numbers in the second column are double the numbers in the first column, the numbers in the fourth column are double the numbers in the second column and the numbers in the fifth column are double the numbers in the fourth column.

3A Multiplication and division facts

Explore 2
Student Book page 70 • Practice Book page 58

Specific learning focus

- Know multiplication/division facts for the 2, 3, 5 and 10 times tables.
- Begin to know the 4 times table.

Global skills

- **Interpersonal skills:** communication

Key vocabulary

- divisible by, divisibility rules, factors

Resources

- mini whiteboards and markers
- 0–9 digit cards

Language support

Ask students to count aloud the sequences that you used at the beginning of the lesson. They should do this individually with you, rather than in front of the whole class.

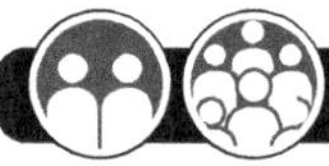 **Introductory activity**

Take two cards at random from the digit cards to form a 2-digit number. Write the number on the board. Give each pair two minutes to write down as many facts about the number as they can, for example it is a multiple of 4, it is even. Take feedback from the group. Try to get a different fact from every pair. Repeat this process twice more.

Then play a game of Fizz-Buzz. Students stand in a circle. The first student says '1' and students count up from 1 around the circle. However, if their number is divisible by 3 the student says 'fizz' instead of their number. If their number is divisible by 5 the student says 'buzz' instead of the number, and if the number is divisible by 3 and 5 the student says 'fizz-buzz'. Any student who makes a mistake sits down. The game continues until only one student is left standing.

 **Main activity**

Introduce students to a True/False game that will prepare them to answer question 2 in the Student Book. You say a statement and students hold up their whiteboards with a 'T' for true or a 'F' for false. Example statements are:

400 is divisible by 10

122 is divisible by 5

123 is divisible by 3.

Students then complete the activities on page 70 of the Student Book in pairs. Move around the class and ask students to 'try out' on you some of the True/False questions they are making up for question 2. They will be giving these questions to their partner to answer so encourage them to redraft questions that are not effective. Make a note of some questions (which you will use in Reflection time).

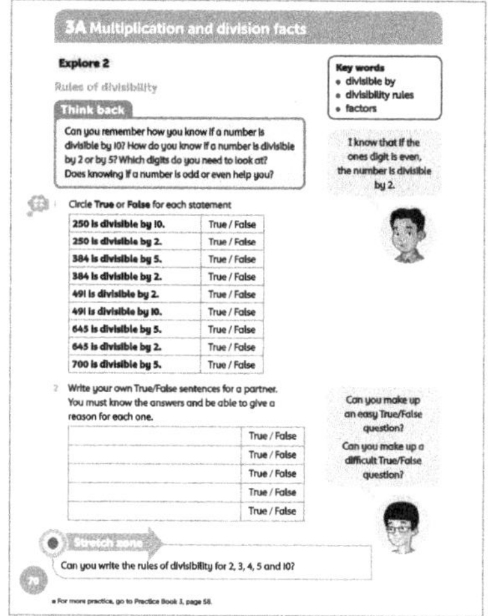

Differentiation

Supporting: Check students' responses and point out errors. *How can you tell that this is incorrect?*

Consolidating: Ask students to explain their thinking and to provide other examples.

Extending: Ask students to explore why the divisibility rules 'work'.

Stretch zone: *Can you write the rules of divisibility for 2, 3, 4, 5 and 10?*

Look to see whether students have written these correctly for each number.

- Any even number divides by 2.
- Any number with a digital sum that is divisible by 3 is itself divisible by 3.
  So 15 327 becomes $1 + 5 + 3 + 2 + 7 = 18$.
  18 is divisible by 3, so 15 327 is too.
- If the last two digits are divisible by 4, then the whole number is too.
- Any whole number with a 0 or 5 in the ones column is divisible by 5.
- If a number has a 0 in the ones column, it is divisible by 10.

 **Reflection time**

Use the list of true/false sentences that you collected during the main activity. Read each sentence to the class. Students should talk to their partner, agree an answer, and hold up their whiteboards with a T for true or an F for false. If there is time, play another game of Fizz-Buzz.

Practice Book: Students complete Practice Book page 58. They can do this directly after the main activity, as homework, or as the focus of a separate mathematics session to help students consolidate their learning and build fluency.

Students complete a table by finding multiples of 2, 4 and 8 using doubling (×2), doubling again (×4) and doubling three times (×8). Encourage students to look for patterns in the numbers when they have completed the table.

Differentiated outcomes	
All students	should understand and apply divisibility rules.
Most students	will use divisibility rules in their True/False questions.
Some students	may devise a range of True/False questions.

Answers

Student Book page 70

1 Table rows 1–5:

True

True

False

True

False

Table rows 6–9:

False

True

False

True

2 Students will write their own number statements to test for divisibility. Check that their partner has given the correct answer for each.

Practice Book page 58

1	16, 32, 64	**7**	64
2	20, 40, 80	**8**	80
3	24, 48, 96	**9**	96
4	30, 60, 120	**10**	120
5	40, 80, 160	**11**	160
6	50, 100, 200	**12**	200

13 Students should notice that the answers to questions 7–12 are the same as the answers in the third column of each row. In other words, multiplying by 8 is the same as doubling, doubling again and doubling again.

Stretch zone: 176, 240

Students might suggest that to multiply a number by 8, they double the number, then double it again and double it again.

3A Multiplication and division facts

Explore 3 — Student Book page 71 · Practice Book page 58

Specific learning focus

- Know multiplication and division facts for the 2 times table and use for doubling and halving larger numbers.

Global skills

- **Creative skills:** exploring

Key vocabulary

- double, multiply, multiply by 4, divide odd, even

Resources

- 100-squares

Language support

Remind students of the terms associated with the 2 times table, such as 'double', 'doubling' and 'even'. As students work on doubling numbers, ask them to say the numbers aloud and listen for patterns in the endings. For example, any number ending in 5 or 0 will end in 0 when doubled, so it will be a tens number ending in '-ty'.

 **Introductory activity**

Ask students to count with you from 0 to 50 in twos. You could use the columns of the 100-square for support. Count on and back. Cover up some of the numbers and repeat the count.

Remind students that these numbers are called multiples of 2 because they are made up of chunks of 2 and are found by multiplying other numbers by 2. Ask students to say what the numbers have in common, for example they are all **even**, or they all end in 0, 2, 4, 6 or 8.

 **Main activity**

Display the 2 times table for students to refer to. Ask a student to choose a multiplication fact from the table, for example 3 × 2 = 6. Now ask: *What is 30 × 2?* To help students calculate this easily, remind them that 3 × 2 has the same answer as 2 × 3, so what has the same answer as 30 × 2? (2 × 30) Students should be able to use their knowledge to **double** 30 to get 60, perhaps calculating 30 + 30.

Repeat with another fact, for example 7 × 2 = 14, so what is 70 × 2?

Now bring in some division facts. For example, give a known division fact such as 25 ÷ 5 = 5 and ask: *What is 250 ÷ 5?*

Give students the opportunity to ask their own questions about multiplication and division facts but explain that they cannot ask a question unless they know the answer.

Students should then work on completing the activities on page 71 of the Student Book.

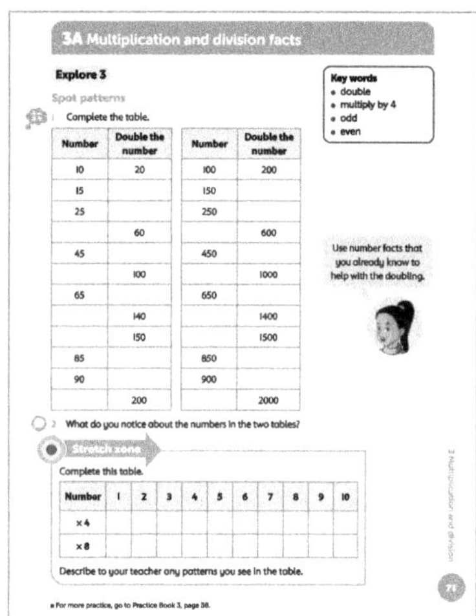

Differentiation

Supporting: Use the 100-square to help students to find the doubles.

Consolidating: Support students in developing a systematic approach to finding all the doubles to 100.

Extending: Students should explore the patterns made by the doubles and extend to investigate doubles beyond 100.

Stretch zone: *Complete this table [the 4 and 8 times tables] Describe to your teacher any patterns you see in the table.*

Notice whether students can show examples of multiples of 8 being double the corresponding multiple of 4.

 **Reflection time**

Write 230 on the board. Ask students whether they can double it. Students should respond with 460, so then ask them to explain how they did it. See whether students use different methods. For example, one might see a pattern and double 23 to get 46, then make it ten times larger, others might calculate it as 200 + 200 + 30 + 30.

Ask students which method they prefer and why. Try the task again by asking them to double 360 and 425.

Practice Book: Students complete Practice Book page 58, if they did not complete it in the last lesson. They can do this directly after the main activity, as homework, or as the focus of a separate mathematics session to help students consolidate their learning and build fluency.

Students complete a table by finding multiples of 2, 4 and 8 using doubling (×2), doubling again (×4) and doubling three times (×8). Encourage students to look for patterns in the numbers when they have completed the table.

Differentiated outcomes	
All students	should double numbers up to 20.
Most students	will double any number to 100.
Some students	may double any number using known doubles.

Answers

Student Book page 71

Number	Double the number		Number	Double the number
10	20		100	200
15	30		150	300
25	50		250	500
30	60		300	600
45	90		450	900
50	100		500	1000
65	130		650	1300
70	140		700	1400
75	150		750	1500
85	170		850	1700
90	180		900	1800
100	200		1000	2000

Practice Book page 58

1 16, 32, 64
2 20, 40, 80
3 24, 48, 96
4 30, 60, 90
5 40, 80, 160
6 50, 100, 200
7 64
8 80
9 96
10 120
11 160
12 200

13 Students should notice that the answers to questions 7–12 are the same as the answers in the third column of each row. In other words, multiplying by 8 is the same as doubling, doubling again and doubling again.

Stretch zone: 176, 240

Students might include in their explanation that to multiply a number by 8, they double the number, then double it again and double it again.

3B Using known facts

Discover 1   Student Book page 72 · Practice Book page 59

Specific learning focus

- Use known multiplication facts and arrays to construct fact families.

Global skills

- **Creative skills:** exploring
- **Interpersonal skills:** communication

Key vocabulary

- array, row, column, fact family

Resources

- counters

Language support

Students could make large colourful versions of the arrays from the Student Book for display and then write the relevant fact families next to each one. They can label their display with the key words: array, row, column, fact family. Refer to the display whenever students use multiplication and division fact families. Listen carefully as students say the fact family they have made. Display some of fact families with a sentence saying 'I know that this number does not belong here because the fact family comes from ____ × ____ = ____!'

Introductory activity

Show students the multiplication fact $4 \times 5 = 20$. Ask them to make an array using counters to represent this fact. Notice how many students have made an array of 4 rows of 5 counters and how many made 5 rows of 4 counters. Discuss how these are the same and how they are different (and how they are both correct). Then repeat for $6 \times 3 = 18$.

Main activity

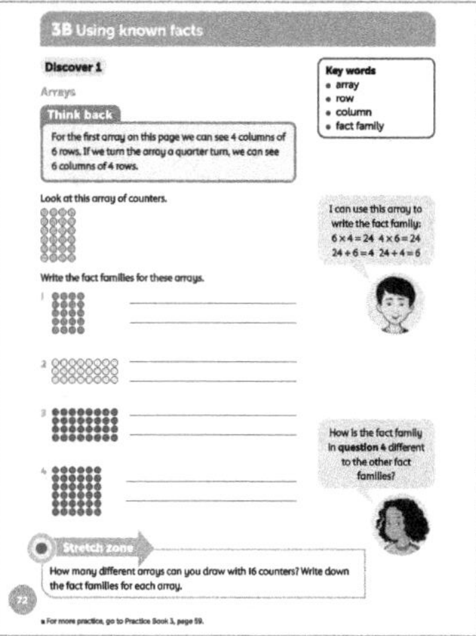

Ask students to work in pairs on the activity in the Student Book on page 72 so that they can explain to each other how they find each **fact family**. While they are working on these activities, move around the pairs and ask questions about the fact families: *How are these the same and how are they different? How else can you represent this fact family, using counters?*

Refer students to the second speech bubble. *Who can explain how this array is different from the others? What other arrays would be similar to this one? Tell me the fact family for those arrays.*

Differentiation

Supporting: Allow students to use counters to make the arrays to find the fact families.

Consolidating: Support students in developing a systematic approach to finding all the facts in a given family for an array. Counters may help with this.

Extending: Students should explore the patterns made by the fact families and extend to investigate ones for larger arrays such as $12 \times 8 = 96$.

Stretch zone: *How many different arrays can you draw with 16 counters? Write down the fact families for each array.*

How do you know you have found them all? Ensure that students work systematically to find all the facts for each array of 16 counters.

 Reflection time

Look back at the fact families that students have produced for each array on page 72 of the Student Book. Ask students to explain their reasoning for recording the multiplication and division facts for each representation of the array.

Show students the multiplication fact $11 \times 6 = 66$. Ask them to say how many rows and columns the array would have for this fact and then to write down the other three facts that belong in this family.

Practice Book: Students complete Practice Book page 59. They can do this directly after the main activity, as homework, or as the focus of a separate mathematics session to help students consolidate their learning and build fluency.

Students continue to draw arrays and write the corresponding fact family for each one. *Which arrays will have four facts as part of their fact family, and which will have only two? How do you know?*

Differentiated outcomes	
All students	should find a multiplication and division fact for an array.
Most students	will find all of the multiplication and division facts for an array.
Some students	may recognise which arrays will have fewer facts and be able to say why.

Answers

Student Book page 72

1 $5 \times 4 = 20$, $4 \times 5 = 20$, $20 \div 5 = 4$, $20 \div 4 = 5$

2 $3 \times 8 = 24$, $8 \times 3 = 24$, $24 \div 8 = 3$, $24 \div 3 = 8$

3 $4 \times 8 = 32$, $8 \times 4 = 32$, $32 \div 8 = 4$, $32 \div 4 = 8$

4 $6 \times 6 = 36$, $36 \div 6 = 6$

Stretch zone: Students should be able to find arrays of 1×16, 2×8 and 4×4, and record the associated fact families for each:

$1 \times 16 = 16$, $16 \times 1 = 16$, $16 \div 1 = 16$, $16 \div 16 = 1$

$2 \times 8 = 16$, $8 \times 2 = 16$, $16 \div 8 = 2$, $16 \div 2 = 8$

$4 \times 4 = 16$, $16 \div 4 = 4$

Practice Book page 59

1 $3 \times 7 = 21$, $7 \times 3 = 21$, $21 \div 3 = 7$, $21 \div 7 = 3$

2 $3 \times 11 = 33$, $11 \times 3 = 33$, $33 \div 3 = 11$, $33 \div 11 = 3$

3 $5 \times 7 = 35$, $7 \times 5 = 35$, $35 \div 5 = 7$, $35 \div 7 = 5$

4 $5 \times 5 = 25$, $25 \div 5 = 5$

5 $2 \times 8 = 16$, $8 \times 2 = 16$, $16 \div 2 = 8$, $16 \div 8 = 2$

Stretch zone: 25 is a square number so the same number (5) is multiplied by itself in the fact family. This only gives one multiplication fact and one division fact, rather than two of each.

3B Using known facts

Discover 2 Student Book page 73 · Practice Book page 60

Specific learning focus

- Understand the idea that multiplication is commutative.
- Understand the relationship between multiplication and division and write connected facts.

Global skills

- **Creative skills:** problem solving

Key vocabulary

- multiplication, division, fact family, inverse

Resources

- mini whiteboards and markers
- counting objects such as cubes, small stones, building blocks

Language support

'Inverse' is a new piece of vocabulary. Work with individuals and write two calculations that are the inverse of each other on their mini whiteboards (e.g. $12 \times 5 = 60$; $60 \div 5 = 12$). Write down the word 'inverse' between the two calculations. Encourage students to say aloud these sentences. 'Multiplication is the inverse of division. Division is the inverse of multiplication.' Explain that each calculation 'undoes' the other.

 **Introductory activity**

This activity is most effective in a large open space or the school hall. Pick 12 students from the class. Ask them to line up in a single line. Then ask them to get into 2 rows. The remaining students should sketch the array made by the 12 students and write $12 = 6 \times 2$ underneath. Repeat this, but now ask the 12 students to get into 3 rows, and then into 4 rows.

Repeat the whole activity with 24 students. This time, let students work out how many different ways they can arrange themselves into rows and columns.

 **Main activity**

Ask individuals for examples of division facts that are the **inverse** of a multiplication and vice versa. For example: *Give me a division fact that is the inverse of: 4×7, 3×9, 8×4.*

Now give me a multiplication fact that is the inverse of: $27 \div 3$, $36 \div 4$, $45 \div 5$.

Look together at page 73 of the Student Book. If you have access to an IWB you could use this. Show students the worked example, which gives the fact family for 8 × 10 = 80. Students should then work in pairs through the activity on page 73 of the Student Book. Their partner can check that they have all the different facts in the fact families. Refer students to the first speech bubble and allow them to decide whether they want to use physical objects or drawings to support them. Remind students of the commutative nature of multiplication and the fact that multiplication and division are inverses.

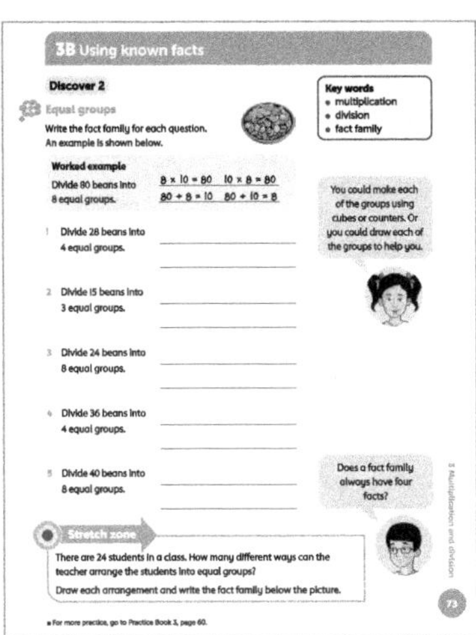

Differentiation

Supporting: Students can create arrays using counting objects and use them to tell you the fact families.

Consolidating: Encourage students to find the full fact family for each array. They may find it useful to use concrete objects to make the arrays, or to draw them.

Extending: Ask students to describe concepts of inverse operations and commutativity to others.

Stretch zone: *There are 24 students in a class. How many different ways can the teacher arrange the students into equal groups? Draw each arrangement and write the fact family below the picture.*

This activity will extend some students to find different factors of 24. *How do you know you have found all the different arrangements?*

 **Reflection time**

Ask one of the students for any multiplication fact they have written down. Write the fact on the board. Ask pairs to write down a division that is the inverse of this fact. Repeat this activity for any division fact students have written down. This time, pairs should write down a multiplication that is the inverse of this fact. Repeat several times.

Practice Book: Students complete Practice Book page 60. They can do this directly after the main activity, as homework, or as the focus of a separate mathematics session to help students consolidate their learning and build fluency.

Students find all the different ways of arranging a number and write the corresponding fact family for each arrangement. *You can use counters (or other objects) to make the arrays.*

Differentiated outcomes	
All students	should find all the fact families for the arrays using counting objects.
Most students	will find all the fact families for the arrays using drawings.
Some students	may know the fact family given one fact.

Answers

Student Book page 73

1 $7 \times 4 = 28, 4 \times 7 = 28, 28 \div 4 = 7, 28 \div 7 = 4$

2 $3 \times 5 = 15, 5 \times 3 = 15, 15 \div 3 = 5, 15 \div 5 = 3$

3 $8 \times 3 = 24, 3 \times 8 = 24, 24 \div 8 = 3, 24 \div 3 = 8$

4 $9 \times 4 = 36, 4 \times 9 = 36, 36 \div 9 = 4, 36 \div 4 = 9$

5 $8 \times 5 = 40, 5 \times 8 = 40, 40 \div 8 = 5, 40 \div 5 = 8$

Stretch zone: Students should draw arrays of 1×24, 2×12, 3×8 and 4×6, and record the associated fact families for each:

$1 \times 24 = 24, 24 \times 1 = 24, 24 \div 1 = 24, 24 \div 24 = 1$

$2 \times 12 = 24, 12 \times 2 = 24, 24 \div 2 = 12, 24 \div 12 = 2$

$3 \times 8 = 24, 8 \times 3 = 24, 24 \div 3 = 8, 24 \div 8 = 3$

$4 \times 6 = 24, 6 \times 4 = 24, 24 \div 4 = 6, 24 \div 6 = 4$

Practice Book page 60

1 $1 \times 24 = 24, 24 \times 1 = 24, 24 \div 1 = 24, 24 \div 24 = 1$

$2 \times 12 = 24, 12 \times 2 = 24, 24 \div 2 = 12, 24 \div 12 = 2$

$3 \times 8 = 24, 8 \times 3 = 24, 24 \div 3 = 8, 24 \div 8 = 3$

$4 \times 6 = 24, 6 \times 4 = 24, 24 \div 4 = 6, 24 \div 6 = 4$

2 $1 \times 12 = 12, 12 \times 1 = 12, 12 \div 1 = 12, 12 \div 12 = 1$

$2 \times 6 = 12, 6 \times 2 = 12, 12 \div 2 = 6, 12 \div 6 = 2$

$3 \times 4 = 12, 4 \times 3 = 12, 12 \div 4 = 3, 12 \div 3 = 4$

3 $1 \times 30 = 30, 30 \times 1 = 30, 30 \div 1 = 30, 30 \div 30 = 1$

$2 \times 15 = 30, 15 \times 2 = 30, 30 \div 2 = 15, 30 \div 15 = 2$

$3 \times 10 = 30, 10 \times 3 = 30, 30 \div 10 = 3, 30 \div 3 = 10$

$5 \times 6 = 30, 6 \times 5 = 30, 30 \div 6 = 5, 30 \div 5 = 6$

Stretch zone: Students may mention that they have found all the factors of 30 (without mentioning the word 'factor' as they will not know this word) or that they have found all the times tables in which 30 appears.

3B Using known facts

Explore 1   Student Book pages 74–75 • Practice Book page 61

Specific learning focus

- Understand the relationship between multiplication and division and write connected facts.

Global skills

- **Creative skills:** exploring
- **Real-world skills:** research/presenting information
- **Interpersonal skills:** communication

Key vocabulary

- double, halve, multiply, divide

Resources

- digit cards 0–9

Language support

Provide a simple speaking frame such as 'I know __________ because ________' to support students to begin to explain how they knew that they had completed all the possible fact families for their chosen numbers. You could model an answer for one number before asking students to use the same structure for a different number. Ask short, closed questions to prompt the next part of a longer response.

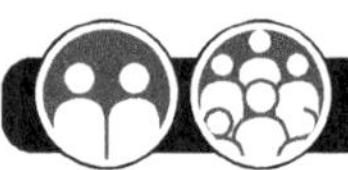

Introductory activity

Remind students that they found the different fact families for given numbers in the previous lessons, either by using an array or by thinking about sharing the number into equal groups. Ask each pair to list quickly all the fact families for 12 (using the arrays 1 × 12, 2 × 6 and 3 × 4). Challenge different pairs of students to tell the class all the fact families for 18. Repeat for other numbers, if there is time.

Main activity

Tell students that they can choose any number from 11–99. They should then work as a pair to find as many of the multiplication and division facts they can for their number. Once they have written as many as they can for their number, they swap books with their partner and check that their partner has found all the fact families for their chosen number correctly. *What other quick facts can we say about a number? How else can we say the fact for ×2 or ÷2?* Remind students that instead of saying 15 × 2 = 30, we can say 'double 15 is 30' and instead of saying 30 ÷ 2 = 15, we can say 'half of 30 is 15'

*How do we **multiply** any number by 10?* Remind students that the digits move one place to the left and that we use a '0' as a place holder in the ones place. *So what is 15 multiplied by 10?*

Look together at page 74 of the Student Book. If you have access to an IWB you could use this. Talk through the completed number bug for 24, looking at all the different facts. Students then complete the number bugs in the Student Book on page 75. *Use your knowledge of fact families, multiplying by 10 and also doubles and halves, when completing your number bugs. What other facts could you include?*

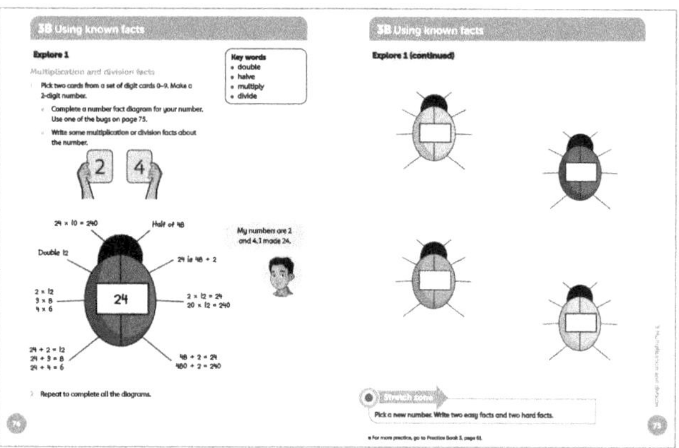

Differentiation

Supporting: You can help write the answers for students and use concrete materials to create fact families by making arrays.

Consolidating: Encourage students to work systematically and always record facts in the same order.

Extending: Students can write down the fact families for a set of numbers between two multiples of 10, say between 40 and 50, and then look for patterns.

Stretch zone: *Pick a new number. Write two easy facts and two hard facts.*

Ask students to explain what they think an easy and a hard fact are. *Why is that an easy fact? Why is that a hard fact?*

 Reflection time

Draw up a list of how many fact families there are for each number from 20 to 40. Ask students whether they can see a pattern. Challenge them to explain the pattern. It may be useful to think about odd and even numbers to help explain the pattern.

Practice Book: Students complete Practice Book page 61. They can do this directly after the main activity, as homework, or as the focus of a separate mathematics session to help students consolidate their learning and build fluency.

Students write different calculations (using multiplication, division, addition or subtraction) with a given number as the answer. They use known facts to help them.

Differentiated outcomes	
All students	should find some members of the fact families but might not complete every fact on each number bug.
Most students	will complete all the fact families and may find some additional facts.
Some students	may find fact families for larger numbers and other facts such as doubles and halves, and multiples of 10.

Answers

Student Book pages 74–75

Answers will vary depending on which numbers students choose. Check that students have completed fact diagrams correctly for the numbers chosen.

3B Using known facts

Explore 2   Student Book page 76 • Practice Book page 62

Specific learning focus

- Understand the relationship between multiplication and division and write connected facts.

Global skills

- **Creative skills:** exploring
- **Interpersonal skills:** communication

Key vocabulary

- multiply, divide, double, halve

Resources

- 1–9 digit cards

Language support

Remind students of 'double' and 'half', and their relationship with the operations of multiplying by 2 and dividing by 2. For example, 'double 4' is the same as 2×4 and 'half of 12 is the same as $12 \div 2$.

 **Introductory activity**

Start by writing 3 on the board. Ask a student to come out and write a multiplication or division calculation that starts with 3 (e.g. $3 \times 5 = 15$). The next student comes out and starts their calculation with 15 (e.g. double 15 = 30).

Practice Book page 61

Answers will vary depending on which numbers students choose. For each number, check that students have written correct facts. For example, for 16:

$8 \times 2 = 16$, $4 \times 4 = 16$, $1 \times 16 = 16$, $2 \times 8 = 16$, $16 \times 1 = 16$ and so on.

$32 \div 2 = 16$, $160 \div 10 = 16$ and so on.

$15 + 1 = 16$, $14 + 2 = 16$, $9 + 7 = 16$ and so on.

$17 - 1 = 16$, $20 - 4 = 16$ and so on.

Continue like this, with each new calculation starting with the answer to the previous one. Students must use multiplication and division only.

 **Main activity**

Following on from the introductory activity, students can now work in pairs on the tasks on page 76 of the Student Book. Each number track is to be built using the same process. Students can either work doing alternate calculations on each number track, checking each other's calculation at each step, or they can write a whole chain each for the partner to check.

As students are working, ask questions to draw out their thinking, for example: *Do you find it easier to start with an odd number or an even number? Why? What is a good calculation to do when the number gets really big? What calculation could you do to 'undo' that one?*

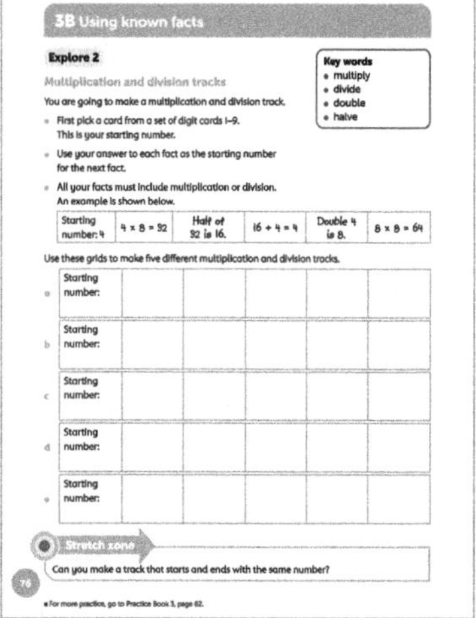

Differentiation

Supporting: Encourage students to use calculations involving the 2, 5 and 10 times tables.

Consolidating: Encourage students to use their knowledge of facts families up to 100.

Extending: Ask students to link the fact families and loop back to the starting number.

Stretch zone: *Can you make a track that starts and ends with the same number?*

This activity can be varied to return to the starting number after a specified number of calculations. Check that they have calculated correctly and used inverse facts from the fact families they have written previously.

 **Reflection time**

Ask different students to explain how they worked out their number chains and offer reasons why they chose the calculations they did. Check the calculations and ask whether they could have used a different calculation to get to the same answer.

Practice Book: Students complete Practice Book page 62. They can do this directly after the main activity, as homework, or as the focus of a separate mathematics session to help students consolidate their learning and build fluency.

Students write multiplication and division chains, some with larger starting numbers. They must use facts they know to work out those they don't know, such as those with larger numbers. *How can you use doubles to work out 35 × 4?*

Differentiated outcomes	
All students	should make a chain of three calculations.
Most students	will complete all the number chains with five calculations.
Some students	may loop their calculations back to the starting number.

 **Answers**

Student Book page 76

Answers will vary because students choose their own starting numbers and calculations. Check that each number track is correct.

Practice Book page 62

Answers will vary because students choose their own starting numbers and calculations. Check that each number track is correct.

Stretch zone: Answers will vary. Check that students' number chains begin on 1 and end on 10 and include all four operations, for example:

$1 \times 100 + (40 - 40) \div 10 = 10$

3C Multiplying mentally

Discover 1   Student Book page 77 • Practice Book page 63

Specific learning focus

- Multiply 2-digit numbers by partitioning tens and ones

Global skills

- **Creative skills:** problem solving

Key vocabulary

- multiply, partition

Resources

- place-value counters

Language support

Support students by reinforcing the language of place value and partitioning. As students work on calculations, help them to say the numbers aloud and describe each part of the number as they partition, for example saying: '45 partitions into 4 tens and 5 ones'.

 **Introductory activity**

Ask a student to choose a digit from 1–9. For example, they choose 4. Write on the board the 2-digit numbers that have a 4 in the tens place: 40, 41, 42, 43, 44, 45, 46, 47, 48, 49.

Model for the class how they are going to say each number in turn and then **partition** it into tens and ones. The sequence will start as follows: '40 is 4 tens and 0 ones, 41 is 4 tens and 1 one, 42 is 4 tens and 2 ones, 43 is 4 tens and 3 ones,…' and so on.

Choose a different tens digit and repeat.

Main activity

Look together at page 77 of the Student Book. If you have access to an IWB you could use this. Ask students to look at the worked example. They can represent the calculation with their own place-value counters as you work through it with them. Ask them to check with their partners that they have set out the counters correctly. When they have formed three rows of 13, they should point to each counter as they count up the total, counting the tens first and then the ones.

Now write on the board 4 × 14. Ask students to form 14 with place-value counters. Then they can make 4 rows of 14. This time, before counting the total, point out that they have more than 10 ones, so they should exchange 10 ones for a tens counter, giving a total of 5 tens and 6 ones, making $4 \times 14 = 56$.

Students now complete the questions on the rest of page 77, remembering to partition the 2-digit numbers, make the right number of identical rows, then exchange if necessary, before counting up the total.

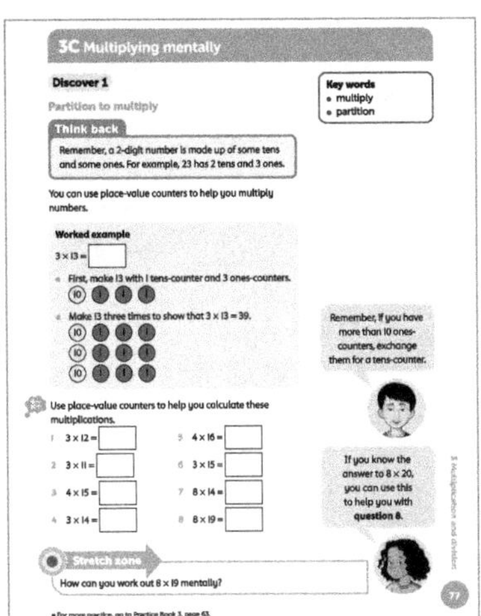

Differentiation

Supporting: Use the place-value counters for support.

Consolidating: Encourage students to use mental methods if possible but use counters where needed.

Extending: Students can work out each calculation mentally, jotting down partial products to help them.

Stretch zone: *How can you work out 8×19 mentally?*

This activity allows students to demonstrate their thinking to partition and multiply where there will be a large number of ones to exchange. You should check that students have formed the partition and carried out the exchanges correctly. Mention that they could use a different mental method and refer them to the second speech bubble on page 77 of the Student Book. For example, they could work out 8×20 and then subtract 8 ($160 - 8 = 152$) to get 8×19.

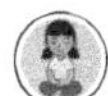 **Reflection time**

Ask students to share their methods for calculating 8×14 and 8×19. How did they arrange the place-value counters? What exchanges did they make? If they solved the calculation mentally, can they describe the process they used?

Invite students to try to multiply 4×23 using the partitioning method. *If you multiply 4×3, how many ones do you have? What do you need to do with the ones?*

Practice Book: Students complete Practice Book page 63. They can do this directly after the main activity, as homework, or as the focus of a separate mathematics session to help students consolidate their learning and build fluency.

Encourage students to draw the place-value counters for each calculation so they can 'see' what is happening each time. *Look at the example. Why are ten of the ones counters circled? What do you need to do with those counters?*

Differentiated outcomes	
All students	should partition 2-digit numbers using place-value counters.
Most students	will partition and multiply using place-value counters.
Some students	may partition and multiply mentally.

Answers

Student Book page 77

1	36	**5**	64
2	33	**6**	45
3	60	**7**	112
4	42	**8**	152

Practice Book page 63

1	42	**4**	64
2	51	**5**	54
3	70		

Stretch zone: Check that students have used a mental strategy. For example, they may suggest multiplying 4 by 20 and then adjusting: $4 \times 20 = 80$; $80 - 4 = 76$.

3C Multiplying mentally

Discover 2 Student Book page 78 • Practice Book page 64

Specific learning focus

- Multiplying teen numbers by 3 and 4.

Global skills

- **Creative skills:** exploring

Key vocabulary

- multiplication, pattern, partition

Resources

- counting stick labelled with the 13 times table (13, 26, 39, 52, 65, 78, 91, 104, 117, 130)

Language support

Encourage students to read their answers aloud so that you can model both the correct use of vocabulary and accurate pronunciation.

 **Introductory activity**

Show students the counting stick and tell them that they are going to learn the multiplication facts for 13. As a class, read the numbers aloud. Then remove the numbers in the order shown below. Each time you remove numbers, recite the 13 times table. Remove:

- 13 and 130 (tell students that these are easy: 13 is obviously in the 13 times table; and we know that $10 \times 13 = 130$)
- 65 (this is half of 130)
- 26 (this is double 13)
- 52 (this is double 26)
- 117 (this is 13 less than 130).

Now try to read the whole table by using the numbers that are left as clues.

 **Main activity**

Draw an array with 4 rows of 13 dots and show students how to use **partitioning** to multiply.

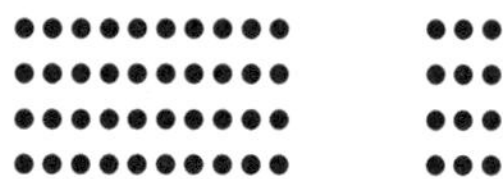

$$10 \times 4 = 40 \quad 3 \times 4 = 12$$

$13 = 10 + 3$ so we can split the array into two smaller arrays and write a multiplication for each. We then add the answers together.

$$13 \times 4 = (10 \times 4) + (3 \times 4) = 40 + 12 = 52$$

Use other examples of multiplying teen numbers by 3 or 4, as necessary, each time, partitioning the teen number into 10 and 'some ones'.

Students should complete Student Book page 78 individually. They use partitioning to multiply the teen numbers and can use place-value counters, or draw arrays for support .

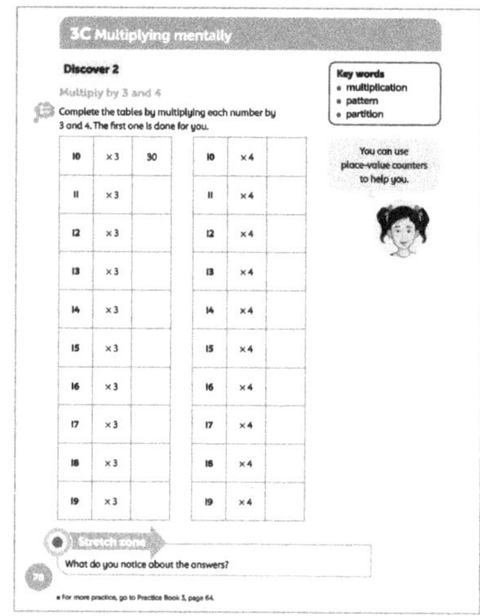

Differentiation

Supporting: Encourage students to explain the way that they are using partitioning to find the answers.

Consolidating: Encourage students to partition the 2-digit numbers to help them complete each table.

Extending: Ask students to extend the tables and multiply numbers up to 30 by 3 and 4 using partitioning.

Stretch zone: *What do you notice about the answers?*

This activity will enable students to see that multiplying by the same number leads to a **pattern** in the ones digits of the answers.

 **Reflection time**

Choose one of the calculations, for example 14×3. Write it on the board and invite students to describe how they calculated this using partitioning. Ask: *Which multiplication facts did you use after you had partitioned the 14?* Repeat for other calculations in the 3 times and the 4 times table.

Practice Book: Students complete Practice Book page 64. (You may choose to leave this until after they have completed Student Book page 79.) They can do this directly after the main activity, as homework, or as the focus of a separate mathematics session to help students consolidate their learning and build fluency.

Students complete a multiplication grid, multiplying teen numbers by 1- and 2-digit numbers. Once completed, they look for patterns in some of the columns of the grid.

Answers

Student Book page 78

$10 \times 3 = 30$	$10 \times 4 = 40$
$11 \times 3 = 33$	$11 \times 4 = 44$
$12 \times 3 = 36$	$12 \times 4 = 48$
$13 \times 3 = 39$	$13 \times 4 = 52$
$14 \times 3 = 42$	$14 \times 4 = 56$
$15 \times 3 = 45$	$15 \times 4 = 60$
$16 \times 3 = 48$	$16 \times 4 = 64$
$17 \times 3 = 51$	$17 \times 4 = 68$
$18 \times 3 = 54$	$18 \times 4 = 72$
$19 \times 3 = 57$	$19 \times 4 = 76$

Practice Book page 64

×	3	6	5	10	12
11	33	66	55	110	132
12	36	72	60	120	144
13	39	78	65	130	156
14	42	84	70	140	168
15	45	90	75	150	180
16	48	96	80	160	192
17	51	102	85	170	204
18	54	108	90	180	216
19	57	114	95	190	228

Stretch zone: Students may notice that the numbers in the 12 column are double the numbers in the 6 column. They may be able to explain that since 12 is double 6, when multiplying any number by 6 and by 12 the products for 12 will always be double those for 6.

3C Multiplying mentally

Discover 3 Student Book page 79 · Practice Book page 64

Specific learning focus

- Multiplying and dividing numbers by 1, 2, 3, 4 and 5.

Global skills

- **Creative skills:** investigating
- **Interpersonal skills:** communication/teamwork

Key vocabulary

- multiplication, pattern, partition

Resources

- 1–5 digit cards and symbol cards for ×, ÷ and =
- a large sheet of paper

Language support

Encourage students to read their answers aloud so that you can model both the correct use of vocabulary and accurate pronunciation.

 **Introductory activity**

Show students the digits 1, 2, 3, 4, 5. Tell them that they can choose two of the digits and multiply them together. How many different answers can they make? (e.g. $2 \times 3 = 6$, $4 \times 1 = 4$, $5 \times 2 = 10$). Give students time to work on this, and then check whether they have found all ten possible products.

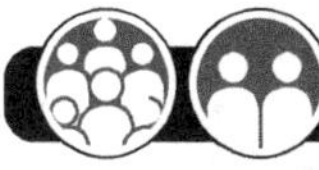 **Main activity**

Ask students whether they can make 10 using more than two of the numbers. They could write $1 \times 2 \times 5 = 10$ or they could write $4 \times 5 \div 2 = 10$, for example. *Can you make 10 in five different ways using more than two numbers?*

Ask students to look at page 79 in the Student Book. Explain how every number from 1 to 20 is now going to be the answer to a calculation using the digits 1–5 and either multiplication or division. Point out the rules highlighted by the two speech bubbles on the Student Book page. For example, show how 2-digit numbers are allowed, as in $55 \div 5 = 11$, and that any digit can be used more than once, so in this example 55 and 5 are the two numbers. Encourage students to work in pairs to find the different answers.

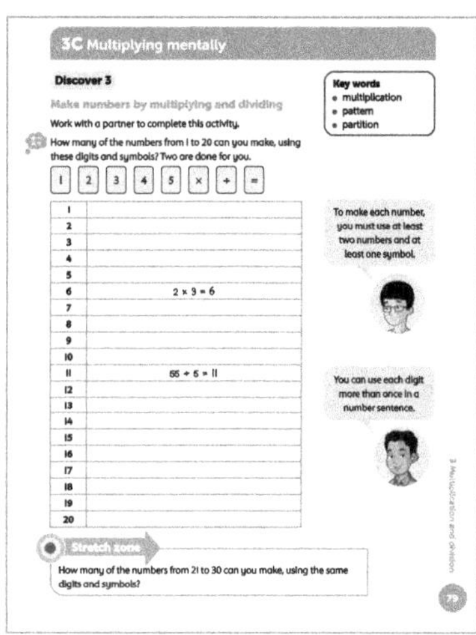

Differentiation

Supporting: Encourage students to use the calculations from the introductory activity to start them off.

Consolidating: Ask students to use times tables to help them write the calculations. For example, if they use 1×5 to make 5, they could use 2×5, 3×5 and 4×5 to make 10, 15 and 20.

Extending: Ask students to write the hardest calculations they can for each number.

Stretch zone: *How many of the numbers from 21 to 30 can you make, using the same digits and symbols?*

This activity will enable students to extend their use of the multiplication and division fact families.

 Reflection time

Write the numbers 1–20 on a large sheet of paper. Walk around the class to find a multiplication or division calculation for each number. If there are several answers for any number, write them all down. Some numbers will be more difficult to make using one operation, so suggest to students that they can use more than one operation. For example, to make 18, they could use $2 \times 3 \times 3$. They can also use an additional operation (e.g. $17 = 5 \times 3 + 2$). Leave the poster on the wall so that they can continue to search for answers in future lessons.

Practice Book: Students complete Practice Book page 64 if they did not complete it in the last lesson. They can do this directly after the main activity, as homework, or as the focus of a separate mathematics session to help students consolidate their learning and build fluency.

Differentiated outcomes	
All students	should complete the answers that use times tables.
Most students	will find a calculation for every number to 20.
Some students	may write more complex calculations to make each number.

Answers

Student Book page 79

There will be varying answers here, but one suggestion for each number is as follows.

$5 \div 5 = 1$	$55 \div 5 = 11$
$1 \times 2 = 2$	$4 \times 3 = 12$
$12 \div 4 = 3$	$52 \div 4 = 13$
$2 \times 2 = 4$	$14 \div 1 = 14$
$15 \div 3 = 5$	$5 \times 3 = 15$
$2 \times 3 = 6$	$4 \times 4 = 16$
$14 \div 2 = 7$	$51 \div 3 = 17$
$4 \times 2 = 8$	$3 \times 3 \times 2 = 18$
$3 \times 3 = 9$	$1045 \div 55 = 19$
$2 \times 5 = 10$	$4 \times 5 = 20$

Practice Book page 64

×	3	6	5	10	12
11	33	66	55	110	132
12	36	72	60	120	144
13	39	78	65	130	156
14	42	84	70	140	168
15	45	90	75	150	180
16	48	96	80	160	192
17	51	102	85	170	204
18	54	108	90	180	216
19	57	114	95	190	228

3C Multiplying mentally

Explore 1
Student Book page 80 · Practice Book page 65

Specific learning focus
- Multiplying numbers by 2, 3, 4 and 5.

Global skills
- **Creative skills:** problem solving/exploring
- **Interpersonal skills:** communication

Key vocabulary
- multiplication, number patterns, partitioning

Resources
- set of 1–9 digit cards
- mini whiteboards and markers
- lollipop in a jar – each lollipop stick has a different student's name on it
- calculators

Language support
Ask students to talk you through the calculations they are doing. Help them to explain their thinking by asking: *What is the first thing that you can do? What number facts do you know that can help you?*

 **Introductory activity**

Select a number from the digit cards and write it down. Select another number and multiply it by the first. For example, if you pick 3 and then 5, write: $3 \times 5 =$

Each pair should write $3 \times 5 =$, and then write the answer. Check all the whiteboards (saying, *Show me*) and ask a pair to explain how they worked this out. Then pick another digit card (e.g. 6) and write: $15 \times 6 = ?$

Again, ask pairs to work this out. They should talk about it with their partner. When all pairs have an answer, discuss as a class the different strategies they have used. An example of a strategy might be to use partitioning:

$6 \times 10 = 60$ and $5 \times 6 = 30$, so $15 \times 6 = 60 + 30 = 90$

Repeat twice more, with different teen numbers.

 **Main activity**

Recap with students how knowing some times table facts can help when multiplying larger numbers. For example: *I know that $4 \times 12 = 48$ so I also know that $40 \times 12 = 480$.* Work through the example given in the Student Book. Encourage the use of known facts to work out the larger calculations. For example, *If you know what 7×3 is, what else do you know? If you know what 8×4 is, what else do you know?*

Look together at page 80 of the Student Book. If you have access to an IWB you could use this. Point to the grid of numbers, the start and finish squares and the directions that the arrows show. Talk through the instructions of the activity. Although students should work through the activities in the Student Book individually, they should share their answers to explore what the biggest and smallest possible totals are. Some students may find lists of multiplication facts helpful.

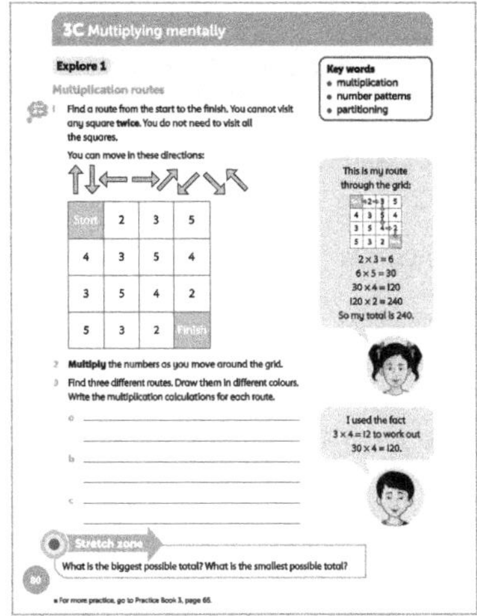

Differentiation
Supporting: Encourage students to use the lists of multiplication facts to support them. They should try to work out the calculations by using facts that they remember.

Consolidating: Ask students to describe the strategy they are using to find the largest and smallest products.

Extending: Students can design 4×4 boards for others to try.

Stretch zone: *What is the biggest possible total? What is the smallest possible total?*

This activity will enable students to see that finding the shortest route with the smallest numbers will give the smallest total, and finding the longest route, using all the numbers, will give the largest total.

 **Reflection time**

Choose an individual by picking a lollipop stick from the jar. Ask them to come to the front and show you the largest total they got. They should try to explain their thinking at each point. If they struggle, support them by asking: *What was the first thing that you did? What number facts did you know that helped you work that out?*

Practice Book: Students complete Practice Book page 65. They can do this directly after the main activity, as homework, or as the focus of a separate mathematics session to help students consolidate their learning and build fluency.

Students practise multiplying 2-digit numbers by 3 and 5, by partitioning the 2-digit number.

Differentiated outcomes	
All students	should find a route through the target board and get individual answers correct.
Most students	will find a route and get the total correct.
Some students	may find the largest and smallest totals by trying different routes.

Answers

Student Book page 80

Answers will vary. Check that students' routes and calculations are correct.

The biggest possible total is:

$4 \times 3 \times 5 \times 3 \times 5 \times 3 \times 2 \times 3 \times 5 \times 5 \times 4 \times 4 \times 2 \times 2 = 25\,920\,000$ (by calculator!)

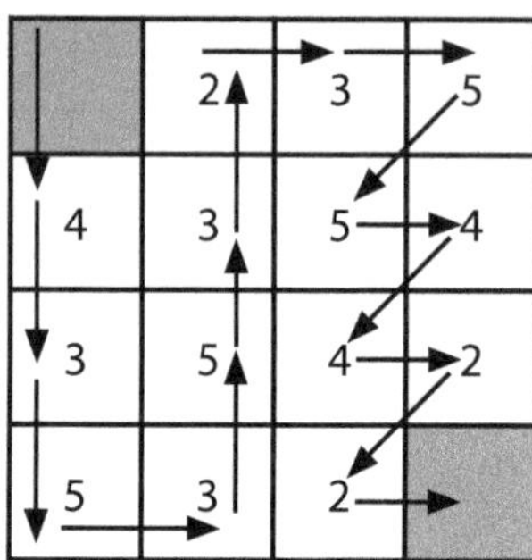

The smallest possible total is:

$3 \times 4 = 12$ (taking a diagonal route from one corner to the other)

Practice Book page 65

Answers will vary depending on the cards students turn over. Check that students' calculations are correct. Ask them to explain the strategies they use when multiplying.

Stretch zone: The product is 12 so the numbers could be 3 and 4, 2 and 6 or 12 and 1.

3C Multiplying mentally

Explore 2 Student Book page 81 · Practice Book page 66

Specific learning focus

- Multiplying numbers by 2, 3, 5 and 10.

Global skills

- **Creative skills:** problem solving/exploring
- **Interpersonal skills:** communication

Key vocabulary

- multiplication, number patterns, partitioning

Resources

- lollipop sticks in a jar – each lollipop stick has a different student's name on it
- calculators

Language support

Ask students to talk you through the calculations that they are doing. Help them to explain their thinking by asking: *What is the first thing that you can do? What number facts do you know that can help you?*

 Introductory activity

Rehearse with students the times tables for 2, 3, 5 and 10. Say them aloud forwards and backwards and then ask random questions from the tables.

Main activity

Recap with students how knowing some times table facts can help when multiplying larger numbers. For example, *I know that $5 \times 12 = 60$ so I also know that $50 \times 12 = 600$.* Encourage the use of known facts to work out the larger calculations. For example, *If you know what 8×3 is, what else do you know? If you know what 9×5 is, what else do you know?*

Although students should work through the activities on page 81 of the Student Book individually, they should share their answers to explore what the biggest and smallest possible totals are. Some students may find lists of multiplication facts helpful.

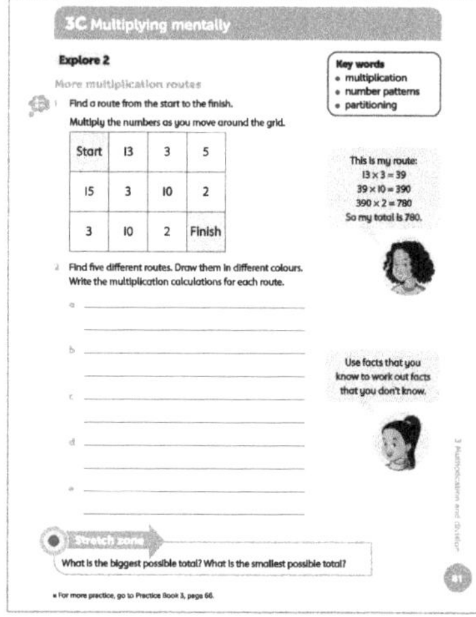

Differentiation

Supporting: Encourage students to use lists of multiplication facts to support them. They should try to work out the calculations first by using facts that they remember.

Consolidating: Ask students to describe the strategy they are using to find the largest and smallest products.

Extending: Students can design multiplication route boards for others to try.

Stretch zone: *What is the biggest possible total? What is the smallest possible total?*

This activity will enable students to see that finding the shortest route with the smallest numbers will give the smallest total, and finding the longest route, using all the numbers, will give the largest total.

Reflection time

Choose an individual by picking a lollipop stick from the jar. Ask them to come to the front and show you the largest total they found. They should try to explain their thinking at each point. If they struggle, support them by asking: *What was the first thing that you did? What number facts did you know that helped you work that out?*

Practice Book: Students complete Practice Book page 66. They can do this directly after the main activity, as homework, or as the focus of a separate mathematics session to help students consolidate their learning and build fluency.

Students complete number chains, by multiplying 1- and 2-digit numbers by 3, 5 and 10. *What times table fact can you use to help you work out that fact?*

Differentiated outcomes	
All students	should find a route through the board and get individual answers correct.
Most students	will find a route and get the total correct.
Some students	may find the largest and smallest totals by trying different routes.

Student Book page 81

Answers will vary. Check that students' routes and calculations are correct.

The biggest possible total is found by visiting each square in turn, and multiplying all the numbers together: $13 \times 3 \times 15 \times 3 \times 10 \times 2 \times 10 \times 3 \times 5 \times 2 = 10\,530\,000$ (by calculator!)

Practice Book page 66

1	4	12	60	600
2	5	15	75	750
3	6	18	90	900
4	10	30	150	1500
5	20	60	300	3000

6 Students may notice that all the products in the 10 row are double the products in the 5 row. And that all the products in the 20 row are double the products in the 10 row and four times the products in the 5 row. They may also notice that all the products in the 4 row are double the products in the 2 row. And that all the products in the 6 row are three times the products in the 2 row.

Stretch zone: 40   120   600   6000

3D Dividing mentally

Discover 1   Student Book page 82 • Practice Book page 67

Specific learning focus

- Divide 2-digit numbers by single-digit numbers.
- Understand that division can leave a remainder (sometimes known as 'left over').

Global skills

- **Creative skills:** problem solving/exploring
- **Interpersonal skills:** communication

Key vocabulary

- division, share, left over, remainder

Resources

- bars of chocolate – enough for at least one chunk per student
- counters or cubes

Language support

Focus on the words 'share', 'left over' and 'remainder', and explain that when division involves sharing we know how many groups we have; we need to find out how many will be in each group. We '**share**' the whole amount equally between the groups. Sometimes we have some left over or a remainder.

 **Introductory activity**

Tell students how many bars of chocolate you have and how many chunks there are in each bar. They should work out how many chunks there are altogether.

Ask a series of questions. Here are two examples.

- If two students share the chocolate equally, how many pieces do they each get?
- If three students share the chocolate equally, how many pieces do they each get?

Make sure some questions leave remainders. Explain that we say '**remainder**' when there is something '**left over**' after division. If necessary, students can use counters or cubes as chocolate so that they can physically 'share'.

Main activity

Encourage students to use their times tables and knowledge of inverses to work on the problems in the Student Book on page 82. They work in pairs. They take it in turns to work out the answer, then the partner checks it. The answer checker should read the answer aloud to their partner to practise the vocabulary of 'share' and 'left over' or 'remainder'. You could write the following on the board as support when students complete and say aloud each problem:

24 shared between ___ people is ____.

24 shared between ___ people is ___ with ___ left over.

24 shared between ___ people is ___ with a remainder of ___.

It may be helpful for some students to have 24 counters or cubes to model the bar of chocolate.

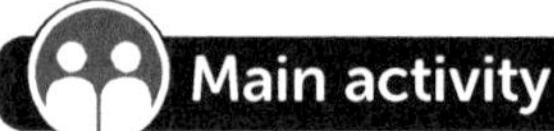

Differentiation

Supporting: Give students counters or cubes to support them.

Consolidating: Ask students which multiplication facts will help them with the calculations.

Extending: Ask students to tell you, before they answer each problem, which problems will give exact answers and which will leave remainders, and how they know.

Stretch zone: *How does knowing your multiplication facts help you divide?*

Discuss this with a partner.

This activity will enable students to see that division is the inverse of multiplication and that knowing their times table facts, with their inverses, is what enabled them to solve the problems.

 **Reflection time**

Ask each pair to make up a word problem that will leave a remainder. For example, *I have 20 apples and I need to share them between 3 bowls. How many apples will be in each bowl?* Select one of the pairs to ask the class to solve their problem. Repeat three times, with different pairs presenting their word problems to the class.

Practice Book: Students complete Practice Book page 67. They can do this directly after the main activity, as homework, or as the focus of a separate mathematics session to help students consolidate their learning and build fluency.

Students work out how to share 30 pieces of fruit between different numbers of people. They write the number that each person gets and any remainder. *Do you think that there will be a reminder when you share the fruit between 6 people/5 people/4 people? How do you know?*

Differentiated outcomes	
All students	should understand the concept of remainders and find answers using counters or cubes.
Most students	will understand the concept of remainders and find answers using known facts.
Some students	may use the idea of multiples to understand, before calculating, which answers will give remainders.

Answers

Student Book page 82

a	3 people – 8	**d**	8 people – 3
b	4 people – 6	**e**	12 people – 2
c	6 people – 4	**f**	24 people – 1

a 7 people – 3 with 3 left over

b 9 people – 2 with 6 left over

c 10 people – 2 with 4 left over

d 11 people – 2 with 2 left over

Practice Book page 67

1 3 people can have exactly 10 pieces of fruit.

2 4 people can have 7 pieces of fruit with 2 left over.

3 5 people can have exactly 6 pieces of fruit.

4 6 people can have exactly 5 pieces of fruit.

5 7 people can have 4 pieces of fruit with 2 left over.

6 8 people can have 3 pieces of fruit with 6 left over.

7   10 people can have exactly 3 pieces of fruit.

8   11 people can have 2 pieces of fruit with 8 left over.

9   12 people can have 2 pieces of fruit with 6 left over.

10   13 people can have 2 pieces of fruit with 4 left over.

11   14 people can have 2 pieces of fruit with 2 left over.

12   15 people can have exactly 2 pieces of fruit.

Stretch zone: 30 can be shared equally between 1, 2, 3, 5, 6, 10, 15 and 30 people, with none left over.

3D Dividing mentally

Discover 2   Student Book page 83 • Practice Book page 68

Specific learning focus

- Divide 2-digit numbers by single-digit numbers.
- Understand that division can leave a remainder (sometimes known as 'left over').

Global skills

- **Creative skills:** problem solving
- **Interpersonal skills:** communication

Key vocabulary

- division, share, left over, remainder

Resources

- counters or cubes

Language support

Focus further on the words 'share', 'left over' and 'remainder', and reinforce that when division involves sharing we know how many groups we have; we need to find out how many will be in each group. We 'share' the whole amount equally between the groups. Sometimes we have some left over or a remainder.

Introductory activity

Ask students to look at the image at the top of page 83 in the Student Book. It shows how many cherries you have. (36) Students are asked to work out how many cherries can be had by different numbers of people. Ask a series of questions to start the activity.

- *If two people share the cherries equally, how many do they each get?*
- *If three people share the cherries equally, how many do they each get?*
- *What if there were 35 cherries. How many would 2 people get? And 3 people?*
- *Tell me a number of people that would leave no reminder if we shared 35 cherries between them.*

Remind students that we say 'remainder' when there is something 'left over' after division. If necessary, students can use counters or cubes to represent the cherries so that they can physically 'share'.

 Main activity

Students should work in pairs on the activity in the Student Book on page 83. They should take it in turns to work out the answer, with the other student checking that the answer is correct. The student that is checking should read the answer aloud to their partner to practise the new vocabulary of 'left over' or 'remainder'. Write the following sentences on the board to support students when they complete and say aloud each problem:

____ people get ___ cherries each with ___ left over.

____ people get ___ cherries each with a reminder of ___.

It may be helpful for some students to have 36 counters or cubes to model the cherries.

Differentiation

Supporting: Give students counters or cubes to support them.

Consolidating: Ask students which multiplication facts will help them with the calculations.

Extending: Ask students to tell you, before they answer each problem, which problems will give exact answers and which will leave remainders, and how they know.

Stretch zone: *How can you tell if there will be a remainder before you divide?*

This activity will enable students to see that knowing their multiplication tables will help them with division problems. They use their knowledge of multiples to know whether an answer will have a reminder or not.

Ask pairs of students to make up a problem that has a reminder of exactly 1 or 2 or 3 and so on. Ask some pairs to say, before they work out the answer, whether they think other pairs are correct or not.

Practice Book: Students complete Practice Book page 68. They can do this directly after the main activity, as homework, or as the focus of a separate mathematics session to help students consolidate their learning and build fluency.

Students work out how to share 24 counters between different numbers of people. They write the number that each person gets and any remainder. *Tell me a number of people that will give a reminder of 1/2/3. How do you know?*

Differentiated outcomes	
All students	should understand the concept of remainder and find answers using counters or cubes.
Most students	will understand the concept of remainder and find answers using known facts.
Some students	may use the idea of multiples to understand, before calculating, which answers will give which remainders.

Answers

Student Book page 83

2 people get 18 cherries each with none left over.

3 people get 12 cherries each with none left over.

4 people get 9 cherries each with none left over.

5 people get 7 cherries each with 1 left over.

6 people get 6 cherries each with none left over.

7 people get 5 cherries each with 1 left over.

8 people get 4 cherries each with 4 left over.

9 people get 4 cherries each with none left over.

10 people get 3 cherries each with 6 left over.

11 people get 3 cherries each with 3 left over.

12 people get 3 cherries each with none left over.

13 people get 2 cherries each with 10 left over.

Practice Book page 68

1 3 people can have 8 counters each with none left over.

2 4 people can have 6 counters each with none left over.

3 6 people can have 4 counters each with none left over.

4 7 people can have 3 counters each with 3 left over.

5 8 people can have 3 counters each with none left over.

6 9 people can have 2 counters each with 6 left over.

7 10 people can have 2 counters each with 4 left over.

8 11 people can have 2 counters each with 2 left over.

9 12 people can have 2 counters each with none left over.

Stretch zone: For there to be 1 counter left over the remaining 23 counters need to be in equal groups. The only ways of grouping them are into 1 group of 23 or 23 groups of 1.

3D Dividing mentally

Explore 1 Student Book page 84 • Practice Book page 69

Specific learning focus

- Divide 2-digit numbers by single-digit numbers.
- Understand that division can leave a remainder (sometimes known as 'left over').

Global skills

- **Creative skills:** problem solving
- **Interpersonal skills:** communication

Key vocabulary

- division, dividend, quotient, remainder, divisor

Resources

- cubes or counters to support students in 'sharing' to solve the problems
- lollipop sticks in a jar – each lollipop stick has a different student's name on it

Language support

Go through these key words with students: 'dividend', 'quotient' and 'divisor'. Provide a division sentence for display with each part of the sentence labelled with the correct terminology, for example: '15 ÷ 5 = 3', labelled to show **'dividend ÷ divisor = quotient'**.

 **Introductory activity**

Place 20 cubes on a table and ask two students to come to the front. They should share the cubes between themselves equally. Write on the board: 20 ÷ 2 = 10

Then ask four students to share the cubes equally. Write: $20 \div 4 = 5$

Repeat with three students. Write: $20 \div 3 = 6$ remainder 2

Repeat with different numbers of students asking the other students to predict the remainder each time before the cubes are shared.

 **Main activity**

Draw 16 counters on the board. Pairs of students should write some division statements using 16 as the dividend (the number to be divided). Some should have exact answers and some should leave remainders.

Students can use cubes or counters to help them work through the questions in the Student Book on page 84. All the division statements should relate to the picture of the 18 bananas on the market stall. Check that students are writing the division statements accurately.

As you move around the class, ask students for the inverse multiplication statements for those answers with no remainder. This reminds them that multiplication is the inverse of division and also acts a check for their answers.

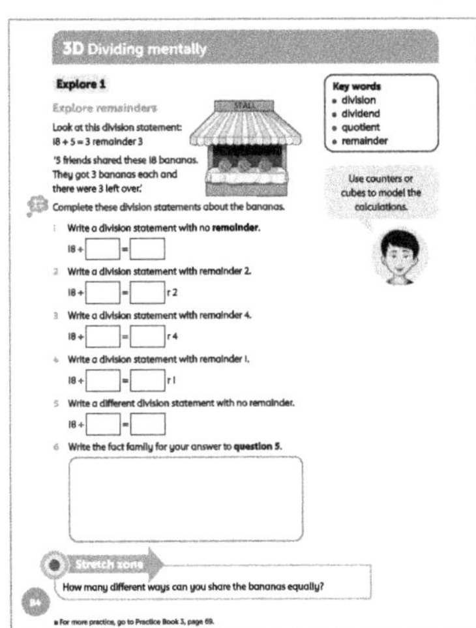

Differentiation

Supporting: Give students 18 counters or cubes to support them.

Consolidating: Ask students which multiplication facts will help them with the calculations.

Extending: Ask students to tell you, before they answer each problem, which problems will give exact answers and which will leave remainders, and how they know.

Stretch zone: How many different ways can you share the bananas equally?

This activity will enable students to find all the factors for 18.

 **Reflection time**

Take a lollipop stick from the jar to select a student. Ask them for the division statement they used as the answer to question 1. Ask them to tell a different 'story' that could go with this statement. For example, if they write '18 ÷ 3 = 6', they might say 'There are 18 pencils altogether. 3 friends shared them, so they had 6 each'.

Repeat for a range of students and different examples.

Practice Book: Students complete Practice Book page 69. They can do this directly after the main activity, as homework, or as the focus of a separate mathematics session to help students consolidate their learning and build fluency.

Students work out how to share 25 cards and then 32 toy cars between different numbers of people. They write the number that each person gets and any remainder.

Differentiated outcomes	
All students	should understand the concept of remainder and find answers using concrete materials.
Most students	will understand the concept of remainder and find answers using known facts.
Some students	may use the idea of multiples to understand, before calculating, which answers will give remainders.

Answers

Student Book page 84

Answers will vary because students choose their own division statements to match the picture and the questions. Answers might include:

1 $18 \div 3 = 6$

2 $18 \div 4 = 4$ remainder 2

3 $18 \div 7 = 2$ remainder 4

4 $18 \div 17 = 1$ remainder 1

5 $18 \div 2 = 9$

6 $18 \div 2 = 9$, $18 \div 9 = 2$, $2 \times 9 = 18$, $9 \times 2 = 18$

Stretch zone: The 18 bananas can be shared equally into groups of 1, 2, 3, 6, 9 and 18.

Practice Book page 69

Answers will vary because students choose their own division statements. Answers might include:

1 I shared the cards equally between 8 people. There was 1 left over. $25 \div 8 = 3$ r 1

2 I shared the cards equally between 3 people. There were 4 left over. $25 \div 3 = 7$ r 4

3 $32 \div 6 = 5$ r 2

  $32 \div 3 = 10$ r 2

  $32 \div 7 = 4$ r 4

Stretch zone: Check students' division sentences all have a reminder of 3. Answers might include $15 \div 4 = 3$ r3, $17 \div 7 = 2$ r3, $19 \div 4 = 4$ r3 and so on.

3D Dividing mentally

Specific learning focus

- Divide 2-digit numbers by single-digit numbers.
- Understand that division can leave a remainder (sometimes known as 'left over').

Global skills

- **Creative skills:** problem solving
- **Interpersonal skills:** communication

Key vocabulary

- division, dividend, remainder

Resources

- cubes or counters to support students in 'sharing' to solve the problems
- lollipop sticks in a jar – each lollipop stick has a different student's name on it

Language support

Reinforce the key words 'dividend', 'quotient', 'divisor' and 'remainder'. Provide further division sentences for display with each part of the sentence labelled with the correct terminology. For example, provide '20 ÷ 3 = 6 remainder 2', labelled to show 'dividend ÷ divisor = quotient and remainder'.

 **Introductory activity**

Place 19 cubes on a table and ask two students to come to the front. They should share the cubes equally. Write on the board '19 ÷ 2 = 9 remainder 1'.

Then ask four students to share the cubes equally. Write '19 ÷ 4 = 4 remainder 3'.

Repeat with three students. Write '19 ÷ 3 = 6 remainder 1'.

Repeat with different numbers of students, asking the other students to predict the remainder each time before the cubes are shared.

 **Main activity**

Draw 23 counters on the board. Pairs of students should write some division statements using 23 as the dividend (the number to be divided). Some statements should have exact answers and some should leave remainders.

Students can use cubes or counters to help them work through the questions in the Student Book on page 85. All the division statements should relate to the pictures given. Check that students are writing the division statements accurately.

As you move around the class, ask students for the inverse multiplication statements for those answers with no remainder. This reminds them that multiplication is the inverse of division and also acts a check for their answers.

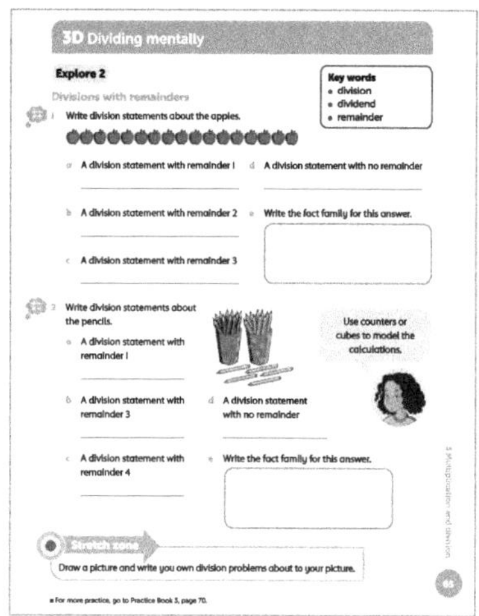

Differentiation

Supporting: Give students counters or cubes to support them.

Consolidating: Ask students which multiplication facts will help them with the calculations.

Extending: Ask students to tell you, before they answer each problem, which problems will give exact answers and which will leave remainders, and how they know this.

Stretch zone: *Draw a picture and write your own division problems about your picture.*

This activity will enable students to create division problems for a context. Check that they can work out the answer correctly.

 **Reflection time**

Take a lollipop stick from the jar to select a student. Ask them for a division statement that they used as an answer to question 1. Ask them to tell the 'story' that goes with this statement. For example, if they write '17 ÷ 3 = 5 remainder 2', they could say 'There are 17 apples altogether. 3 friends shared them, so they had 5 each and 2 were left over'.

Repeat for a range of students and different examples. You could then ask students to come up with a different number story, using a different context and to tell you some more division statements.

Practice Book: Students complete Practice Book page 70. They can do this directly after the main activity, as homework, or as the focus of a separate mathematics session to help students consolidate their learning and build fluency.

Students draw pictures or diagrams to represent each division statement, including the remainders. *What object will you draw? How can you group them? How will you represent the remainder?*

<table>
<tr><td colspan="2">Differentiated outcomes</td></tr>
<tr><td>All students</td><td>should understand the concept of remainders and find answers using concrete resources.</td></tr>
<tr><td>Most students</td><td>will understand the concept of remainders and find answers using known facts.</td></tr>
<tr><td>Some students</td><td>may use the idea of multiples to understand, before calculating, which answers will give remainders.</td></tr>
</table>

Answers

Student Book page 85

Answers may vary. Possible answers include:

1 a $17 \div 8 = 2$ remainder 1

  b $17 \div 5 = 3$ remainder 2

  c $17 \div 7 = 2$ remainder 3

  d $17 \div 17 = 1$

  e $17 \times 1 = 17, 1 \times 17 = 17, 17 \div 17 = 1, 17 \div 1 = 17$

2 a $25 \div 8 = 3$ remainder 1

  b $25 \div 11 = 2$ remainder 3

  c $25 \div 7 = 3$ remainder 4

  d $25 \div 5 = 5$

  e $5 \times 5 = 25, 25 \div 5 = 5$

Practice Book page 70

Check that students have drawn pictures to match each division calculation correctly.

Stretch zone: Answers will vary. For example, students may suggest 24, which can be divided into equal groups of 1, 2, 3, 4, 6, 8 and 24 with no remainder.

3E Written methods for multiplication

Discover Student Book page 86 • Practice Book page 71

Specific learning focus

- Multiply 2-digit numbers by single-digit numbers using base-10 equipment.

Global skills

- **Creative skills:** problem solving
- **Interpersonal skills:** communication

Key vocabulary

- multiply, product, tens, ones

Resources

- base-10 equipment
- 0–9 digit cards

Language support

As students work with the base-10 equipment, model the language they need to carry out the multiplication and help them to understand the words in the problems that indicate what the calculation should be. Remind students about partitioning when setting up their numbers in base-10 equipment and encourage them to talk about the 'product' when they find the answer.

 Introductory activity

Ask a student to choose a digit card at random (e.g. 3). Firstly, count on in 3s to 30, then count back in 3s. Now ask students how much is 10×3? (30) This time, using knowledge of the 3 times table, count on and back in 30s to 300: 30, 60, 90, 120, 150, 180, 210, 240, 300.

Choose different multiples of 10 and repeat.

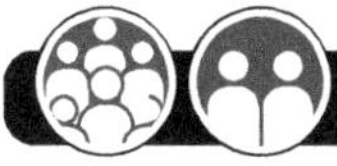 Main activity

Look together at page 86 of the Student Book. If you have access to an IWB you could use this. Explain that students are going to multiply bigger numbers using base-10 equipment. Ask students to look at the worked example in the Student Book. Read the problem to students. *What calculation do we need to do?* Agree that you need to work out 32×3. They can represent the calculation with their own base-10 equipment as you work through it with them. Ask them to check with their partners that they have both set out the base-10 equipment correctly. When they have formed three rows of 32, they should work out the total, pointing to each piece as they count, counting the tens first and then the ones.

Now write on the board 4×23. Ask students to form 23 with base-10 equipment. Then they can make 4 rows of 23. This time, before counting the total, point out that they have more than 10 ones, so they should exchange 10 ones-cubes for a tens-rod, giving a total of 9 tens and 2 ones, making $4 \times 23 = 92$.

Students should now complete the questions on the rest of page 86 of the Student Book. They need to remember to partition the 2-digit numbers, make the right number of identical rows, then exchange if necessary, before working out the total.

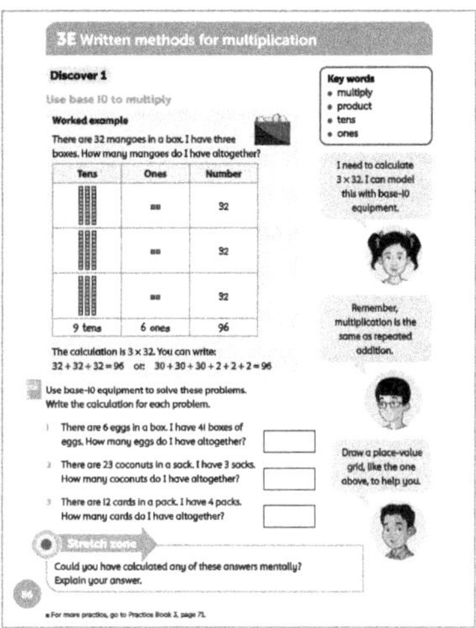

Differentiation

Supporting: Use the base-10 equipment for support.

Consolidating: Encourage students to use mental methods if possible but to use base-10 where needed.

Extending: Students can work out each calculation mentally, jotting down partial products in a place-value grid to help them.

Stretch zone: *Could you have calculated any of these answers mentally? Explain your answer.*

This activity will enable students to demonstrate the extent to which they understand when they can use a mental method, and when to use a different method to multiply a 2-digit number by a single digit. For example, they may choose to solve question 3 using times tables facts, and they may be able to solve questions 1 and 2 using partitioning mentally, first multiplying the tens and then the ones, and then adding them together.

 **Reflection time**

Ask students to share their methods for answering each question. How did they arrange the base-10 equipment? What exchanges did they make? If they solved it mentally, can they describe the process they used?

Invite students to try to multiply 7×37 using the same method. Ask one student to talk through the process they used and the rest of the students can follow their instructions.

Practice Book: Students complete Practice Book page 71. They can do this directly after the main activity, as homework, or as the focus of a separate mathematics session to help students consolidate their learning and build fluency.

Students draw the base-10 equipment they use to solve 2-digit × 1-digit calculations, then solve the calculation. *Do you have to exchange any ones-cubes for a tens-rod? How will you draw this?*

Differentiated outcomes	
All students	should partition 2-digit numbers using base-10 equipment.
Most students	will partition and multiply using base-10 equipment or with jottings.
Some students	may partition and multiply mentally or with jottings.

Answers

Student Book page 86

1 246   **2** 69   **3** 48

Practice Book page 71

Check that students have drawn out the correct base-10 equipment for each of the calculations. The answers should be 52 and 74.

Stretch zone: Students may suggest that they multiply the tens and the ones separately and then add them together:

$30 \times 2 = 60, 7 \times 2 = 14$

$60 + 14 = 74$

3E Written methods for multiplication

Discover 2   Student Book page 87 • Practice Book page 72

Specific learning focus

- Multiply 2-digit numbers by single-digit numbers using the grid method.

Global skills

- **Creative skills:** exploring

Key vocabulary

- multiply, product, grid method, tens, ones

Resources

- mini whiteboards and markers

Language support

As students use the grid method, model the language they need to carry out the multiplication and help them to understand the words in the problems that indicate what the calculation should be. Remind students about partitioning when writing their numbers in the grid and encourage them to talk about the 'product' when they find the answer.

Introductory activity

Practise counting on and back in multiples of 20: 20, 40, 60, 80, 100, 120, 140, 160, 180, 200. Then count forward and back in 30s to 300: 30, 60, 90, 120, 150, 180, 210, 240, 300. Repeat for multiples of 40, 50, 60, 70, 80, 90.

Main activity

Look together at page 87 of the Student Book. If you have access to an IWB you could use this. Explain that students are going to multiply bigger numbers using a **grid method**. Ask students to look at the worked example in the Student Book. First read the problem to students. *What calculation so we need to do?* Agree that you need to work out 32×3. They can draw the calculation on their whiteboards as you work through it with them.

Now write on the board 4×23. Ask students to use the grid method to set out 4×23 using the example on page 87 of the Student Book as a template. Talk through with them the process of partitioning the 2-digit number, calculating mentally the separate multiples and then adding to get the total.

Students should now complete the questions on the rest of page 87, remembering to partition the 2-digit numbers correctly, multiply each partition separately, before working out the total.

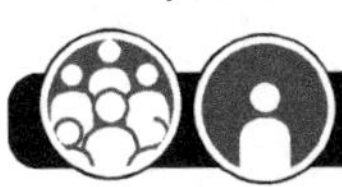

Differentiation

Supporting: Support students in partitioning the numbers correctly and writing them in the grid.

Consolidating: Encourage students to use the grid method to solve the multiplication problems. Check that their grids match their partner's grid before they complete each part of it.

Extending: Encourage students to work independently, using the grid method to help them.

Stretch zone: *Could you have calculated any of these answers mentally? Explain your answer.*

This activity will enable students to demonstrate the extent to which they understand when they can use a mental method, and when they need to use a different method to multiply a 2-digit number by a single digit.

 Reflection time

Ask students to share their methods for answering each question. How did they decide what to write in each part of the grid? If they solved it mentally, can they describe the process they used?

Invite students to try to multiply 7×37 using the grid method. Ask one student to talk through the process they used and the rest of the students can follow their instructions. Compare this method with the partitioning of base-10 equipment in the previous lesson and ask students which method they find easier and why.

Practice Book: Students complete Practice Book page 72. They can do this directly after the main activity, as homework, or as the focus of a separate mathematics session to help students consolidate their learning and build fluency.

Students use the grid method to solve 2-digit × 1-digit calculations. *Do you prefer this method or using base-10 equipment? Why?*

Differentiated outcomes	
All students	should partition 2-digit numbers correctly for the grid.
Most students	will partition and multiply using the grid method.
Some students	may partition and multiply mentally, using the grid method if needed.

Answers

Student Book page 87

1 306

2 88

3 248

Practice Book page 72

1 52

2 74

3 136

4 141

Stretch zone: Answers will vary. Check that students have written a suitable explanation for their preferred method.

3E Written methods for multiplication

Explore 1   Student Book page 88 · Practice Book page 73

Specific learning focus

- Multiply 2-digit numbers by single-digit numbers using base-10 equipment.

Global skills

- **Creative skills:** problem solving
- **Interpersonal skills:** communication

Key vocabulary

- multiply, product

Resources

- base-10 equipment

Language support

Focus on the language of multiplication, reminding students about partitioning, 'multiply' and 'product'. Ask students to talk through a calculation, describing what they are doing to show you that they are using the correct vocabulary.

 **Introductory activity**

Ask a student to choose three digit cards at random (e.g. 1, 3 and 6). Ask the class to form different multiplication calculations using the digits (e.g. 3 × 16). *How many different ones can you make?* (1 × 36, 1 × 63, 3 × 16, 3 × 61, 6 × 13, 6 × 31) *Which ones can you do mentally?*

Choose three different digits and repeat. For example, 2, 4, and 9 would make:

2 × 49, 2 × 94, 4 × 29, 4 × 92, 9 × 24 and 9 × 42.

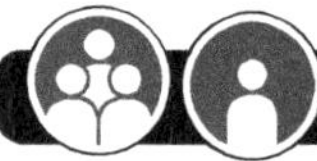 **Main activity**

Look together at page 88 of the Student Book. If you have access to an IWB you could use this. Talk through the worked example in the Student Book to show students how you expect them to set out their work (drawing a place-value grid to record the tens and the ones separately). Choose three of the examples from the introductory activity and, as a class, work through these three questions using base-10 equipment and recording in a place-value grid, as shown. Ensure that you select one example question that requires exchanges (e.g. 3 × 16 or 4 × 29).

Students then complete the activity in the Student Book on page 88. They can use base-10 equipment for support

or, if they are confident, they can partition mentally and then record in the place-value grid.

Encourage students to work in pairs to find the different answers. They need to remember to partition the 2-digit numbers, make the right number of identical rows if using base-10 equipment, then exchange if necessary before working out the total.

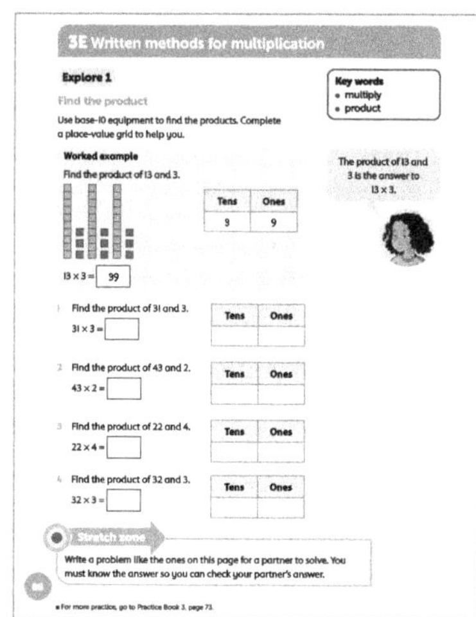

Differentiation

Supporting: Support students with the exchanges they need to make, using the base-10 equipment for support.

Consolidating: Encourage students to work in pairs, using base-10 equipment where needed.

Extending: Students can work out each calculation mentally, jotting down partial products to help them.

Stretch zone: *Write a problem like the ones on page 88 for a partner to solve. You must know the answer so you can check your partner's answer.*

Can students challenge their partner by including the need to exchange ones for a ten?

 **Reflection time**

Ask students to share their methods for answering each question. *How did you arrange the base-10 equipment? What exchanges did you make?* If students solved it mentally, can they describe the process they used?

Invite students to try to multiply 9 × 87 using the same method. Ask one student to talk through the process they used and the rest of the students can follow their instructions.

Practice Book: Students complete Practice Book page 73. They can do this directly after the main activity, as homework, or as the focus of a separate mathematics session to help students consolidate their learning and build fluency.

Students solve 2-digit × single-digit multiplications. Allow students the option of using base-10 equipment and drawing a place-value grid or using a grid method. *Which method do you find easier? Why?*

Differentiated outcomes	
All students	should partition 2-digit numbers using base-10 equipment.
Most students	will partition and multiply using base-10 equipment or by using jottings.
Some students	may partition and multiply mentally or by using jottings.

Answers

Student Book page 88

1 93

2 86

3 88

4 96

Practice Book page 73

1 160

2 128

3 168

4 306

Stretch zone: Students may suggest that they multiply the tens and the ones separately and then add them together.

3E Written methods for multiplication

Explore 2 Student Book page 89 • Practice Book page 74

Specific learning focus

- Multiply 2-digit numbers by single-digit numbers using the grid method.

Global skills

- **Creative skills:** problem solving

Key vocabulary

- multiply, product, grid method, tens, ones

Resources

- 100-square

Language support

As students use the grid method, model the language they need to carry out the multiplication and help them to understand the words and numbers in the problems that indicate what the calculation should be. Remind students about partitioning when writing their numbers in the grid and encourage them to talk about the 'product' when they find the answer.

 **Introductory activity**

Practise counting on and back in harder multiples (e.g. 13), which encourages students to partition mentally and count on in the tens and then the ones. You could use a 100-square to help with this.

 **Main activity**

Ask students to look back at the worked example on page 87 of the Student Book. They can use this layout to help them set out the questions in the Student Book on page 89. Remind them to think about how to solve word problems, using RUCSAC to help them:

Read the question – what is the important information?

Understand the question – what do you need to find out?

Choose the correct method of calculation and operation(s)

Solve the problem – make sure that you follow the steps.

Answer the question – what were you meant to find out?

Check your answer – use the inverse to check your working out.

Students should now complete the questions on page 89 of the Student Book, remembering to partition the 2-digit numbers correctly, then multiply each partition separately, before working out the total. They may do some of the calculation mentally. Students should also check that the answer makes sense in the context of the problem.

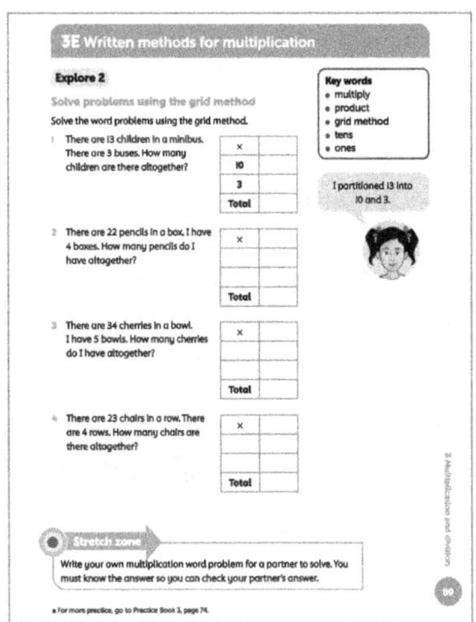

Differentiation

Supporting: Support students in partitioning the numbers correctly and writing them in the grid.

Consolidating: Encourage students to use the grid method to solve the multiplication problems. Check that their grids match their partner's grid before they complete each part of it.

Extending: Ask students to work independently, using the grid method to help them.

Stretch zone: *Write your own multiplication word problem for a partner to solve. You must know the answer so you can check your partner's answer.*

What context will students use? How can they make their problem more challenging?

 **Reflection time**

Ask students to share their methods for answering each question. How did they decide what to write in each part of the grid? If they solved it mentally, can they describe the process they used?

Invite students to try to multiply 9×98 using the grid method. Ask one student to talk through the process they used and the rest of the students can follow their instructions.

Practice Book: Students complete Practice Book page 74. They can do this directly after the main activity, as homework, or as the focus of a separate mathematics session to help students consolidate their learning and build fluency.

Students solve further word problems involving multiplication. They use the grid method to solve them. Check that they are choosing the correct numbers from the word problems and that that they are partitioning them correctly when writing them in the grid.

Differentiated outcomes	
All students	should partition 2-digit numbers correctly for the grid.
Most students	will partition and multiply using the grid method.
Some students	may partition and multiply mentally, using the grid method if needed.

Answers

Student Book page 89

1 39 children

2 88 pencils

3 170 cherries

4 92 chairs

Practice Book page 74

1 84 stickers

2 104 students

3 115 pencils

4 138 children

3F Written methods for division

Discover 1 Student Book page 90 • Practice Book page 75

Specific learning focus

- Divide 2-digit numbers by single-digit numbers using base-10 equipment.

Global skills

- **Creative skills:** exploring
- **Interpersonal skills:** communication

Key vocabulary

- divide, equal groups, regroup, exchange

Resources

- base-10 equipment

Language support

As students learn the process of division, focus on using the correct vocabulary when talking about the stages of the calculations. Students should see and hear the words 'divide', 'division', equal groups', 'regroup', 'exchange'. These words can be made up as a wall poster for students to see at all times.

 **Introductory activity**

Count on in threes to 36 and back. Now ask students some division questions from the 3 times table. For example, ask: *How many threes are there in 24? How many threes are there in 15?*

Repeat for other times tables.

 **Main activity**

Look together at page 90 of the Student Book. If you have access to an IWB you could use this. Explain that students are going to divide bigger numbers using base-10 equipment. Ask them to look at the worked example in the Student Book. They can represent the calculation with their own base-10 equipment as you work through it with them. Ask them to check with their partners that they have both set out the base-10 equipment correctly and when they have formed 56, they should point to each piece as they count.

Explain to students that as they are dividing by 4, they need 4 tens-rods, but they have 5 to start with. One of the tens-rods needs to be **exchanged** for 10 ones-cubes, so they will then have 4 tens-rods and 16 ones-cubes. Share the rods into four equal groups and then share the cubes into four **equal groups**. Each group will have a rod and 4 cubes, making 14.

Students should now complete the questions on the rest of page 90 in the Student Book, remembering to make the right number of ten-rods to share equally, **regrouping** any extra rods, then sharing the rods and cubes into the same number of equal groups as the number being divided by.

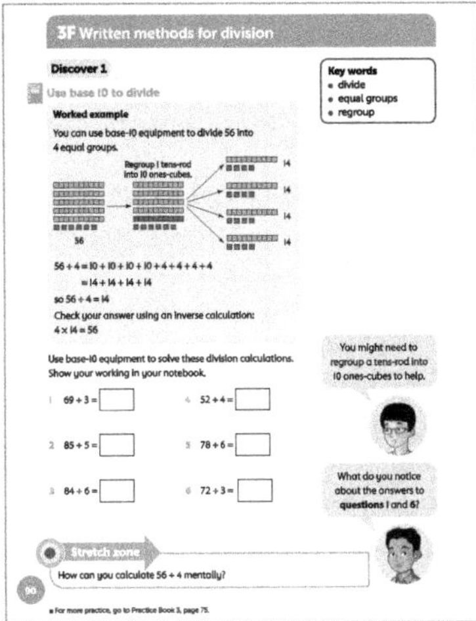

Differentiation

Supporting: Use the base-10 equipment for support.

Consolidating: Encourage students to use jottings and mental methods if possible but use base-10 where needed.

Extending: Students can work out each calculation mentally, using jottings to help them.

Stretch zone: *How can you calculate 56 ÷ 4 mentally?*

This task will enable students to see whether they can carry out the division process mentally, using a strategy such as halving and halving again instead of using base-10 equipment.

 **Reflection time**

Ask students to share their methods for answering each question. *How did you arrange the base-10 equipment? What regrouping did you do?* If they solved it mentally, can they describe the process they used?

Invite students to try to divide 91 by 7 using the same method, inviting one student to talk through the process while the others follow.

Practice Book: Students complete Practice Book page 75. They can do this directly after the main activity, as homework, or as the focus of a separate mathematics session to help students consolidate their learning and build fluency.

Students use base-10 equipment to divide. If they are doing this activity at home, and do not have base-10 equipment, ask them to draw base-10 equipment in their notebook. Point out the note in the Practice Book – they do not have to draw the rods and cubes accurately; they can use a line for a ten-rod and a dot for a ones-cube.

Differentiated outcomes	
All students	should regroup 2-digit numbers using base-10 equipment.
Most students	will regroup and divide using base-10 equipment.
Some students	may regroup and divide mentally, using base-10 if needed.

Answers

Student Book page 90

1	23	**4**	13
2	17	**5**	13
3	14	**6**	24

Practice Book page 75

1	18	**4**	11
2	14	**5**	12
3	13	**6**	18

Stretch zone: Answers will vary but students might suggest that they use their knowledge of multiplication tables to solve mentally: 72 ÷ 6 = 12 and 44 ÷ 4 = 11

3F Written methods for division

Discover 2 — Student Book page 91 • Practice Book page 76

Specific learning focus

- Divide 2-digit numbers by single-digit numbers using a written method.

Global skills

- **Creative skills:** problem solving

Key vocabulary

- divide, equal groups, bus-stop method

Resources

- mini whiteboards and markers

Language support

As students use the bus-stop method, model the language they need to carry out the division. Remind students about 'regrouping' when writing their numbers in the bus-stop.

 **Introductory activity**

Count on in fours up to 480 and back. Now ask students some division questions from the four times table. For example, ask: *How many fours are there in 24? How many fours are there in 32?*

Repeat for other tables.

 **Main activity**

Look together at page 91 of the Student Book. If you have access to an IWB you could use this. Explain that students are going to divide bigger numbers using a written method called short division or the '**bus-stop method**'. Ask students to look at the worked example. They can write the calculation on their whiteboards as you work through it with them.

Now write on the board 72 ÷ 4. Ask students to use the bus-stop method to set out 72 ÷ 4 using the layout shown on page 91 of the Student Book. Talk through with them the process of grouping the tens in the 2-digit number and exchanging leftover tens for ones.

Students should now complete the questions on the rest of page 91.

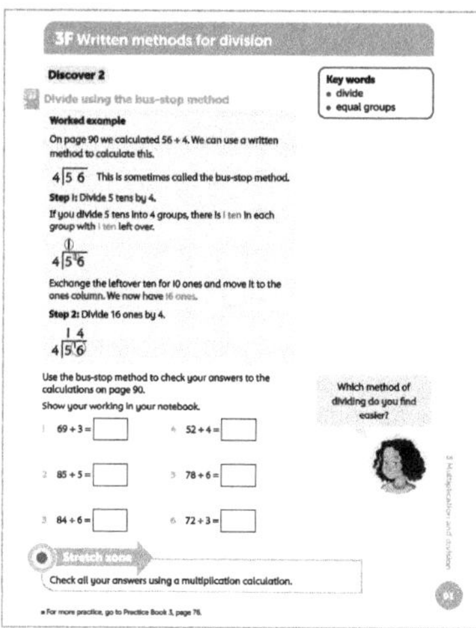

Differentiation

Supporting: Help students to set out the calculations correctly, using the bus-stop layout. They could also use base-10 equipment as they go through each step of the division, but they record it using the bus-stop method.

Consolidating: Encourage students to use mental methods if possible but use base-10 where needed. They record the division using the bus-stop method.

Extending: Encourage students to work out each calculation mentally and record accurately using the bus-stop method.

Stretch zone: *Check all your answers using a multiplication calculation.*

This activity will enable students to see that multiplying is the inverse of dividing. *Which method will you use for multiplying? Which method is quicker/easier?*

 **Reflection time**

Ask students to share their methods for answering each question. How did they decide what to write in each part of the bus-stop? If they solved it mentally, can they describe the process they used?

Invite students to try to calculate 96 ÷ 8 using the bus-stop method, inviting one student to talk through the process while the others follow.

Practice Book: Students complete Practice Book page 76. They can do this directly after the main activity, as homework, or as the focus of a separate mathematics session to help students consolidate their learning and build fluency.

Students solve division calculations using the bus-stop method. Ensure that they are setting out the calculations correctly. Refer them to the example at the top of the page, which shows them how to set out the calculations.

Differentiated outcomes	
All students	should regroup 2-digit numbers for the bus-stop.
Most students	will regroup and divide using the bus-stop method.
Some students	may regroup and divide mentally, or using the bus-stop method.

Answers

Student Book page 91

1	23	**4**	13
2	17	**5**	13
3	14	**6**	24

Practice Book page 76

1	18	**4**	11
2	14	**5**	12
3	13	**6**	18

3F Written methods for division

Explore 1 Student Book page 92 · Practice Book page 77

Specific learning focus

- Divide 2-digit numbers by single-digit numbers using base-10 equipment to solve a word problem.

Global skills

- **Creative skills:** problem solving
- **Real-world skills:** interpreting information

Key vocabulary

- divide, equal groups, regroup

Resources

- base-10 equipment

Language support

Encourage students to read the word problem carefully and pick out any words or phrases that indicate that the calculation needed is a division and which numbers need to be used. Once the answers are found, ask students to state the answer in a sentence to fit the context of the problem.

 **Introductory activity**

Use the following word problem to model the use of RUCSAC.

There are 72 students in a school year, spread across 3 classes equally. How many students are in each class?

Read the question – the important information is the number of students (72), the number of classes (3) and that there are equal class sizes.

Understand the question – you need to find out how many students are in each class.

Choose the correct method of calculation and operation(s) – divide 72 by 3.

Solve the problem – use base-10 equipment to do the division.

Answer the question – there are 24 students in each class.

Check your answer – $24 \times 3 = 72$.

 **Main activity**

Look at question 1 on page 92 of the Student Book. Discuss with students how RUCSAC will help them solve the word problem. Ask students to identify the words and numbers that they need to help them solve the problem. *What do we need to find out? Which words might tell us what we need to do? Which numbers should be in the calculation?* Students then complete question 1 and the remaining questions on page 92. As they work through each problem, remind them to follow the stages in the RUCSAC method. For the calculations, they can use base-10 equipment and make jottings in their notebooks.

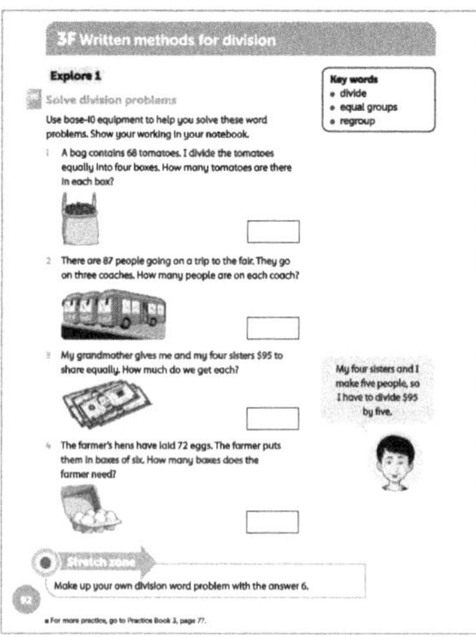

Differentiation

Supporting: Work through more examples talking through each step of the RUCSAC method.

Consolidating: Encourage students to work through each problem using RUCSAC.

Extending: Challenge students to solve problems in ways that make sense to them.

Stretch zone: *Make up your own division word problem with the answer 6.*

Prompt students by getting them to think about how they would use multiplication to check their answer.

 ### Reflection time

Talk through with students their methods for solving the word problems. Ask them to describe the process they used and ask some of them to come to the front and write or draw their calculation on the board for other students to see and discuss.

Practice Book: Students complete Practice Book page 77. They can do this directly after the main activity, as homework, or as the focus of a separate mathematics session to help students consolidate their learning and build fluency.

Students use base-10 equipment to divide. If they are doing this activity at home, and do not have the base-10 equipment, ask them to draw the base-10 equipment in their notebook. Point out the note in the Practice Book – they do not have to draw the tens-rods and ones-cubes accurately; they can use a line for a rod and a dot for a cube.

Differentiated outcomes	
All students	should be able to follow RUCSAC, perhaps with support.
Most students	will be able to solve the problems using RUCSAC.
Some students	may be able to solve word problems in different ways.

Answers

Student Book page 92

1	17 tomatoes	**3**	$19
2	29 people	**4**	12 boxes

Practice Book page 77

1	13 apples	**3**	$24
2	19 people	**4**	15 boxes

3F Written methods for division

Explore 2   Student Book page 93 · Practice Book page 78

Specific learning focus

- Divide 2-digit numbers by single-digit numbers using the bus-stop method to solve a word problem.

Global skills

- **Creative skills:** problem solving
- **Real-world skills:** interpreting information

Key vocabulary

- divide, equal groups, regroup

Resources

- none needed

Language support

Encourage students to read the word problem carefully and pick out any words or phrases that indicate that the calculation needed is a division and which numbers need to be used. While working, ask students to talk through the calculations and use the language of division and regrouping. Once the answers are found, ask students to state the answer in a sentence to fit the context of the problem.

 ### Introductory activity

Give students the following problem.

When a bag of sweets was shared between 4 children, they each received 18 sweets. How many sweets were in the bag at the start?

Ask students how they would calculate the answer. Encourage them to think about fact families and inverse calculations. *Can you describe the connection between division and multiplication?*

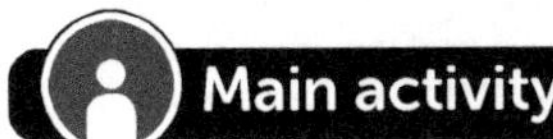

Main activity

Students can be reminded of the RUCSAC process as in the previous lesson, then apply it to solve the problems on page 93 of the Student Book. For each problem, students should identify the important information in the word problem, then use the bus-stop method to calculate the answer. *What do we need to find out? Which words might tell us what we need to do? Which numbers should be in the calculation?* Remind students to interpret the answer in relation to the problem. For example, an answer should say 24 boxes, not just 24 (although students are not asked to do this in the Student Book).

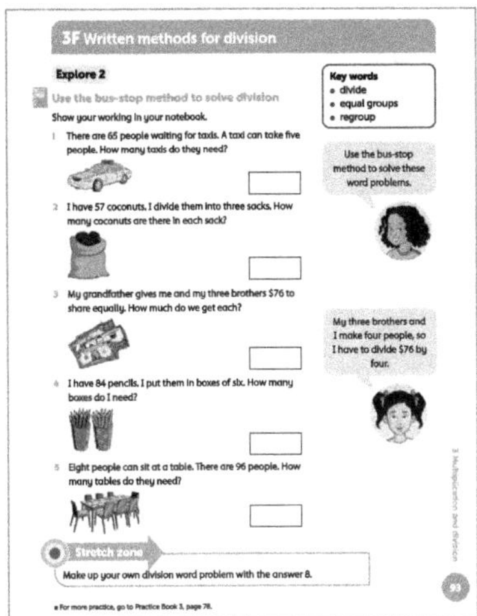

Differentiation

Supporting: Encourage students to explain the way that they are using the bus-stop method to find the answers.

Consolidating: Ask students to estimate answers before they calculate them.

Extending: Ask students to make their own division word problems.

Stretch zone: *Make up your own division word problem with the answer 8.*

Prompt students by getting them to think about how they would use multiplication to check their answer.

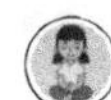

Reflection time

Talk through with students their methods for solving the word problems. Ask them to describe the process they used and ask some of them to come to the front and write or draw their calculation on the board for other students to see and discuss.

Practice Book: Students complete Practice Book page 78. They can do this directly after the main activity, as homework, or as the focus of a separate mathematics session to help students consolidate their learning and build fluency.

Remind students how to set out the bus-stop method to solve the calculation. You could refer them back to the example on page 76 of the Practice Book.

Differentiated outcomes	
All students	should use RUCSAC to help them set out the bus-stop method for doing division.
Most students	will also solve the problems using division by the bus-stop method.
Some students	may make their own division word problems to solve using the bus-stop method.

Answers

Student Book page 93
1 13 taxis

2 19 coconuts

3 $19

4 14 boxes

5 12 tables

Practice Book page 78
1 17 rickshaws

2 32 pineapples

3 23 boxes

4 16 glasses

3G Problem solving with multiplication and division

Discover 1 Student Book page 94 • Practice Book page 79

Specific learning focus

- Multiply and divide 2-digit numbers by single-digit numbers.
- Understand that division can leave a remainder (sometimes known as 'left over').

Global skills

- **Creative skills:** problem solving

Key vocabulary

- multiplication, division, word problem, inverse

Resources

- set of 1–9 digit cards
- lollipop sticks in a jar – each stick has a different student's name on it

Language support

You may need to suggest possible stories for some students. Focus on the vocabulary of multiplication and division with these students. For example, focus on 'multiply', 'divide', 'product', 'quotient', 'remainder'.

 **Introductory activity**

Ask students to suggest words and phrases that they know tell them to do a particular calculation. Make a list of all these words for display. They might suggest, for example, 'share equally', 'they each'. Once the list is created, sort the words or phrases into two columns, one for multiplication and one for division. Students can refer to this as they work on the word problems on page 94 of the Student Book.

 **Main activity**

Talk through the example in the Student Book on page 94. Ask students, for example, *How many are in each row? How many are in each column? What could these represent?* Refer students to the speech bubble for one suggestion. Can they think of another? *What other number story could you think of if the picture was not there? What is the context for your story? What are the objects in your story? How can you multiply or divide those objects?* Students should work on the activity in the Student Book as individuals. The number stories can all be illustrated by a picture, although question 5 does not make a complete array.

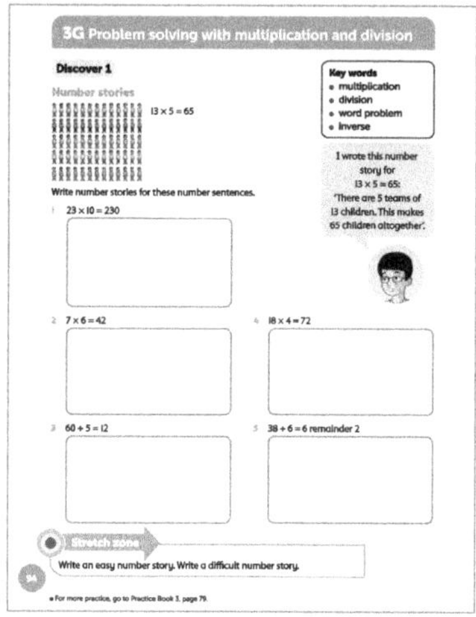

Differentiation

Supporting: Ask students to tell you their number stories and help write the stories in their books.

Consolidating: Encourage students to think of different stories for each number sentence.

Extending: Encourage students to describe how their number story can include using inverse operations.

Stretch zone: *Write an easy number story. Write a difficult number story.*

This activity will enable students to consider what numbers might make easier or more difficult calculations from number stories, for example creating two-step problems.

 **Reflection time**

Select a student by taking a lollipop stick from the jar. They should read out one of their number stories. Each pair then writes down the family of multiplication and division facts that link to the story. Specify that they should choose a number story that does not involve a remainder. Repeat three times.

Practice Book: Students complete Practice Book page 79. They can do this directly after the main activity, as homework, or as the focus of a separate mathematics session to help students consolidate their learning and build fluency.

Students write more number stories to match a multiplication or division number sentence. Two examples are given. *Can you think of a different context for each of your stories?*

Differentiated outcomes	
All students	should tell you their number stories.
Most students	will write down appropriate number stories.
Some students	may write challenging number stories, including two-step number stories.

Student Book page 94

Answers will vary because students choose their own number stories for the multiplication and division number sentences given. Check that the number stories are appropriate for the number sentences.

Practice Book page 79

Answers will vary because students choose their own multiplication and division questions. Check that the stories match the given answers.

3G Problem solving with multiplication and division

Discover 2 Student Book page 95 • Practice Book page 80

Specific learning focus

Use multiplication and division of 2-digit numbers by single-digit numbers to make loop cards with multiplication and division calculations.

Global skills

- **Interpersonal skills:** communication

Key vocabulary

- multiplication, division, multiple

Resources

- A4 paper

Language support

In making loop card sets, model for students the idea of giving a statement about a number and then asking a question to generate a new number. Reinforce the question style: 'Who is___' does not refer directly to a person, but to the person who has that answer.

 **Introductory activity**

Allocate each student in the class a number from 1 to however many students there are, including yourself (e.g. 36). Write the numbers from 1–36 on the board. You start by saying 'I am (your number, e.g. 36)', then look at the list on the board and choose another number from the list. Ask, for example, *Who is 3 × 5?* The student with the number that correctly answers your calculation comes out, crosses off their number from the list and makes up a calculation with one of the remaining numbers as the answer.

The game continues with each student taking a turn as their number is called and they come out to choose a new number from those remaining and ask a question. Eventually, each student will have come out and the final student will be left with 36, which is the teacher's number.

 **Main activity**

Look together at page 95 of the Student Book. If you have access to an IWB you could use this. Show how a loop card game is written, using the first few completed cards in the example given. Students then work individually to make their own loop card game on a piece of A4 paper, using the structure of the game played in the introductory activity. Remind students to use a mixture of multiplication and division problems, and to include single- and 2-digit numbers. Make sure that the questions and answers flow from one card to the next.

When everyone is finished, form small groups and use one set of loop cards to play the game. Each student in the group will have a small number of the cards and so will have to check through their cards each time a question is asked.

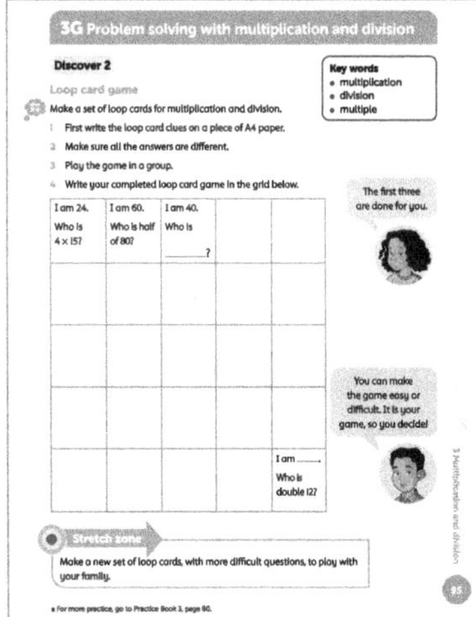

Differentiation

Supporting: Model the questions and responses for students as they play the game in their group.

Consolidating: Suggest that students use 2- and 3-digit numbers.

Extending: Encourage students to explore a range of question types, for example: 'What is half of 8 × 3?'

Stretch zone: *Make a new set of loop cards, with more difficult questions, to play with your family.*

Ask students to give feedback on how their cards worked at home.

 **Reflection time**

Ask students to recall their reasoning as they built their sets of loop cards and describe how they chose some of their calculations. *What was their hardest calculation? What was the easiest?*

Choose one or two students and use their sets of cards with the whole class.

Practice Book: Students complete Practice Book page 80. They can do this directly after the main activity, as homework, or as the focus of a separate mathematics session to help students consolidate their learning and build fluency.

Students make another loop card game. This can be the one used in the Stretch zone, to be used at home, or as an additional game for the group activity.

Differentiated outcomes	
All students	should complete a set of loop cards, perhaps with support.
Most students	will make a set of loop cards without support.
Some students	may make a set of more complex calculations for their loop cards.

3G Problem solving with multiplication and division

Explore 1 Student Book page 96 · Practice Book page 81

Specific learning focus

- Share groups of objects equally with no remainder.

Global skills

- **Creative skills:** problem solving, investigating

Key vocabulary

- multiplication, division, multiple

Resources

- cubes or counters

Language support

As students work on the activities, encourage them to talk about multiples and whether one number is divisible by another. Help them to describe their sets of objects in arrays, referring to the rows and columns and the number of objects in each.

 **Introductory activity**

Show students a pile of 24 cubes. Ask them first to estimate how many there are and how they estimated the number. Now arrange the cubes into an array of 4 rows of

Answers

Student Book page 95

Each set of loop cards will vary. Students should check as they play that each calculation links to the next card until returning to the first card.

Practice Book page 80

Answers will all vary because students make their own sets of loop cards. They should check as they play that the questions and answers match up.

6 cubes. *Now how many cubes are there? How do you know?* Students may say that they counted one row and then multiplied by the number of rows.

Ask students whether the cubes could have been arranged in a different array. They may suggest 6×4, and you can show them by turning the array so that it is the same as 4×6. They may offer other suggestions such as 8×3, 3×8, 2×12, 12×2.

 **Main activity**

Arrange the 24 cubes into a single straight line. Ask students whether they think this is an array. They may say that it is not as there is only a single row, but they may suggest (or you can point out) that you have made 1 row of 24 (1×24) and turning around it can be 24 rows of 1 (24×1).

Explain that every number can be put into an array with 1 row. Now ask them to look at the start of question 1 on page 96 of the Student Book. This shows an array of 5×6 for 30 counters. Discuss with them what other array and multiplication fact they can make by turning this one around (6×5). *What other arrays can you think of for 30 counters?* As they suggest each one, they can draw it to complete question 1.

Students should now work individually to complete the remaining questions. As they work, ask: *How do you know that you have found all of the possible arrays for each number?* Have they done the reverse of each one?

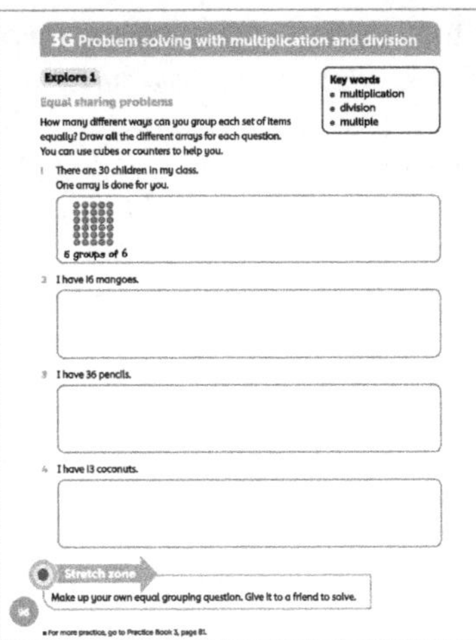

Differentiation

Supporting: Encourage students to explain the way that they are using their knowledge of multiplication and division facts to find the answers.

Consolidating: Ask students to predict how many arrays they will find for each number.

Extending: Ask students to find all the possible arrays for each number.

Stretch zone: *Make up your own equal grouping question. Give it to a friend to solve.*

This activity will enable students to show that they understand something about divisibility.

 **Reflection time**

Ask students to describe the arrays they found for 36 pencils. *How many different arrays did you find? Why do you think there are so many for 36?* Then ask how many they found for 13 coconuts. *Why do you think there were so few for 13?*

Practice Book: Students complete Practice Book page 81. They can do this directly after the main activity, as homework, or as the focus of a separate mathematics session to help students consolidate their learning and build fluency.

Students complete a table to show which numbers can be divided equally (with no remainder) into different-sized groups. They can use cubes or counters to support them, if needed.

Differentiated outcomes	
All students	should make at least one array for any number.
Most students	will use known facts to help make more arrays.
Some students	may describe the arrays and explain how they know they have found them all.

Answers

Student Book page 96

1 5 × 6, 6 × 5, 10 × 3, 3 × 10, 2 × 15, 15 × 2, 30 × 1, 1 × 30

2 1 × 16, 16 × 1, 2 × 8, 8 × 2, 4 × 4

3 1 × 36, 36 × 1, 2 × 18, 18 × 2, 3 × 12, 12 × 3, 4 × 9, 9 × 4, 6 × 6

4 1 × 13, 13 × 1

Practice Book page 81

Number	1 group	2 groups	3 groups	4 groups	5 groups
9	✓	✗	✓	✗	✗
10	✓	✓	✗	✗	✓
11	✓	✗	✗	✗	✗
12	✓	✓	✓	✓	✗
13	✓	✗	✗	✗	✗
14	✓	✓	✗	✗	✗
15	✓	✗	✓	✗	✓
16	✓	✓	✗	✓	✗
17	✓	✗	✗	✗	✗
18	✓	✓	✓	✗	✗
19	✓	✗	✗	✗	✗
20	✓	✓	✗	✓	✓

Stretch zone: 23. The tick will be in the 1 group column.

3G Problem solving with multiplication and division

Explore 2
Student Book page 97 · Practice Book page 82

Specific learning focus

- Solve word problems using multiplication and division calculations.

Global skills

- **Creative skills:** problem solving
- **Interpersonal skills:** communication

Key vocabulary

- multiplication, multiple, division, divisible

Resources

- cubes or counters

Language support

As students work on word problems, help them to understand what is being asked by using the RUCSAC acronym. Help them to interpret the words in the problems that imply what calculations are needed and to make sense of the answer by referring back to the wording in the problem.

Introductory activity

Select two single-digit cards and use them to make a 2-digit number. Then write down this number multiplied by 10. For example, if you pick 3 and then 7, you could write: $37 \times 10 = 370$.

Make up a number story. For example: *I run for 37 minutes every day. In ten days, I have run for 370 minutes.* Repeat for two single-digit numbers multiplied together. This time ask pairs of students to make up a story.

Main activity

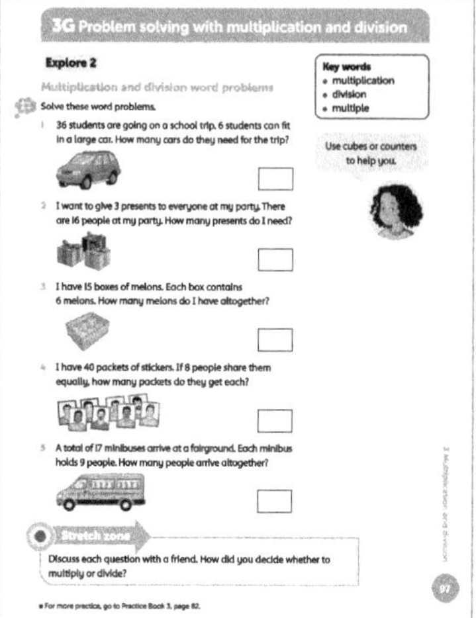

Ask students to look at the first problem on page 97 of the Student Book. Work on this as a class, using RUCSAC and identify what type of calculation is needed and which words tell us this. You could refer students to the list of words made in the Discover 1 lesson on page 94.

Which numbers tell us the calculation we need? Once the calculation has been identified, students can complete the calculation and then reread the problem to check what it is asking. *Does your answer make sense? What calculation can you do to check that it is correct?*

Students then go on to work individually on the remaining word problems on page 97 of the Student Book.

Differentiation

Supporting: Model the use of RUCSAC and help students to find which calculation to use.

Consolidating: Ask students to solve the problems with the help of RUCSAC.

Extending: Encourage students to solve the problems using any method and check the answers.

Stretch zone: *Discuss each question with a friend. How did you decide whether to multiply or divide?*

This activity will enable students to focus on the way the vocabulary is used to help them decide. They can use the class list of words to help them.

 Reflection time

Talk through with students their methods for solving the word problems. Ask them to describe the process they used. Ask individual students to come to the front and write or draw their calculation on the board for other students to see.

Practice Book: Students complete Practice Book page 82. They can do this directly after the main activity, as homework, or as the focus of a separate mathematics session to help students consolidate their learning and build fluency.

Encourage students to write out the answer in full. For example, in question 1 they write: 'They need 7 cars. Not all the cars are full.'

Differentiated outcomes	
All students	should identify the calculations needed in word problems.
Most students	will be able to carry out the calculations.
Some students	may be able to solve word problems using a range of methods.

Student Book page 97

1 6 cars

2 48 presents

3 90 melons

4 5 packets

5 153 people

Practice Book page 82

1 7 cars. Not all the cars are full.

2 6 cherries with 2 left over

3 60 eggs

4 7 boxes. Not all the boxes are full.

5 56 people

3 Multiplication and division

Connect Student Book page 98

Big idea

- I can use multiplication facts, place-value facts, and the relationship between multiplication and division, to help me multiply and divide.

Global skills

- **Creative skills:** investigating
- **Interpersonal skills:** communication

Key vocabulary

- multiples, multiplication, multiplication facts

Resources

- cubes (enough for at least 32 per pair)

Language support

Students may need support with the language of dimension when looking at 3D shapes. You could label a picture of a cuboid showing the length, width and height of the shape and leave this on display throughout this lesson.

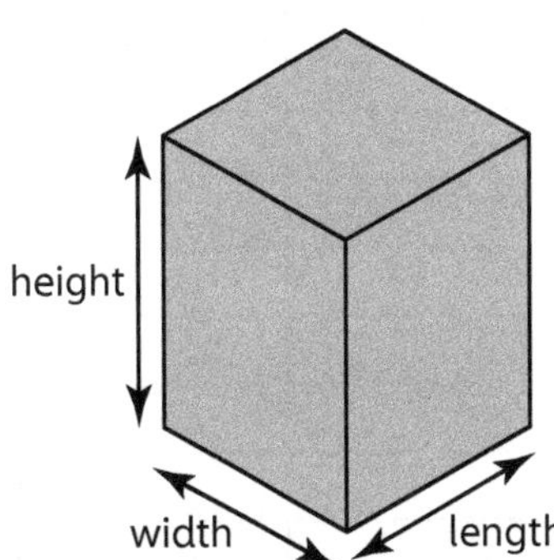

 Introductory activity

Ask students to look at the different rows and columns of various cuboids. Explain that you can relate this to multiplication. Look together at page 98 of the Student Book. Use an IWB, if possible. Refer students to the photograph of the cube building, which is the Pompidou Centre in Malaga, Spain.

How many cubes are there on the top layer of the building? How do you know? There are 36 because there are 6 rows and 6 columns. *Six multiplied by six is thirty-six.* You can also remind students of the terms 'arrays', 'rows' and 'columns'.

Main activity

Students should work in mixed-attainment groups on this activity. Students should go straight into the investigation without an introduction. This will encourage them to think for themselves. Make sure that they have enough cubes per pair so that they can make the cuboids.

As they begin to find different cuboids, encourage them to come up with ways of recording the different cuboids they find. They may choose to draw them or they may come up with a way of recording them by dimensions.

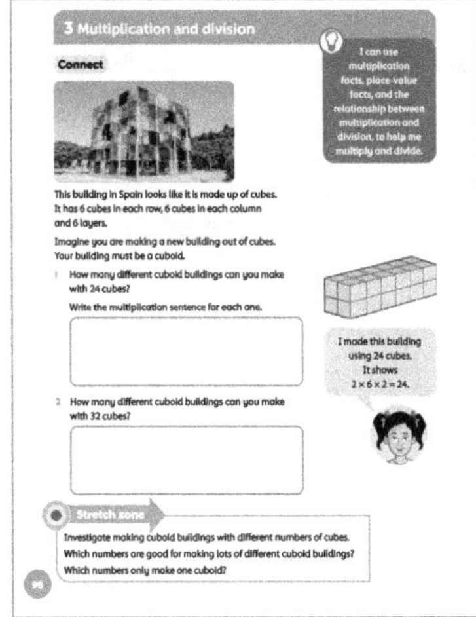

Differentiation

All students should work in mixed-attainment groups for this activity. You may have to help write the answers for those less confident in writing but all students should join in the discussion.

Stretch zone: *Investigate making cuboid buildings with different numbers of cubes.*

Which numbers are good for making lots of different cuboid buildings?

Which numbers only make one cuboid?

Check that students have found numbers that make several different cuboids, such as 18, 24 or 36, and numbers that only make one cuboid (one row of single cubes) such as 2, 3, 5, 7, 11, 13, 17 and so on.

 ### Reflection time

Still within their pairs from the main activity, students should present their solutions. To help draw out some of their understanding, ask, for example: *How did you decide how to arrange the 24 cubes in the first part? How did the arrangement help you write the number sentence? Could you write the number sentence in different ways from the same cube arrangement?* Ask students to compare the different arrangements with different numbers of cubes. *Why are some numbers better for making lots of different cuboid buildings?* It may be appropriate to remind students of the language of 'multiples'. For example, 24 is a multiple of 2, 3, 4, 6, 8 and 12, which might help with arranging the cubes into a cuboid.

Differentiated outcomes	
All students	should work in mixed-attainment groups for this activity. You may have to help write the answers for those less confident in writing, but all students should join in the discussion.

3 Multiplication and division

Review Student Book page 99 · Practice Book page 83

Global skills

Self-development skills: reflecting on learning

Student Book

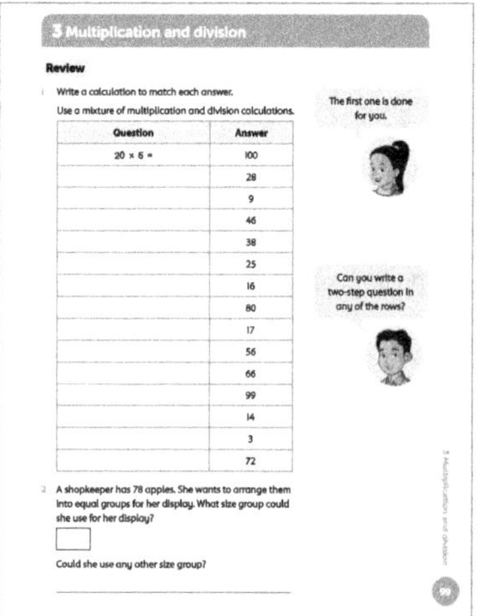

With young children, assessment activities are most effective when carried out as an everyday classroom activity. Students should be able to work on developing their knowledge and understanding of multiplication and division facts and then use them in calculations. It is important for students to learn all their times table facts,

Student Book page 98

1 24 cubes in the cuboid

$1 \times 1 \times 24$

$1 \times 2 \times 12$

$1 \times 3 \times 8$

$1 \times 4 \times 6$

$2 \times 2 \times 6$

$2 \times 3 \times 4$

2 32 cubes in the cuboid

$1 \times 1 \times 32$

$1 \times 2 \times 16$

$1 \times 4 \times 8$

$2 \times 2 \times 8$

$2 \times 4 \times 4$

including the division facts, and use these in different areas of mathematics such as fractions, ratio and proportion, and more complex multiplication and division calculations with larger numbers. Watch as students increase their knowledge and use it to solve more complex problems using multiplication, division and fractions.

Answers

1 Check that students have written correct multiplication and division sentences for each answer. For example:

Question	Answer
20×5	100
4×7	28
$27 \div 3$	9
23×2	46
$76 \div 2$	38
5×5	25
$2 \times 2 \times 4$	16
$2 \times 5 \times 8$	80
17×1	17
$112 \div 2$	56
11×6	66
9×11	99
$28 \div 2$	14
1×3	3
$6 \times 3 \times 4$	72

2 The shopkeeper could display the apples in groups of 2, 3, 6, 13, 26 or 39.

Practice Book

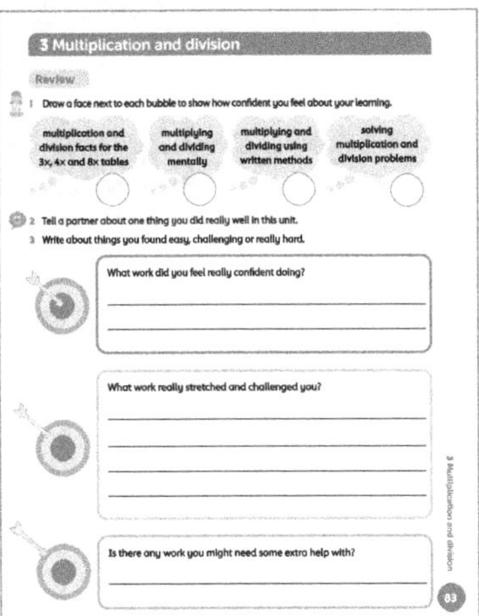

It is appropriate to complete this Practice Book review as a whole-class discussion. You may choose to keep a record of the class discussion or a copy of the review page for your own records. The review provides an opportunity for students to reflect on their learning from the unit, to discuss any areas of mathematics that they feel went particularly well, and any areas that they feel less confident about. Ensure that all students have a copy of the Student Book as a reminder of the areas of mathematics that they have worked on in this unit.

Allow students plenty of time for discussion before asking them to complete the Practice Book page individually, and then, if appropriate, to share their responses with the rest of the class. If students complete this self-assessment at home, encourage them to discuss this with adults. Make a note of areas that students still feel unsure about. For example, how do students feel about the different methods they have learned? Are they more confident using mental methods or written methods? Are they more confident when using concrete materials to support either mental or written methods? Are they confident choosing a suitable method when solving problems? Which methods do they feel they still need help with?

Additional material

There are additional end-of-unit assessments available on the *Oxford Owl for School* website.

4 Fractions

Overview

Big idea

Students have previously been introduced to fractions in terms of halves and quarters of shapes and amounts. The big idea in Year 3 is that there is more to fractions than just halves and quarters; that fractions come in all sorts of shapes and sizes and we can write them in different ways. For example, using tenths, students count in tenths, comparing the fraction and decimal equivalents and dividing whole numbers by 10 to get tenths.

In this unit, students are introduced to calculating with fractions for the first time, as they add and subtract fractions with the same denominator. Their previous work on equivalent fractions is developed, as they continue recognising and finding equivalent fractions, and begin to compare and order fractions.

Look out for

- **Students who are unable to recognise equivalent fractions, such as $\frac{5}{10}$ and $\frac{1}{2}$ or $\frac{2}{3}$ and $\frac{4}{6}$.** To overcome this, display a fraction wall showing fractions of different denominators. Explain to students how to read equivalent fractions from the fraction wall by identifying where vertical lines on different rows line up.
- **Students who use the numerator as the division calculation when finding non-unit fractions of amounts.** Talk through the steps of finding, for example, $\frac{3}{5}$ of ten objects, first finding $\frac{1}{5}$, by dividing the ten objects by 5, and then using that to find 2 or more fifths by multiplying the numerator by the answer.
- **Students who find it difficult to halve numbers with odd tens digits (e.g. 58).** Help students with partitioning to halve the tens and ones separately. So, half of 58 is half of 50 plus half of 8, or $25 + 4 = 29$.

Possible misconceptions

- **Students think that the bigger the denominator, the bigger the fraction.** To help with this, divide a rectangle into thirds and another, the same size, into fifths to show that the greater the number of parts, the smaller the unit fraction.
- **Students think that they need to add the numerators *and* the denominators when adding fractions.** Model adding (and subtracting) fractions using concrete resources such as number rods to demonstrate that the denominator stays the same when adding and subtracting.
- **Students think that you can't halve odd numbers.** Show students that halving an odd number leaves a remainder of 1 which can be halved to make two halves. For example, half of 7 is 3 remainder 1, or $3\frac{1}{2}$.

Key vocabulary and language structures

- one whole, one half, two halves
- one quarter, two quarters, three quarters, four quarters
- halves, thirds, quarters, eighths, tenths
- halve (verb); one third, two thirds; one tenth two tenths, ..., nine tenths
- decimal, decimal point, equivalent, equivalent fraction
- fifths, sixths, sevenths, ninths, twelfths
- share, amount, same, add, equal
- numerator, denominator, number line
- greater than (>), less than (<), more than, smaller than, order
- unit fraction, non-unit fraction

Learning objective	E	4A	4B	4C	4D	4E	C	R
Count up and down in tenths; recognise that tenths arise from dividing an object into 10 equal parts and in dividing one-digit numbers or quantities by 10.		✓						
Recognise, find and write fractions of a discrete set of objects: unit fractions and non-unit fractions with small denominators.			✓					✓
Recognise and use fractions as numbers: unit fractions and non-unit fractions with small denominators.			✓					✓
Recognise and show, using diagrams, equivalent fractions with small denominators.	✓				✓			✓
Add and subtract fractions with the same denominator within one whole $\left(\text{for example: } \frac{5}{7} + \frac{1}{7} = \frac{6}{7}\right)$.				✓				
Compare and order unit fractions, and fractions with the same denominators.						✓		
Solve problems that involve all of the above.							✓	✓

4 Fractions

Engage Student Book page 100

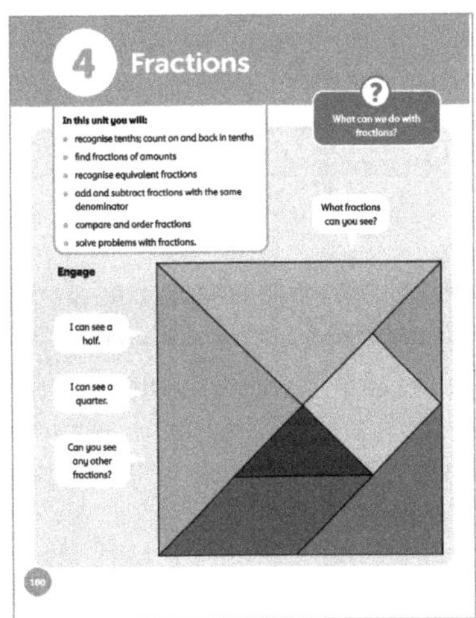

Big question

- What can we do with fractions?

Global skills

- **Creative skills:** exploring
- **Interpersonal skills:** communication

Key vocabulary

- equivalent fractions, one whole, half, halves, third, quarter, fifth, sixth, eighth, tenth, sixteenth

Resources

- mini whiteboards and markers
- tangram sets, one per pair

Language support

Use the tangram patterns that students make to create a poster that shows the following key vocabulary: part, equal parts, fraction, one whole, quarter, half, third, eighth, tenth. Display this in the classroom for students to reference throughout this unit. Students or the teacher can add new fraction-relevant terms and diagrams to the poster, which should remain visible and added to as students work through the rest of the unit.

 Introductory activity

In pairs, students list on their whiteboards all the fractions they know. They should write each fraction and draw an image to illustrate each fraction. Share all the fractions and images as a whole class. Decide which is the smallest fraction and which is the largest. *How do you know that is the smallest? How do you know that is the largest?*

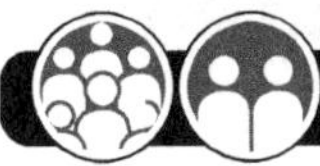 Main activity

Look together at page 100 of the Student Book. If you have access to an IWB you could use that. Ask students to look at the tangram image in the Student Book. Hold a discussion about what a tangram is. Ask, for example: *Have you seen a tangram before? How many pieces does it have? What shapes can you see? Which is the biggest piece? Can you see any pieces that are the same shape but different sizes? What about the same size but different shapes? How do you know these are the same size?*

Ask students in pairs to use their own tangram pieces, and to discuss the answers to these questions.

- *What fractions can you see?*
- *Can you see any other fractions?*
- *If the tangram is 'one whole', can you tell me what fraction of the whole this shape is?*
- *How many of the smaller pieces can fit over the larger ones? So what fraction of the whole shape is the smaller piece?*

Differentiation

Supporting: Help students to identify the simpler fractions that are visible, for example the green and orange quarters.

Consolidating: Ask students what fraction each piece of the tangram represents. They should understand the relationship between quarters and eighths and sixteenths.

Extending: Ask students how many different fractions they can make using their set of tangram pieces. Encourage students to describe them using a wide range of fractions.

 **Reflection time**

There are many tangram activities that students may enjoy, which you can search for on the internet. Handling the pieces is a fun way to develop spatial awareness and careful questioning can focus on fraction work. Some research on the internet will give you many different tangram puzzles. For example, one such puzzle is to use the pieces of the tangram to make this camel:

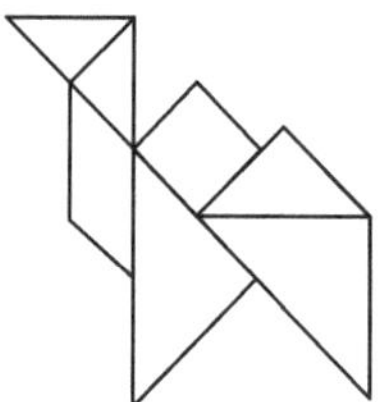

Ask students in pairs to make the camel using their pieces. *Can you design your own animal shape using the pieces? Can you describe it using fractions?* For example, the camel's head (the two smaller triangles) takes up $\frac{1}{8}$ of the whole camel.

4A Tenths

Discover Student Book page 101 • Practice Book page 84

Specific learning focus

- Count in tenths and express tenths as fractions or decimals.

Global skills

- **Real-world skills:** interpreting information

Key vocabulary

- tenths, decimal, decimal point

Resources

- counting stick

Language support

Add to the fraction display using several shapes including circles, squares and triangles. For example, cut one large shape in half. Label one '$\frac{1}{2}$' and the other 'half'. Display the two halves together to show one whole. Add a label saying 'Two halves make one whole. Both halves must be the same size or they are not halves.'

Underneath, display other shapes divided into, for example, tenths, with each one labelled and the sentence 'Ten tenths make one whole. Each of the tenths are the same size.'

 **Introductory activity**

Show students a bar of chocolate with ten equal pieces. *How many pieces are there? Are they all the same size?* Explain that as there are ten equal pieces, each piece is $\frac{1}{10}$ of the chocolate bar.

Break the chocolate into its individual pieces and then hold up different numbers of pieces in turn (e.g. seven pieces). *What fraction of the bar do I have?* $\left(\frac{7}{10}\right)$ Repeat with different numbers of pieces, asking for the fraction of the whole each time.

 **Main activity**

Hold up the counting stick marked into ten equal sections. Tell students that one end is 0 and the other end is 1. Point to halfway along the stick and ask what fraction this is. $\left(\frac{1}{2}\right)$ Now count the **tenths** from the 0 end until you get to halfway. *How many tenths have you counted?* (5) *This shows that $\frac{5}{10}$ is the same as $\frac{1}{2}$.*

Start from 0 and count along the stick together in tenths: one tenth, two tenths and so on. Then count back from ten tenths: nine tenths, eight tenths and so on

Draw some column headings on the board to show hundreds, tens, ones and then tenths. Write 0.1 under the correct column heading and explain to students that this represents one tenth. Explain that the **decimal point** separates the whole numbers from the **decimal fractions**.

Now count up and back on the counting stick in decimal tenths: 0.1, 0.2, 0.3 and so on to 1.0, then 1.0, 0.9, 0.8 and so on.

Students should then complete the counting activities on page 101 of the Student Book.

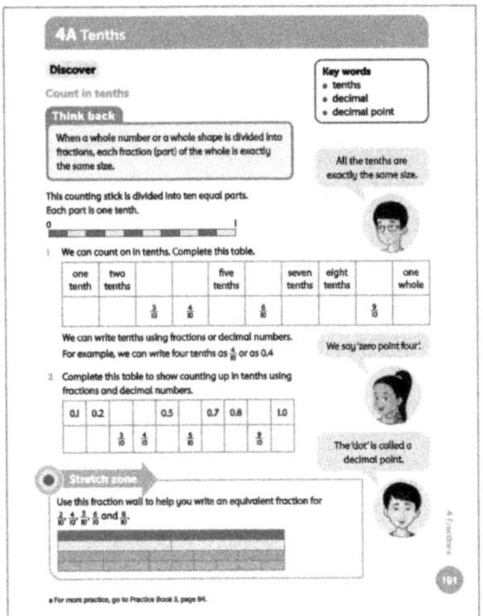

Differentiation

Supporting: Use concrete materials to support counting in tenths. Use base-10 equipment to consolidate fractions or decimals in tenths up to 1, using a tens-rod as a whole one and the ones-cubes as tenths.

Consolidating: Ask students to spot number bonds to 1 in tenths (e.g. 0.4 + 0.6 = 1.0).

Extending: Encourage students to count in tenths beyond 1.0.

Stretch zone: *Use this fraction wall to help you write an equivalent fraction for* $\frac{2}{10}, \frac{4}{10}, \frac{5}{10}, \frac{6}{10}$ *and* $\frac{8}{10}$.

Students should be able to recognise that $\frac{2}{10} = \frac{1}{5}$, $\frac{4}{10} = \frac{2}{5}$, $\frac{5}{10} = \frac{1}{2}$, $\frac{6}{10} = \frac{3}{5}$ and $\frac{8}{10} = \frac{4}{5}$.

 **Reflection time**

Check with students that they completed the tables in the Student Book correctly with tenths in words, fractions and decimals. Ask them to describe to you what is the same and what is different about a fraction and a decimal.

Practice Book: Students complete Practice Book page 84. They can do this directly after the main activity, as homework, or as the focus of a separate mathematics session to help students consolidate their learning and build fluency.

They need to recognise different numbers of tenths in diagrams and write the value as a fraction and a decimal.

Ask questions about equivalents: *What is another way to write* $\frac{5}{10}$? *What about* $\frac{2}{10}$? *And* $\frac{4}{10}$?

Differentiated outcomes	
All students	should be able to count in tenths up to 1.
Most students	will count in tenths as fractions or decimals.
Some students	may count beyond 1 in tenths as fractions or decimals.

Answers

Student Book page 101

1

one tenth	two tenths	three tenths	four tenths	five tenths	six tenths	seven tenths	eight tenths	nine tenths	one whole
$\frac{1}{10}$	$\frac{2}{10}$	$\frac{3}{10}$	$\frac{4}{10}$	$\frac{5}{10}$	$\frac{6}{10}$	$\frac{7}{10}$	$\frac{8}{10}$	$\frac{9}{10}$	$\frac{10}{10}$

2

0.1	0.2	0.3	0.4	0.5	0.6	0.7	0.8	0.9	1.0
$\frac{1}{10}$	$\frac{2}{10}$	$\frac{3}{10}$	$\frac{4}{10}$	$\frac{5}{10}$	$\frac{6}{10}$	$\frac{7}{10}$	$\frac{8}{10}$	$\frac{9}{10}$	$\frac{10}{10}$

Practice Book page 84

1 $\frac{6}{10}$ 0.6

2 $\frac{3}{10}$ 0.3

3 $\frac{9}{10}$ 0.9

4 $\frac{2}{10}$ 0.2

5 $\frac{5}{10}$ 0.5

Stretch zone:

Same: the numerator of the fraction is the same digit as the digit in the tenths place in a decimal.

Different: we write 10 as the denominator in the fraction to tell us these are tenths, but in decimals it is the place value of the digit that tells us they are tenths.

4A Tenths

Specific learning focus

- Recognise the equivalence between the decimal fraction and vulgar fraction forms of halves, quarters, tenths and hundredths.

Global skills

- **Creative skills**: problem solving

Key vocabulary

- tenths, decimal, decimal point

Resources

- mini whiteboards and markers
- set of 1–9 digit cards
- base-10 equipment

Language support

Explain to students that you write a zero in front of decimal numbers less than 1. For example: '0.4' not '.4'. It Is easy to miss the decimal point, so the zero makes it clear. We read this as 'zero point four'. For decimals with two places (e.g. 0.25), make sure that students are saying 'zero point two five' and not 'zero point twenty-five'.

Introductory activity

Pick two digit cards. Create a number with one decimal place with each card. So, for example, if you pick a 2 and a 7, write 0.2 and 0.7. Use the cards to make an addition calculation: 0.2 + 0.7 =.

Students should write the answer on their whiteboards and show you. Ask for the strategy each time. Focus particularly on the additions that bridge 1 (e.g. 0.6 + 0.5 = 1.1). Point out how our knowledge of adding single-digit numbers can help us when adding decimals.

Repeat several times.

Main activity

Explain to students that they will use base-10 equipment to represent decimal fractions. Say that each cube will represent one tenth. Ask them to tell you what fraction is represented by 3 cubes $\left(\frac{3}{10}\right)$, 7 cubes $\left(\frac{7}{10}\right)$ and 5 cubes $\left(\frac{5}{10} = \frac{1}{2}\right)$, for example.

Now show ten cubes and ask what number they think this represents. Some students may say ten tenths, which is correct, but then remind students how we can exchange 10 of the smaller unit to make 1 of the next larger unit. 10 tenths is the same as 1 whole, so exchange the 10 cubes for 1 rod and then this represents 1 whole.

Now show 12 cubes. *What number is this?* Students may say 12 tenths, but then exchange 10 cubes for a rod to get 1 whole and 2 tenths. Write on the board 1.2 and explain how the whole number part is written to the left of the decimal point and the fraction part is written to the right of the decimal point.

Discuss some other decimal equivalents for fractions, for example: $2\frac{1}{2} = 2.5$, $3\frac{4}{10} = 3.4$, $1\frac{7}{10} = 1.7$.

Ask students to count up from 0 in tenths to 3 wholes, and back to zero. Make sure that they are clear about how to cross the ones boundaries: …, 0.8, 0.9, 1.0, 1.1, … Students should then work to complete the activities in the Student Book on page 102.

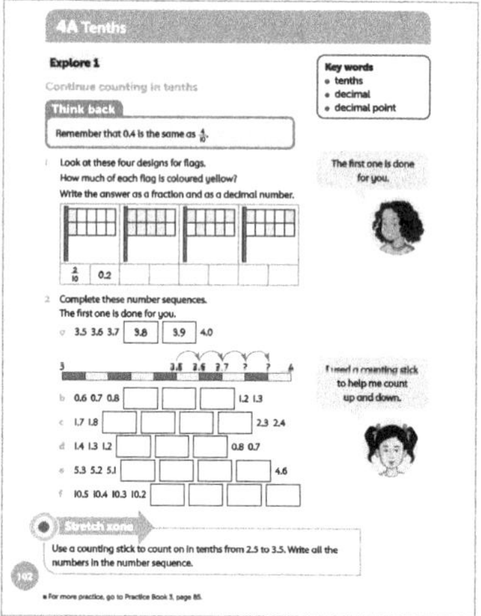

Differentiation

Supporting: Encourage students to read the number sequences aloud in question 2 so that they can hear the pattern.

Consolidating: Encourage students to make up sequences of increasing fractions of their own.

Extending: Ask students to share their strategies for completing the missing number sequences with others.

Stretch zone: *Use a counting stick to count up in tenths from 2.5 to 3.5. Write all the numbers in the number sequence.*

Encourage students to look for the pattern when counting in tenths. *Can you go beyond 3.5? Can you tell me the next five numbers?*

 **Reflection time**

Choose a student to share their answer to each of the parts of question 2. Encourage them to be clear about the strategy that they used. For example, in part e they might be counting in tenths (52 tenths, 51 tenths, 50 tenths, 49 tenths,…) or they may be counting in wholes and tenths (five point 2, five point 1, five point zero, four point nine,…). Choose confident students for this, whom you have identified during the activity.

Practice Book: Students complete Practice Book page 85. They can do this directly after the main activity, as homework, or as the focus of a separate mathematics session to help students consolidate their learning and build fluency.

Students complete decimal number sequences with larger numbers. *Look carefully at the numbers in each sequence to see if the sequences are counting on or counting back.*

Differentiated outcomes	
All students	should count on or back in tenths.
Most students	will recognise the equivalence between tenths as fractions and decimals.
Some students	may complete sequences of fractions in tenths and share their strategies.

Answers

Student Book page 102

1 $\frac{2}{10}$ 0.2 $\frac{8}{10}$ 0.8 $\frac{5}{10}$ 0.5 $\frac{3}{10}$ 0.3

2 **b** 0.9, 1.0, 1.1

   c 1.9, 2.0 (or 2), 2.1, 2.2

   d 1.1, 1.0 (or 1), 0.9

   e 5.0 (or 5), 4.9, 4.8, 4.7

   f 10.1, 10.0 (or 10), 9.9, 9.8

Stretch zone:

2.5, 2.6, 2.7, 2.8, 2.9, 3, 3.1, 3.2, 3.3, 3.4, 3.5

Practice Book page 85

	Start number	Sequence						
	1.0	1.1	1.2	1.3	1.4	1.5	1.6	1.7
1	1.2	1.3	1.4	1.5	1.6	1.7	1.8	1.9
2	2.2	2.3	2.4	2.5	2.6	2.7	2.8	2.9
3	2.4	2.5	2.6	2.7	2.8	2.9	3.0	3.1
4	3.5	3.4	3.3	3.2	3.1	3.0	2.9	2.8
5	5.5	5.4	5.3	5.2	5.1	5.0	4.9	4.8
6	15.5	15.4	15.3	15.2	15.1	15.0	14.9	14.8
7	21.5	21.6	21.7	21.8	21.9	22.0	22.1	22.2
8	22.5	22.4	22.3	22.2	22.1	22.0	21.9	21.8
9	25.7	25.8	25.9	26.0	26.1	26.2	26.3	26.4
10	29.6	29.7	29.8	29.9	30.0	30.1	30.2	30.3

In rows 11 and 12 students choose their own sequence. Check that they have completed the sequence correctly.

Stretch zone: Answers will vary but will include any decimal, with one decimal place, from 3.2 to 4.0. (Students may also include decimals with more than one decimal place, although they have not yet learned about hundredths or thousandths and beyond.)

4A Tenths

Explore 2
Student Book page 103 • Practice Book page 86

Specific learning focus
- Moving digits one place to the right when dividing by 10.

Global skills
- **Creative skills:** problem solving
- **Real-world skills:** interpreting information
- **Interpersonal skills:** communication

Key vocabulary
- tenths, decimal, decimal point

Resources
- mini whiteboards and markers
- place-value grids

Language support
Encourage students to say the number names of decimals correctly. For example, they should say 'zero point five' and not 'nought point five'. Use the language of place value, asking: *How many tenths are there in 0.4? How many tenths are there in 1.3?*

 **Introductory activity**

Ask students to count together in tenths, starting from $\frac{1}{10}$ and continuing up to 2 wholes, then count back down in tenths from 2 wholes to 0.

Ask questions about tenths, for example: *How many tenths in 0.7?, How many tenths in 1.3?, What is the total of $\frac{7}{10}$ and $\frac{4}{10}$?*

 **Main activity**

Ask each student to draw a place-value grid on their whiteboard showing tens, ones and tenths.

Tens	Ones	•	Tenths

Ask students to write the number 8 in the place-value grid, in the ones column. Explain that when we divide a number by 10, the digits move one place to the right. Explain that they will need to use 0 as a place holder in the ones column, so dividing 8 by 10 gives us 0.8.

Now repeat with the number 3 in the ones column and write what you get when you divide 3 by 10. (0.3)

This time write 24 in the place-value grid, with 2 in the tens and 4 in the ones column. Model how, when each digit moves one column to the right, the 2 tens become 2 ones and the 4 ones become 4 tenths and there is no need to use 0 as a place holder. 24 ÷ 10 = 2.4

Students should work with a partner and look at the table in question 1 on page 103 of the Student Book. They describe to each other what they notice about how the value of digits change when divided by 10. Listen to their discussions and ask questions to check their understanding. They should then complete question 2.

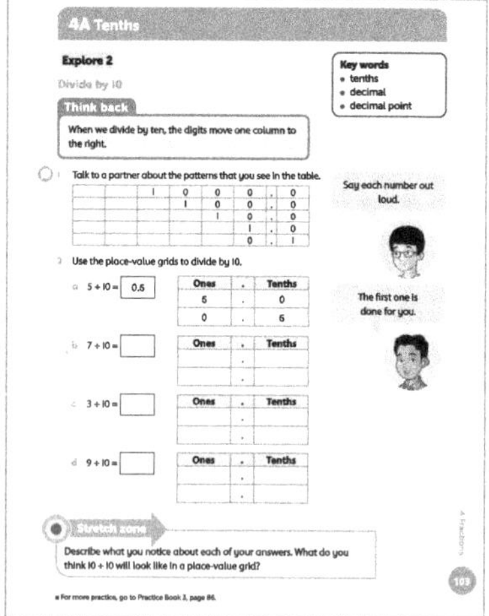

Differentiation

Supporting: Continue to use place-value grids to support division by 10. Base-10 equipment can help consolidate bonds to 1 in tenths and help with small addition of tenths by using ones-cubes as tenths and exchanging 10 ones-cubes for a tens-rod which represents 1 whole.

Consolidating: Ask students to fill in grids of 10 squares to show different numbers of tenths (e.g. $\frac{7}{10}$).

Extending: Ask students to divide 3-digit numbers by 10 (e.g. 234 ÷ 10 = 23.4).

Stretch zone: *Describe what you notice about each of your answers. What do you think 10 ÷ 10 will look like in a place-value grid?*

Check that students understand the effect of dividing 10 by 10 using the place-value grid. *What happens to the zero? Do we need to include it in the answer?*

 **Reflection time**

Ask some students to share with the class the answers they found for the Student Book activity on page 103. If they have completed the Practice Book activity at this stage, ask them to explain how they found the answers and describe what happened to each of the digits in 2-digit numbers when they divide by 10.

Write 632 on the board and ask a confident student whether they can divide it by 10 using the place-value grid. Ask other students to check the answer.

Repeat with other 3-digit numbers, including some with a zero in the tens or ones place.

Practice Book: Students complete Practice Book page 86. They can do this directly after the main activity to focus on dividing 2-digit numbers, as homework, or as the focus of a separate mathematics session to help students consolidate their learning and build fluency.

Differentiated outcomes	
All students	should divide a single digit by 10 with support.
Most students	will divide single digits by 10.
Some students	may divide any 2- or 3-digit number by 10.

Student Book page 103

1 a 0.5 **b** 0.7 **c** 0.3 **d** 0.9

Practice Book page 86

1 0.4

2 0.6

3 0.1

4 1.5

5 2.5

6 2.0

Stretch zone: 0.6, 0.1

4B Finding fractions of quantities

Discover Student Book page 104 · Practice Book page 87

Specific learning focus

- Find halves, quarters, thirds, fifths, eighths and tenths of objects and numbers.
- Relate finding fractions to division.

Global skills

- **Creative skills:** exploring
- **Interpersonal skills:** communication

Key vocabulary

- fraction, share, amount

Resources

- interlocking cubes

Language support

In this lesson, students are finding factors. You can ask questions such as these.

- *Does 2 divide exactly into halves?*
- *Does 3 divide exactly into thirds?*
- *Does 4 divide exactly into quarters with no remainder?*

Introductory activity

Explain that this lesson looks at **fractions** of sets of objects and numbers. Ask eight students to stand at the front of the class. Ask them to divide into halves. Ask

them to divide into quarters. *Can you split into thirds? Can you split into fifths? Why not?* Repeat with 12 students, 15 students and 24 students. Explore which fractions are possible with these numbers.

Main activity

Ask eight students to split into quarters again. *If one quarter of 8 students is 2, what is three quarters of 8 students?* Repeat with 12 and 24 students.

Students should work in pairs to answer the questions on page 104 of the Student Book. Encourage students to use the cubes to find the answers to question 1, if necessary, before they start recording. As they are working on the remaining questions, ask them additional questions, making the link between finding fractions and division: *How do you find an eighth/a tenth/a half/a quarter? How could you find three quarters?*

Students should discuss their thinking with a partner before they agree on an explanation and write it in the Student Book.

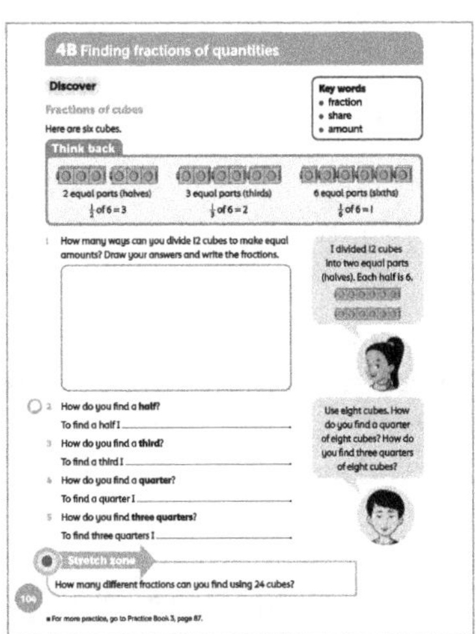

Differentiation

Supporting: Model for students how to divide numbers of cubes into halves and quarters.

Consolidating: Ask students to show how to find different fractions of numbers using cubes.

Extending: Ask students to find different fractions of 16.

Stretch zone: *How many different fractions can you find using 24 cubes?*

Students can explore dividing 24 into different groupings and reinforce the connection between fractions and division.

 Reflection time

Look at the answers to questions 2–5 together. Draw on the more-confident students to explain their thinking to the rest of the group.

Ask pairs to work together on finding fractions of 36. Write their responses on the board and ask pairs to explain how they found each fraction.

Practice Book: Students complete Practice Book page 87. They can do this directly after the main activity, as homework, or as the focus of a separate mathematics session to help students consolidate their learning and build fluency.

Differentiated outcomes	
All students	should be able to split sets of cubes into halves and quarters.
Most students	will divide sets of cubes into a range of fractions and explain their thinking.
Some students	may divide sets of cubes into a range of fractions, including non-unit fractions, and explain their thinking.

4B Finding fractions of quantities

Explore Student Book page 105 • Practice Book page 88

Specific learning focus

- Understand halves, thirds, quarters, fifths, eighths and tenths and draw diagrams to represent them.
- Identify simple fractions with a total of 1.

Global skills

- **Creative skills:** problem solving
- **Real-world skills:** interpreting information

Answers

Student Book page 104

1 Students should find that 12 divides equally into 1, 2, 3, 4 and 6 equal parts.

2 To find a half I divide the number into 2 equal groups.

3 To find a third I divide the number into 3 equal groups.

4 To find a quarter I divide the number into 4 equal groups.

5 To find three quarters I divide the number into 4 equal groups, then count 3 of them.

Practice Book page 87

1	12	**6**	10
2	6	**7**	6
3	18	**8**	14
4	8	**9**	4
5	16	**10**	12

Stretch zone: Check students' answers. For example, $\frac{1}{2}$ of 8 is 4, $\frac{1}{4}$ of 8 is 2, $\frac{3}{4}$ of 8 is 6.

Key vocabulary

- parts, fractions, equal

Resources

- interlocking cubes

Language support

When students are using the language of fractions, check their pronunciation, for example: 'fifths', 'sixths', 'eighths'.

Introductory activity

Ask students to count forward and back from 0 to 1 in tenths: $0, \frac{1}{10}, \frac{2}{10}, \frac{3}{10}, \frac{4}{10}, \frac{5}{10}, \frac{6}{10}, \frac{7}{10}, \frac{8}{10}, \frac{9}{10}, \frac{10}{10}, \frac{9}{10}, \frac{8}{10}, \frac{7}{10}, \frac{6}{10}, \frac{5}{10}, \frac{4}{10}, \frac{3}{10}, \frac{2}{10}, \frac{1}{10}, 0$.

Ask students to recall any equivalent fractions they can from 4A Discover (e.g. $\frac{4}{10} = \frac{2}{5}$). Now count again, this time replacing the tenths with an equivalent fraction if there is one:

$0, \frac{1}{10}, \frac{1}{5}, \frac{3}{10}, \frac{2}{5}, \frac{1}{2}, \frac{3}{5}, \frac{7}{10}, \frac{4}{5}, \frac{9}{10}, 1.$

Main activity

Show students a line of 10 connected cubes. *How many cubes can you see?* (10) Now break the line of cubes into two pieces. Ask students to say what fraction each piece is of the whole line. For example, if the break is into 3 cubes and 7 cubes, students can say $\frac{3}{10}$ and $\frac{7}{10}$. Write on the board $\frac{3}{10} + \frac{7}{10} = 1$ whole.

Join the cubes together again and repeat for a different break point (e.g. 4 cubes and 6 cubes). Students might say $\frac{4}{10}$ and $\frac{6}{10}$, some might say $\frac{2}{5}$ and $\frac{3}{5}$. Write the addition on the board: $\frac{4}{10} + \frac{6}{10} = 1$ whole, or $\frac{2}{5} + \frac{3}{5} = 1$ whole.

Look together at page 105 of the Student Book. If you have access to an IWB you could use this. Discuss with students the worked example and explain that this is how you would like them to record their answers to the activity. Make sure that they can describe the cubes in and out of the bag accurately as a number and as a fraction in tenths.

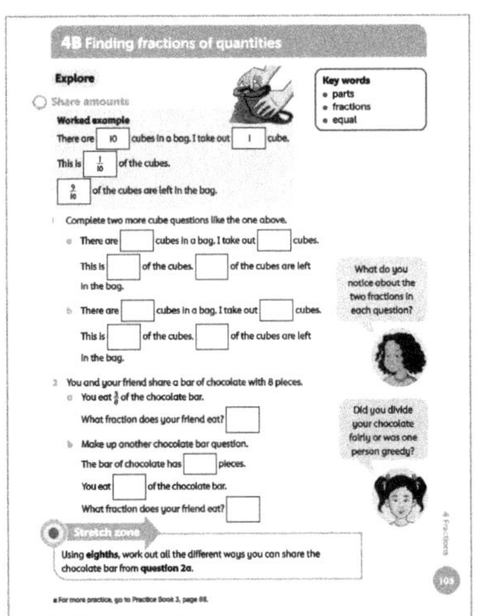

Differentiation

Supporting: Help students count the cubes and form the fractions.

Consolidating: Ask students to describe the numbers of cubes as fractions.

Extending: Students could make up their own addition calculations using the different numbers of cubes that can be taken out of the bag.

Stretch zone: *Using eighths, work out all the different ways you can share the chocolate bar from question 2a.*

Check that students have found all the possible ways and represented them correctly.

Reflection time

Show students an array of 12 cubes arranged in 2 rows of 6 cubes. *What fraction of the whole array is each cube?*

$(\frac{1}{12})$ Now split the array into two parts and ask students to describe the numbers of cubes in each part, and what fraction this represents. For example, the break might be into 8 cubes and 4 cubes, which would be $\frac{8}{12}$ and $\frac{4}{12}$, or some students might recognise this as $\frac{2}{3}$ and $\frac{1}{3}$ of the total.

Practice Book: Students complete Practice Book page 88. They can do this directly after the main activity, as homework, or as the focus of a separate mathematics session to help students consolidate their learning and build fluency.

Students find different fractions of amounts for 2-digit numbers. *How do you know you have found all the possible fractions for each number?*

Differentiated outcomes	
All students	should identify fractions from a number of cubes.
Most students	will describe the number of cubes and what fraction it represents.
Some students	may describe the fraction of cubes taken out and the fraction remaining and know that these total 1 whole.

Answers

Student Book page 105

1 **a** There are 10 cubes in a bag. I take out 2 cubes. This is $\frac{2}{10}$ of the cubes. $\frac{8}{10}$ of the cubes are left in the bag.

   b There are 10 cubes in a bag. I take out 3 cubes. This is $\frac{3}{10}$ of the cubes. $\frac{7}{10}$ of the cubes are left in the bag.

   (Other examples are possible, using 10 cubes or a different number of cubes.)

2 **a** $\frac{5}{8}$

   b For example, the bar of chocolate has 12 pieces. You eat $\frac{5}{12}$ of the chocolate bar. What fractions does your friend eat? $\frac{7}{12}$.

Practice Book page 88

1 There are various possibilities including: $\frac{1}{40} = 1$, $\frac{1}{20} = 2$, $\frac{1}{10} = 4$, $\frac{1}{8} = 5$, $\frac{1}{5} = 8$, $\frac{1}{4} = 10$, $\frac{1}{2} = 20$, $\frac{3}{4} = 30$, $\frac{2}{5} = 16$, $\frac{3}{5} = 24$.

2 Again, there are many possibilities for the chosen numbers, so check that students have completed the diagrams correctly for the numbers they chose.

Stretch zone: Students may mention that they know they have found all the fractions because they have found all the factors of their chosen number (without mentioning the word 'factor' as they will not know this word) or that they have found all the times tables in which their number appears.

 **Unit 4 Fractions**   **131**

4C Adding and subtracting fractions

Discover 1 Student Book page 106 • Practice Book page 89

Specific learning focus
- Add fractions with the same denominator.

Global skills
- **Creative skills:** investigating
- **Real-world skills:** interpreting information
- **Interpersonal skills:** communication

Key vocabulary
- fraction wall, add, unit fraction, numerator, denominator, equivalent fractions

Resources
- fraction wall showing one whole, halves, thirds, quarters, sixths and twelfths (Resource sheet 4.1)

Language support

Remind students of the words **'numerator'** and **'denominator'** and add a labelled diagram to their fraction poster. Also add a diagram of an addition fraction sentence with 'add the numerators' pointing to the numerators and 'leave the denominators' pointing to the denominators. Help them to understand that when adding or subtracting, the denominators are the same and the answer has the same denominator.

$$\frac{1}{5} + \frac{2}{5} = \frac{3}{5}$$

Add the numerators

Leave the denominators

Introductory activity

Ask students to look at the **fraction wall**. They could use Resource sheet 4.1. If you have access to an IWB, you could use it to display the resource sheet. Ask questions about the fraction wall to start a discussion, for example: *Can someone come to the fraction wall I have displayed and show me one quarter? Three quarters? Two sixths? What other fraction can you see that is the same as $\frac{2}{6}$? Can you see fifths on this fraction wall?* Focus on **equivalent fractions**.

Main activity

Look together at page 106 of the Student Book. If you have access to an IWB you could use this. Ask students to look at the fraction wall and, in pairs, find examples of fractions that they can add together to make one whole.

They should record them as an addition sentence and be able to illustrate the sentence by pointing to the fraction

wall. After five minutes, record all the examples on the board (e.g. $\frac{1}{2} + \frac{1}{2} = 1$, $\frac{1}{3} + \frac{2}{3} = 1$, $\frac{1}{4} + \frac{3}{4} = 1$).

Talk through the worked example in question 2 with students. Ask them to say what they notice about the numerators and the denominators in the calculation. Write on the board the following fraction calculations.

$$\frac{1}{2} + \frac{1}{2} \qquad \frac{1}{4} + \frac{1}{4} \qquad \frac{2}{3} + \frac{1}{3} \qquad \frac{3}{6} + \frac{1}{6}$$

Ask students to discuss, in pairs, how they would find the answers to these. Encourage them to look at the fraction wall to see whether they can work out the answers. Choose students to give the answers and say what they notice. *Are the answers smaller than or larger than 1?* Use this as an opportunity to explain that fractions do not have to add up to 1.

Ask students, in pairs, to work on the activities in the Student Book on page 106. Ask students to explain their thinking to their partner. This gives them additional practice at using the vocabulary in context. Listen to these explanations carefully as you support pairs by using the fraction wall or the bar model.

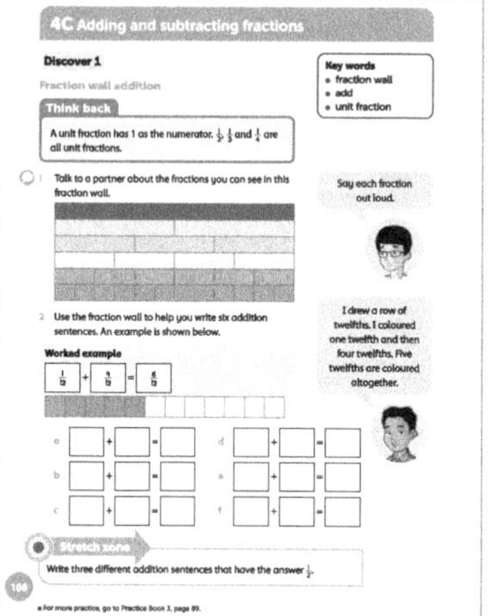

Differentiation

Supporting: Model for students how to use the fraction wall, bar model or concrete resources. Print individual copies for students to use.

Consolidating: Ask students to explain their reasoning clearly. Ask them to record the fraction number sentence for each example.

Extending: Ask students to find as many different fractions as they can that total 1. Can they also record more complex additions, using more than two fractions with the same denominator?

Stretch zone: *Write three different addition sentences that have the answer $\frac{1}{2}$.*

Check that students have written correct sentences (e.g. $\frac{1}{4} + \frac{1}{4}, \frac{2}{6} + \frac{1}{6}, \frac{4}{12} + \frac{2}{12}$).

 Reflection time

Ask students to share their solutions. Ask them to work in pairs to find further fractions that total a given fraction, such as $\frac{3}{4}$, $\frac{1}{2}$ or 1. Check their answers and understanding.

Practice Book: Students complete Practice Book page 89. They can do this directly after the main activity, as homework, or as the focus of a separate mathematics session to help students consolidate their learning and build fluency.

The Practice Book activity includes subtraction as well as addition so you may choose to leave this until students have completed the Discover 2 lesson.

Differentiated outcomes	
All students	should find the answers to fraction additions using the fraction wall, and record these.
Most students	will write and solve fraction additions and explain their thinking clearly.
Some students	may find fractions that total 1 or another given total.

Answers

Student Book page 106

1 As students discuss the fractions they can see, move around the room to listen and help correct misconceptions if they arise.

2 Answers may vary because students choose their own numbers. Check that their addition sentences are correct.

Practice Book page 89

1 Check that each line of the fraction wall is correct: The first line is 1 whole, each successive line is $\frac{1}{2}$, $\frac{1}{3}$, $\frac{1}{4}$, $\frac{1}{6}$, $\frac{1}{12}$ in each box for that line.

2 $\frac{1}{3}$

3 $\frac{2}{3}$

4 $\frac{3}{6}$ or $\frac{1}{2}$

5 $\frac{5}{6}$

6 $\frac{4}{12}$ or $\frac{2}{6}$ or $\frac{1}{3}$

7 $\frac{7}{12}$

8 $\frac{2}{4}$ or $\frac{1}{2}$

9 $\frac{3}{4}$

10 $\frac{4}{12}$ or $\frac{2}{6}$ or $\frac{1}{3}$

Stretch zone: No, she is wrong. She has added the numerators and the denominators. When adding fractions it is only the numerators that are added; the denominators remain unchanged.

4C Adding and subtracting fractions

Discover 2 Student Book page 107 · Practice Book page 90

Specific learning focus

- Subtract simple fractions with the same denominator.

Global skills

- **Creative skills:** problem solving
- **Interpersonal skills:** communication

Key vocabulary

- fraction wall, subtraction, unit fraction

Resources

- fraction wall showing halves, thirds, quarters, sixths and twelfths (Resource sheet 4.1)

Language support

When students are using the language of fractions, check their pronunciation, for example, 'fifths', 'sixths', 'twelfths'. Help them understand that when adding or subtracting, the denominators are the same and the answer has the same denominator.

 Introductory activity

Ask students to look at the fraction wall. They could use Resource sheet 4.1 or if you have access to an IWB you could use it to display the resource sheet.

Introduce subtracting from one whole. For example, tell students to look at the third row of the fraction wall. *This shows the thirds. How many thirds are there in this row?* Cover up one third. *How many are left?* Repeat this with other rows. Record the fraction number sentences (e.g. $1 - \frac{1}{3} = \frac{2}{3}$).

They should record the sentences as **subtraction** sentences and be able to illustrate the sentence by pointing to the fraction wall. After five minutes, record all the examples on the board.

Main activity

Look together at page 107 of the Student Book. If you have access to an IWB you could use this. Ask pairs of students to use the fraction wall to find examples of fractions they could use to make a subtraction. Their subtractions do not have to start with 1 whole. Talk through the worked example to show students how to record the fraction number sentence for subtraction. For example, in thirds use $\frac{2}{3} - \frac{1}{3} = \frac{1}{3}$, or in sixths $\frac{5}{6} - \frac{3}{6} = \frac{2}{6}$. Students can then find and write down more examples.

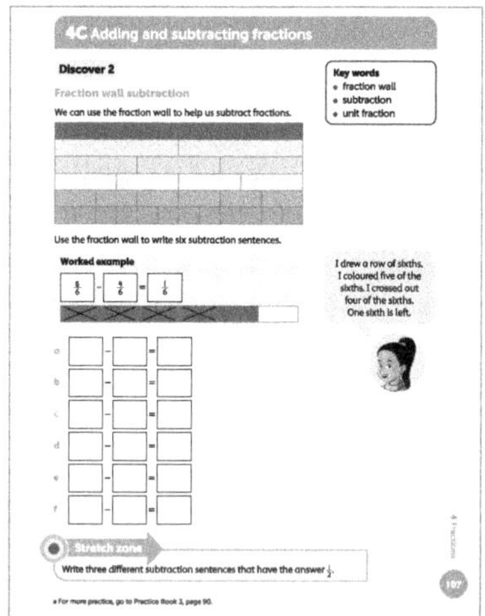

Ask students to work in pairs on the activities in the Student Book on page 107. Ask students to explain their thinking to their partner. This gives them additional practice at using the vocabulary in context. Listen to these explanations carefully as you support students by drawing a bar model as in the worked example, or by referring back to the fraction wall.

Differentiation

Supporting: Model for students how to use the fraction wall, a bar model or concrete resources to support them. Print individual copies for students to use.

Consolidating: Ask students to explain their reasoning clearly. *How do you know what fraction remains after the subtraction?* Ask students to record the correct fraction number sentence.

Extending: Ask students to find as many different fraction subtractions as they can that have a given answer, for example $\frac{1}{4}$ or $\frac{1}{3}$, using the fraction wall.

Stretch zone: *Write three different subtraction sentences that have the answer $\frac{1}{2}$.*

Check that students have written correct sentences.

 **Reflection time**

Ask students to share their solutions. Ask them to work in pairs to find further fraction subtractions using the fraction wall. Check their answers and understanding. Students can draw diagrams to represent their solutions and add these to the class poster.

Practice Book: Students complete Practice Book page 90. They can do this directly after the main activity, as homework, or as the focus of a separate mathematics session to help students consolidate their learning and build fluency.

This activity focuses on both addition and subtraction of fractions. *Read the question carefully so you know whether you need to add or subtract to find the answer.* Encourage students to use the fraction wall to answer the calculations.

Differentiated outcomes	
All students	should subtract two fractions with the same denominator.
Most students	will explain their thinking clearly as they subtract fractions and record the number sentences accurately.
Some students	may find fractions that differ by 1 or another given total.

Answers

Student Book page 107

Answers may vary because students choose their own numbers. Check that their subtraction sentences are correct.

Practice Book page 90

Some variation in answers is possible, but these are example solutions.

1 $\frac{2}{10} + \frac{1}{10} = \frac{3}{10}$

2 $\frac{4}{10} - \frac{1}{10} = \frac{3}{10}$

3 $\frac{2}{8} + \frac{1}{8} = \frac{3}{8}$

4 $\frac{7}{8} - \frac{4}{8} = \frac{3}{8}$

5 $\frac{1}{3} + \frac{1}{3} = \frac{2}{3}$

6 $1 - \frac{1}{3} = \frac{2}{3}$

7 $\frac{1}{5} + \frac{1}{5} = \frac{2}{5}$

8 $\frac{3}{5} - \frac{1}{5} = \frac{2}{5}$

9 $\frac{5}{12} + \frac{4}{12} = \frac{9}{12}$

Check students' answers. For example, $\frac{1}{9} + \frac{4}{9} = \frac{5}{9}$ and $\frac{8}{9} - \frac{3}{9} = \frac{5}{9}$.

4C Adding and subtracting fractions

Specific learning focus

- Understand halves, thirds, quarters, fifths, eighths, ninth and tenths and draw diagrams to represent adding them.

Global skills

- **Creative skills:** problem solving

Key vocabulary

- add, fraction, total

Resources

- fraction wall showing from halves through to twelfths (Resource sheet 4.2)

Language support

Support students by referring to fractions as being a fraction of the whole, and encourage them to talk about, for example, $\frac{2}{3}$ of the whole.

 ## Introductory activity

Display a fraction wall, showing up to twelfths, on an IWB if possible, or use Resource sheet 4.2. Invite students to focus on just one row of the fraction wall, for example the row showing sixths. Write on the board the calculation $\frac{1}{6} + \frac{3}{6}$. Draw a bar model divided into 6 equal pieces, just like the sixths row of the fraction wall.

How many sections represent $\frac{1}{6}$? Colour one of the sections to represent $\frac{1}{6}$.

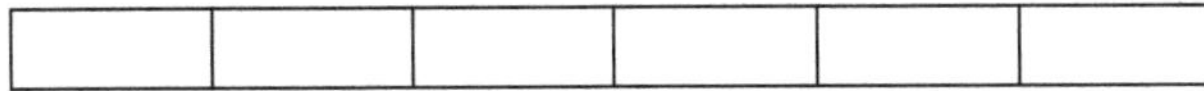

How many sections represent $\frac{3}{6}$? Colour three more sections to show this.

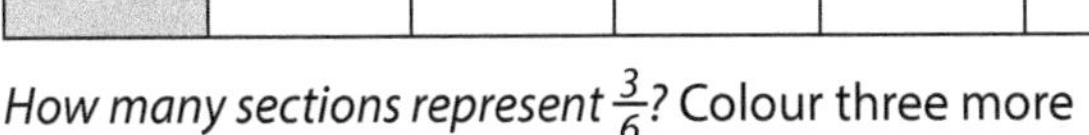

How many sections have been coloured in total? (4)

What fraction does this show? $\left(\frac{4}{6}\right)$

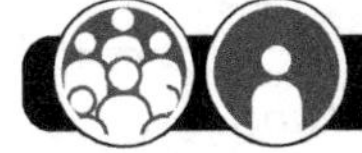 ## Main activity

Look together at page 108 of the Student Book. If you have access to an IWB you could use this. Refer students to the worked example and compare it with the example from the introductory activity. *What is the same and what is different about the two examples?* Students should notice that the number of equal pieces on the bar model represents the denominator of the fractions being calculated. In both cases, the colouring represents the fractions being added.

Students should complete the activity on page 108 of the Student Book, using each bar model to help them calculate the fraction additions by colouring the respective fractions.

Refer students to the speech bubble and share responses. *Can you explain why this happens?*

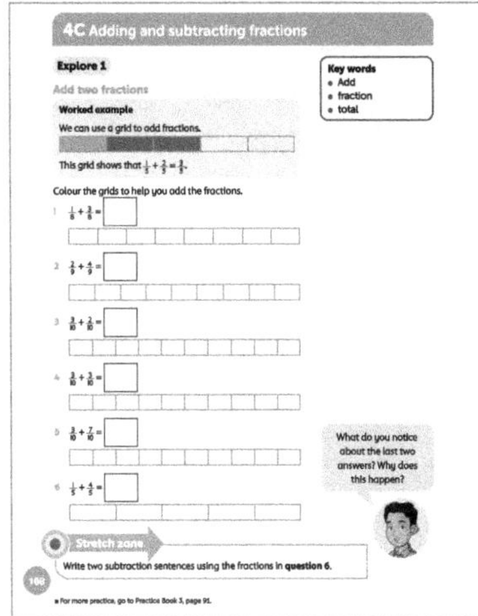

Differentiation

Supporting: Ask students to count the number of equal sections in the bar model and check that it matches the denominators.

Consolidating: These students should be able to describe how the bar model matches the numbers being added.

Extending: Students who complete the additions could extend by exploring fractions that add to more than one whole (e.g. $\frac{4}{5} + \frac{3}{5}$) and draw a diagram to show this. *How would you represent the whole? How would you represent the fraction part?*

Stretch zone: *Write two subtraction sentences using the fractions in question 6.*

Check that students have written subtraction number sentences correctly: $1 - \frac{1}{5} = \frac{4}{5}$ and $1 - \frac{4}{5} = \frac{1}{5}$.

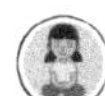 ## Reflection time

Ask students to share their strategies for the additions in the Student Book on page 108. Try adding the fractions in a different order to remind students that we can add numbers (including fractions) in any order. Ask students to work in pairs to develop a word problem using fractions, for example:

'I ate $\frac{3}{10}$ of a chocolate bar and my friend ate $\frac{1}{10}$. What fraction did we eat altogether?'

Practice Book: Students complete Practice Book page 91. They can do this directly after the main activity, as homework, or as the focus of a separate mathematics session to help students consolidate their learning and build fluency.

Students write fraction additions that all have a **total** of 1. Remind students that the denominators must be the same when we add fractions.

Differentiated outcomes	
All students	should colour in the fractions on a bar model.
Most students	will use the coloured bar model to find the total.
Some students	may draw and use bar models for additions of fractions that total more than 1.

Answers

Student Book page 108

Check that students have correctly coloured the diagrams.

1 $\frac{4}{8}$

2 $\frac{6}{9}$

3 $\frac{5}{10}$

4 $\frac{6}{10}$

5 $\frac{10}{10}$

6 $\frac{5}{5}$

Practice Book page 91

1 $\frac{7}{10}$

2 $\frac{3}{5}$

3 $\frac{2}{3}$

4 $\frac{3}{4}$

5 $\frac{1}{6}$

6 $\frac{1}{8}$

7 $\frac{3}{10}$

8 $\frac{5}{8}$

9 $\frac{8}{12}$

For questions 10–13 answers may vary.

10 $\frac{3}{4} + \frac{1}{4} = 1$

11 $\frac{1}{2} + \frac{1}{2} = 1$

12 $\frac{9}{10} + \frac{1}{10} = 1$

13 $\frac{3}{7} + \frac{4}{7} = 1$

Stretch zone: Check that students have written a correct rule. For example, If the denominator is 8 then the numerators of the two fractions must total 8.

4C Adding and subtracting fractions

the vocabulary of numerators and denominators accurately when describing the process.

Explore 2 — Student Book page 109 • Practice Book page 92

Specific learning focus

- Understand halves, thirds, quarters, fifths, eighths, ninths and tenths and draw diagrams to represent subtracting them.

Global skills

- **Creative skills:** problem solving

Key vocabulary

- subtract, fraction

Resources

- fraction wall showing one whole then all fractions through to twelfths (Resource sheet 4.2)

Language support

Support students by referring to fractions as being a fraction of the whole, and encourage them to talk about, for example, $\frac{2}{3}$ of the whole. When subtracting fractions, encourage students to use

Introductory activity

Display a fraction wall, showing up to twelfths, on an IWB if possible, or use Resource sheet 4.2. Invite students to focus on just one row of the fraction wall, for example the row showing fifths. Write on the board the calculation $\frac{4}{5} - \frac{1}{5}$. Draw a bar model divided into five equal pieces, just like the fifths row of the fraction wall.

How many sections represent $\frac{4}{5}$? Colour four of the sections to represent $\frac{4}{5}$.

How many sections represent $\frac{1}{5}$? Strike through one section to show this.

How many of the coloured-in sections are left uncrossed? (3)

What fraction does this show? $\left(\frac{3}{5}\right)$

Look together at page 109 of the Student Book. If you have access to an IWB you could use this. Refer students to the worked example and compare it with the example from the introductory activity. *What is the same and what is different about the two examples?* Students should notice that the number of equal pieces on the bar model represents the denominator of the fractions being calculated. In both cases, the colouring represents the starting fraction.

Students should complete the activity on page 109 of the Student Book, using each bar model to help them calculate the fraction subtractions by colouring and crossing through the respective fractions.

Refer students to the second speech bubble and share responses. *Can you explain why this happens?* Students can use the fraction wall to see that $\frac{4}{10}$ and $\frac{2}{5}$ are equivalent fractions.

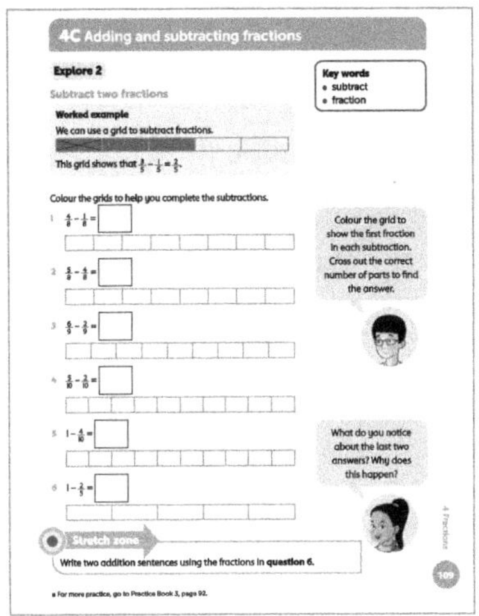

Differentiation

Supporting: Ask students to count the number of equal sections in the bar model and check that it matches the denominators.

Consolidating: Ask these students to describe how the bar model matches the numbers being subtracted.

Extending: Encourage students who complete the subtractions to extend by exploring fractions that subtract from more than 1 (e.g. $1\frac{1}{2} - \frac{5}{8}$) and draw a diagram to show this.

Stretch zone: *Write two addition sentences using the fractions in question 6.*

Check that students have written addition number sentences correctly: $\frac{3}{5} + \frac{2}{5} = 1$ and $\frac{2}{5} + \frac{3}{5} = 1$.

 **Reflection time**

Ask students to share their strategies for the subtractions in the Student Book on page 109. Then ask them to work in pairs to develop a word problem using fractions, for example:

'My friend and I ate $\frac{5}{8}$ of a whole chocolate bar between us. What fraction of the bar was left?'

Practice Book: Students complete Practice Book page 92. They can do this directly after the main activity, as homework, or as the focus of a separate mathematics session to help students consolidate their learning and build fluency.

Students complete diagrams to represent addition and subtraction of fractions. *Read the question carefully so you know whether you need to shade more parts of the diagram, or cross out some to find the answer.*

Differentiated outcomes	
All students	should colour in the fractions on a bar model.
Most students	will use the coloured bar model to find the difference.
Some students	may draw and use bar models for subtractions of a fraction from more than 1.

Student Book page 109

Check that students have correctly coloured the diagrams.

1 $\frac{3}{8}$

2 $\frac{1}{8}$

3 $\frac{4}{9}$

4 $\frac{3}{10}$

5 $\frac{6}{10}$

6 $\frac{3}{5}$

Practice Book page 92

1 $\frac{6}{8}$

2 $\frac{3}{9}$

3 $\frac{7}{9}$

4 $\frac{1}{10}$

5 $\frac{8}{10}$

6 $\frac{5}{10}$

7 $\frac{1}{5}$

8 $\frac{5}{8}$

Stretch zone: When adding a pair of fractions with the same denominator just add the numerators.

4D Equivalent fractions

Discover
Student Book page 110 · Practice Book page 93

Specific learning focus
- Find equivalent fractions.

Global skills
- **Creative skills:** problem solving
- **Interpersonal skills:** communication

Key vocabulary
- equal, equivalent, fraction wall

Resources
- mini whiteboards and markers
- sheets of A4 paper

Language support

Listen to each student saying the fractions out loud. Ask questions to monitor students' understanding, and to model phrases, for example:

- … is equivalent to …
- in fractions equivalent to a half, the top number is half the bottom number.

 **Introductory activity**

Ask students to draw eighths on a number line on their whiteboards. The ends of the line should be marked 0 and 1. If they need help, suggest that they draw halves, quarters and finally eighths. Ask them to draw tenths: they mark halfway and then visualise dividing each half into five. Ask them to compare the two number lines so they see that the divisions on the tenths number line are closer together than the divisions on the eighths line (so they can see that $\frac{1}{10}$ is smaller than $\frac{1}{8}$).

 **Main activity**

Complete the paper folding exercise together (question 1 in the Student Book on page 110). Then ask students to work in pairs on question 2. As they work, ask, *What is happening to the top number each time? What is happening to the bottom number each time?* Discuss with students how to use multiplication (and division) to find equivalent fractions by multiplying (or dividing) the numerator and denominator by the same number. For example, start with $\frac{1}{4}$. Multiply the numerator and denominator by 5. $1 \times 5 = 5$ and $4 \times 5 = 20$, so $\frac{1}{4} = \frac{5}{20}$.

Students complete question 3 individually. As they complete the fraction wall, ask them to talk about the equivalent fractions they know. *Give me an equivalent fraction to $\frac{1}{4}$. How do you know?, What do you notice about the relationship between the top and bottom numbers?* and so on.

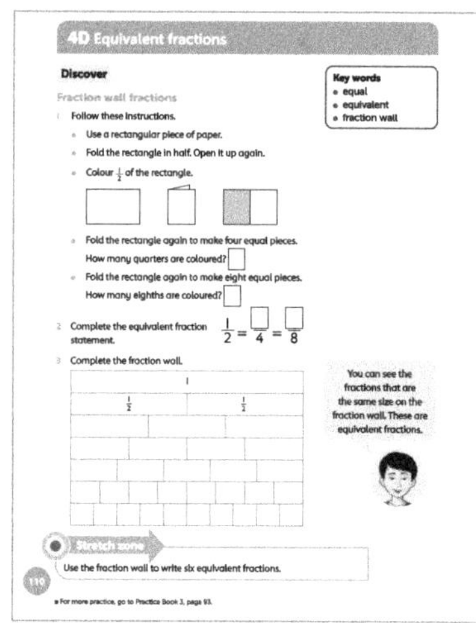

Differentiation

Supporting: Work with students to find equivalent fractions on the fraction wall.

Consolidating: Ask students to say a rule for finding equivalent fractions.

Extending: Ask students to explore other fractions that can be made by folding a piece of paper and see whether they can mark those fractions on a number line.

Stretch zone: *Use the fraction wall to write six equivalent fractions.*

Check that students have written equivalent fractions, using a good range of denominators.

 **Reflection time**

Ask pairs to discuss what the rule is to create equivalent fractions. Take feedback from pairs and agree a class 'rule'. Record this on the class fractions poster and display it for the rest of unit.

Practice Book: Students complete Practice Book page 93. They can do this directly after the main activity, as homework, or as the focus of a separate mathematics session to help students consolidate their learning and build fluency.

Students write pairs of equivalent fractions, using a fraction wall to help them. *Can you write another equivalent fraction for each pair? How did you work it out?*

Differentiated outcomes	
All students	should find equivalent fractions using the fraction wall.
Most students	will understand how to create equivalent fractions using multiplication facts.
Some students	may find a wide range of equivalent fractions by using the rule of multiplying or dividing the numerator and denominator by the same number.

Answers

Student Book page 110

1 $\frac{2}{4}$ and $\frac{4}{8}$

2 $\frac{1}{2} = \frac{2}{4} = \frac{4}{8}$

3 1

$\frac{1}{2}$

$\frac{1}{3}$

$\frac{1}{4}$

$\frac{1}{5}$

$\frac{1}{8}$

$\frac{1}{10}$

Practice Book page 93

Answers may vary as students find their own pairs of equivalent fractions. Check they are correct.

Stretch zone: In equivalent fractions, the numerator and the denominator have been multiplied (or divided) by the same number. For example, in $\frac{2}{3}$ and $\frac{4}{6}$ both the numerator and denominator have been multiplied by 2.

4D Equivalent fractions

Explore Student Book pages 111–112 · Practice Book page 94

Specific learning focus

- Find equivalent fractions using fractions of shapes.

Global skills

- **Creative skills:** problem solving

Key vocabulary

- equivalent, same, fraction

Resources

- strips and circles of paper, large square of paper
- fraction wall (Resource sheet 4.1)

Language support

Support students by using the language of equivalence. Reinforce the use of fraction vocabulary, including 'numerator', 'denominator', 'equivalent' and add these to the poster display so that the vocabulary can be matched to the fractions being used.

 **Introductory activity**

Show students a large square of paper. *This square of paper represents one whole.* Now fold the paper in half to form a rectangle. Open it up and ask students what fraction each piece is. $\left(\frac{1}{2}\right)$ Now fold in half again, in the other direction to make a second fold. *What shape you have made?* (square) *What fraction of the whole is that square?* $\left(\frac{1}{4}\right)$.

Now make a third fold in half again, to make a smaller rectangle. *What fraction is it of the whole?* $\left(\frac{1}{8}\right)$. If students can't work out why, unfold the square of paper to show the eighths that have been formed by the folds.

 **Main activity**

Give each student a circle of paper and ask them to fold it in half, then in half again and then in half a third time. Ask them to estimate how many sections there will be when they unfold it. Then they can unfold the circle and count the sections to find that they have made eighths.

Using your circle, tell me how many eighths make $\frac{1}{4}$? How many eighths make $\frac{1}{2}$? What is $\frac{3}{4}$ in eighths?

Students now complete pages 111–112 of the Student Book, using paper strips and circles, which they can fold, as practical supports.

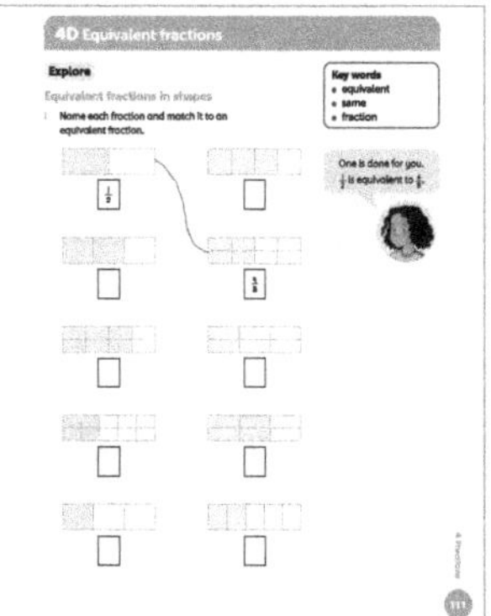
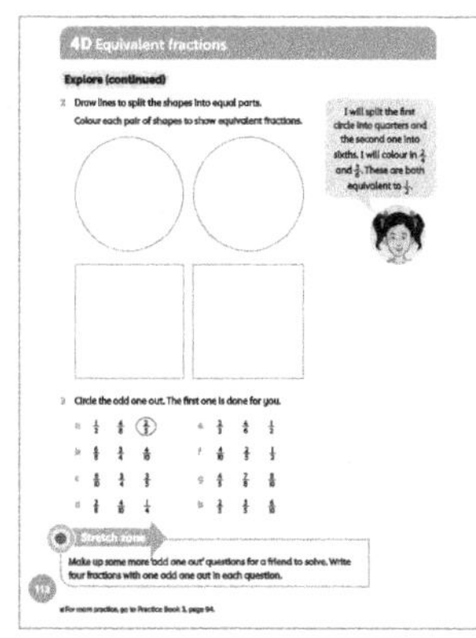

Differentiation

Supporting: Ask students to fold paper strips and circles to form the equivalent fractions.

Consolidating: Ask students to draw equal sections on rectangles to find equivalent fractions.

Extending: Before students complete the Student Book activities, encourage them to use circles or rectangles and try to fold them to create a wider range of fractions and to write equivalents.

Stretch zone: *Make up some more 'odd one out' questions for a friend to solve. Write four fractions with one odd one out in each question.*

Check that students have written suitable questions for a partner.

 ### Reflection time

Ask students to share their responses to question 3 on page 112. *How did you decide which fraction was the odd one out? Did you fold paper strips to help?* For those who used paper folds, ask them to share the folded paper and talk through how it helped them find the odd one out in each set of fractions.

Practice Book: Students complete Practice Book page 94. They can do this directly after the main activity, as homework, or as the focus of a separate mathematics session to help students consolidate their learning and build fluency.

Students shade different grids to show equivalent fractions. *Can you remember the rule for finding equivalent fractions? What do you need to do to the denominator if you multiply the numerator by 2? What if you divide the numerator by 2?*

Differentiated outcomes	
All students	should find equivalent halves, quarters and eighths using folded paper.
Most students	will draw diagrams with equal sections to find equivalent fractions.
Some students	may explore other denominators by making folded paper strips.

Answers

Student Book pages 111–112

1 $\frac{1}{2} = \frac{4}{8}$; $\frac{2}{3} = \frac{4}{6}$; $\frac{6}{8} = \frac{3}{4}$; $\frac{4}{10} = \frac{2}{5}$; $\frac{1}{3} = \frac{2}{6}$

2 Various answers are possible. Check that students' responses are correct.

3 **a** $\frac{2}{5}$ **b** $\frac{4}{10}$ **c** $\frac{3}{4}$ **d** $\frac{4}{10}$ **e** $\frac{1}{2}$ **f** $\frac{1}{2}$ **g** $\frac{7}{8}$ **h** $\frac{2}{3}$

Practice Book page 94

Check that students have shaded the diagrams correctly.

4E Comparing and ordering fractions

Discover 1 Student Book page 113 • Practice Book page 95

Specific learning focus

- Use equivalence to compare and order fractions.

Global skills

- **Creative skills:** exploring
- **Real-world skills:** interpreting information
- **Interpersonal skills:** communication

Key vocabulary

- denominator, numerator, greater than (>), less than (<)

Resources

- mini whiteboards and markers
- fraction wall (Resource sheet 4.2)
- tenths fraction cards
- eighths fraction cards

Language support

Model the language and pronunciation of fractions with unusual denominators, for example 'sevenths', 'ninths', 'elevenths'.

 ### Introductory activity

Discuss the **greater than (>)** and **less than (<)** symbols. (The larger number is always at the open end of the symbol.) Use the set of tenths fraction cards. Ask a student to choose two cards.

Ask students to name the fractions and compare them on their whiteboards. For example, they write:

$$\frac{4}{10} < \frac{7}{10}$$

Repeat using the eighths cards.

Ask students a series of questions that they can answer using Resource sheet 4.2. *Tell me a fraction greater than $\frac{2}{5}$. Tell me two fractions less than $\frac{7}{8}$* and so on.

 ### Main activity

Look together at page 113 of the Student Book. If you have access to an IWB you could use this. Ask students to refer to the fraction wall. Now use the ideas from the introductory activity to find fractions that meet two conditions at the same time. For example, ask students to give you a fraction greater than $\frac{1}{3}$ and less than $\frac{4}{5}$. There are several possible answers but encourage students to make sure that their answer fits both the conditions that it must be greater than $\frac{1}{3}$ and less than $\frac{4}{5}$.

Then ask students to give you two fractions that go either side of a given one. For example, if you give them $\frac{3}{4}$, they must name a fraction less than $\frac{3}{4}$ and another that is greater than $\frac{3}{4}$.

Allow students some time to ask a partner similar questions, then direct them to complete the questions on page 113 of the Student Book.

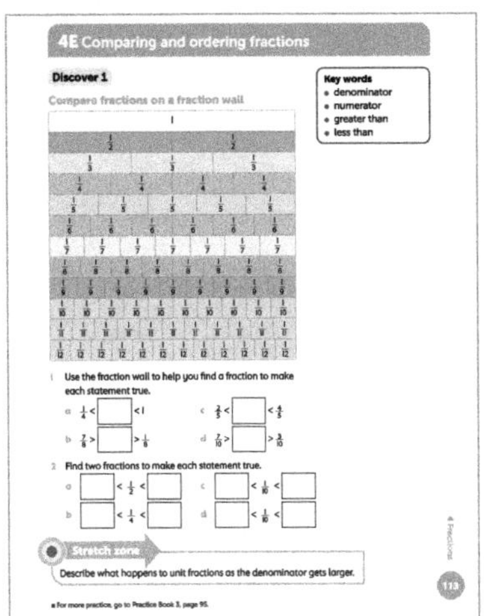

Differentiation

Supporting: Students should continue to use the fraction wall in the Student Book for each question to check that they have found the correct comparisons.

Consolidating: Ask students to begin to choose fractions without using the fraction wall.

Extending: Ask students to choose missing fractions from decreasing intervals. For example, if they can name a fraction between $\frac{1}{4}$ and $\frac{1}{2}$, can they then name one between $\frac{1}{4}$ and $\frac{3}{8}$ and so on?

Stretch zone: *Describe what happens to unit fractions as the denominator gets larger.*

Check that students understand that a larger denominator means a smaller fraction.

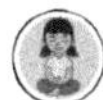 **Reflection time**

Students can share some of their answers to the questions in the Student Book on page 113. Ask them to explain their strategy for finding fractions within a given range, or for finding two fractions around a given one. *Can anyone tell me another fraction that we could use here instead? And another?*

Practice Book: Students complete Practice Book page 95. They can do this directly after the main activity, as homework, or as the focus of a separate mathematics session to help students consolidate their learning and build fluency.

Students continue to look at fractions on a fraction wall to recognise the relative size of fractions. They then order a set of fractions, which all have different denominators. *Which is the smallest fraction? Which is the largest fraction? How do you know? Explain how you know that fraction is smaller than that one but larger than that one?*

Differentiated outcomes	
All students	should be able to compare two fractions using a fraction wall.
Most students	will find fractions that fit in an interval of two other fractions.
Some students	may know which fractions are greater than or less than others without using the fraction wall.

Answers

Student Book page 113

1 Answers may vary but these are some possible examples.

   a $\frac{1}{4} < \frac{1}{2} < 1$

   b $\frac{7}{8} > \frac{1}{4} > \frac{1}{8}$

   c $\frac{2}{5} < \frac{3}{5} < \frac{4}{5}$

   d $\frac{7}{10} > \frac{5}{8} > \frac{3}{10}$

2 Answers may vary but these are some possible examples.

   a $\frac{1}{4} < \frac{1}{2} < \frac{3}{5}$

   b $\frac{1}{10} < \frac{1}{4} < \frac{5}{6}$

   c $\frac{1}{11} < \frac{1}{10} < \frac{1}{2}$

   d $\frac{1}{12} < \frac{1}{10} < \frac{3}{10}$

Practice Book page 95

1 $\frac{2}{3}$

2 $\frac{6}{10}$

3 $\frac{3}{5}$

4 $\frac{3}{8}$

5 $\frac{1}{2}$

6 $\frac{5}{6}$

7 $\frac{1}{4}$

Check that the rows are shaded correctly.

8 $\frac{1}{6}$ $\frac{3}{8}$ $\frac{1}{2}$ $\frac{2}{3}$ $\frac{9}{10}$

Stretch zone: $\frac{9}{10}, \frac{1}{6}$

4E Comparing and ordering fractions

Discover 2
Student Book page 114 • Practice Book page 96

Specific learning focus

- Use equivalences to order fractions on number lines.

Global skills

- **Creative skills:** problem solving
- **Real-world skills:** presenting information

Key vocabulary

- number line, denominator, numerator

Resources

- fraction wall (Resource sheet 4.2)

Language support

Create number lines marked with different fractions, labelled with the denominator as a word, for example 'thirds', 'fifths'. Add these to the class display poster.

 **Introductory activity**

Ask students to write down as many fractions as they can equivalent to $\frac{1}{2}$. They might suggest $\frac{2}{4}, \frac{3}{6}, \frac{4}{8}$ and so on. Ask them to describe what they notice about the fractions. They may notice that the numerator is always half of the denominator, or that the denominators are all even numbers.

How many twentieths would equal $\frac{1}{2}$. (10) What about fiftieths? (25) Can you write a fraction with as large a denominator as you can that is equivalent to $\frac{1}{2}$?

 **Main activity**

Draw a **number line** marked in quarters from 0 to 1. Under each mark, ask students what fraction can be written there. For example, under $\frac{1}{4}$ they could write $\frac{2}{8}$ or $\frac{3}{12}$ and so on. Give them time to work with a partner to find different equivalents to write under each marking.

Now write on the board $\frac{1}{6}, \frac{1}{8}, \frac{1}{9}, \frac{1}{10}$. Ask students to write down next to each fraction any equivalents they know or can find using a fraction wall (e.g. $\frac{1}{6} = \frac{2}{12}$).

Students complete the number line activities on page 114 of the Student Book. Remind them to use their knowledge of finding equivalents to write the fractions in the correct place on the number lines. They can use Resource sheet 4.2 for support.

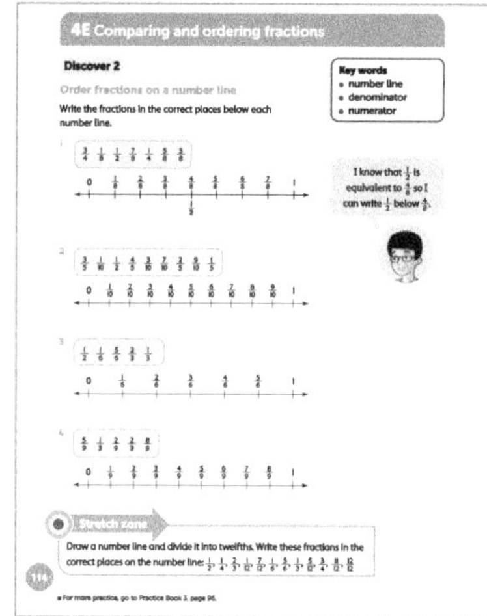

Differentiation

Supporting: Model for students how to locate where fractions should be placed on the number lines.

Consolidating: Ask students to write fractions equivalent to those given, to help them place these correctly on the number lines.

Extending: Ask students to describe the strategies they are using to order the fractions on the number lines.

Stretch zone: *Draw a number line and divide it into twelfths. Write these fractions in the correct places on the number line: $\frac{1}{2}, \frac{1}{4}, \frac{2}{3}, \frac{1}{12}, \frac{7}{12}, \frac{1}{6}, \frac{5}{6}, \frac{1}{3}, \frac{5}{12}, \frac{3}{4}, \frac{11}{12}, \frac{12}{12}.$*

Check that students have understand the equivalent fractions to twelfths and placed the fractions correctly.

 **Reflection time**

Ask students to explain how they were able to order each set of fractions and label the number lines. Compare strategies and see whether any students have found what they might regard as interesting equivalents.

Practice Book: Students complete Practice Book page 96. They can do this directly after the main activity, as homework, or as the focus of a separate mathematics session to help students consolidate their learning and build fluency.

Students write fractions, in the correct place, on a number line. The number lines are already divided and partly labelled. *What equivalent fraction will help you to work out where that fraction goes on that number line? How do you know that you have written that fraction in the correct place on the number line?*

Differentiated outcomes	
All students	should place fractions with matching denominators on the number lines.
Most students	will place fractions with a range of denominators on the number lines.
Some students	may find equivalent fractions to the ones given using multiplication and division.

Answers

Student Book page 114

1 $\frac{1}{8}, \frac{1}{4}, \frac{3}{8}, \frac{1}{2}, \frac{5}{8}, \frac{3}{4}, \frac{7}{8}$

2 $\frac{1}{10}, \frac{1}{5}, \frac{3}{10}, \frac{2}{5}, \frac{1}{2}, \frac{3}{5}, \frac{7}{10}, \frac{4}{5}, \frac{9}{10}$

3 $\frac{1}{6}, \frac{1}{3}, \frac{1}{2}, \frac{2}{3}, \frac{5}{6}$

4 $\frac{2}{9}, \frac{1}{3}, \frac{5}{9}, \frac{2}{3}, \frac{8}{9}$

Practice Book page 96

1 $\frac{1}{10}, \frac{1}{2}, \frac{3}{5}, \frac{4}{5}, \frac{9}{10}$

2 $\frac{1}{4}, \frac{1}{3}, \frac{3}{6}, \frac{9}{12}, \frac{5}{6}$

3 $\frac{3}{4}, 1\frac{1}{4}, 1\frac{1}{2}, 1\frac{3}{4}, 2\frac{1}{2}$

4 $\frac{1}{2}, \frac{7}{10}, 1\frac{1}{10}, 1\frac{1}{2}, 1\frac{4}{5}$

Stretch zone: For example, $\frac{3}{10}, \frac{1}{7}$

4E Comparing and ordering fractions

Explore 1   Student Book page 115 • Practice Book page 97

Specific learning focus

- Use equivalences to compare fractions.

Global skills

- **Creative skills:** problem solving

Key vocabulary

- fraction wall, order, smaller than, less than, greater than, more than

Resources

- fraction wall

Language support

Remind students that '<' means 'less than' or '**smaller than**' and '>' means '**more than**' or 'greater than' and that both can be used when comparing fractions. Encourage students to use the terms 'numerator' and 'denominator' when talking about fractions.

Introductory activity

Show students some equivalent fractions, for example: $\frac{1}{2} = \frac{3}{6}, \frac{1}{4} = \frac{2}{8}, \frac{2}{5} = \frac{6}{15}$. Ask them to examine each pair and see whether they can describe how to get from one to the other using multiplication or division.

Now ask students to use multiplication to make an equivalent fraction for each of the following fractions: $\frac{2}{3}$, $\frac{5}{8}$ and $\frac{7}{10}$.

Main activity

Ask students to look at the fraction wall on page 115 of the Student Book. Ask questions about finding equivalent fractions involving fifths and tenths. *Can you give me an equivalent fraction for $\frac{1}{5}$? $\frac{2}{5}$? $\frac{3}{5}$? What about $\frac{2}{10}$? How many fifths is $\frac{6}{10}$ equivalent to? How many equivalent fifths and tenths pairs can you find?* $\left(\frac{1}{5} = \frac{2}{10}, \frac{2}{5} = \frac{4}{10}, \frac{3}{5} = \frac{6}{10}, \frac{4}{5} = \frac{8}{10}, \frac{5}{5} = \frac{10}{10}\right)$

Now ask students to look for fractions greater than $\frac{3}{4}$. $\left(\frac{4}{5}, \frac{7}{8}, \frac{8}{10}, \frac{9}{10}\right)$ *Can you find all the fractions on the wall that are less than $\frac{1}{4}$?* $\left(\frac{1}{5}, \frac{1}{8}, \frac{1}{10}, \frac{2}{10}\right)$

Students complete the activities in the Student Book on page 115. Remind them to use the fraction wall on the page to help them answer the questions, if they need to.

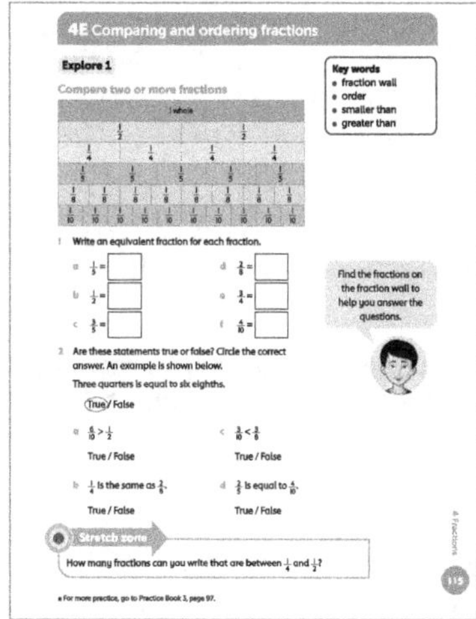

Differentiation

Supporting: Ask students to write pairs of equivalent fractions using quarters and eighths.

Consolidating: Ask students to look for examples of three equivalent fractions that appear on the wall, for example $\frac{1}{2} = \frac{2}{4} = \frac{5}{10}$.

Extending: Ask students to find equivalent fractions by multiplying or dividing the numerator and denominator by the same number.

Stretch zone: *How many fractions can you write that are between $\frac{1}{4}$ and $\frac{1}{2}$?*

Ask students to explain how they worked these out and how they know that the fractions are in the correct range.

Reflection time

Students share the results of their activities. Ask students to explain how they worked out whether one fraction was greater or smaller than another in question 2. *Did you find an equivalent fraction to check?*

Practice Book: Students complete Practice Book page 97. They can do this directly after the main activity, as homework, or as the focus of a separate mathematics session to help students consolidate their learning and build fluency.

Students write inequality statements using pairs of fractions with different denominators. Remind students what the < and > symbols mean. *Look carefully at which symbol is used in each question so that you are clear about is being asked of you.*

Differentiated outcomes	
All students	should find a pair of equivalent fractions on the fraction wall.
Most students	will find several pairs of equivalent fractions.
Some students	may find equivalent fractions that are not on the fraction wall.

Answers

Student Book page 115

1 a $\frac{2}{10}$

   b $\frac{2}{4}, \frac{3}{6}, \frac{4}{8}, \frac{5}{10}$

   c $\frac{6}{10}$

   d $\frac{1}{4}$

   e $\frac{6}{8}$

   f $\frac{2}{5}$

2 **a–d** All are true.

Practice Book page 97

Other answers are possible so check for accuracy.

1 $\frac{5}{8}$

2 $\frac{7}{10}$

3 $\frac{11}{12}$

4 $\frac{5}{6}$

5 $\frac{2}{3}$

6 $\frac{3}{5}$

7 $\frac{3}{8}$

8 $\frac{4}{5}$

4E Comparing and ordering fractions

Explore 2 Student Book page 116 · Practice Book page 98

Specific learning focus

- Order unit and non-unit fractions.

Global skills

- **Creative skills:** problem solving
- **Interpersonal skills:** communication

Key vocabulary

- unit fraction, non-unit fraction, denominator, numerator

Resources

- fraction wall
- number lines

Language support

Add some unit and non-unit fractions to the class poster display. For example, write some unit fractions: $\frac{1}{2}, \frac{1}{3}, \frac{1}{4}, \frac{1}{5}$ and label the line with 'unit fractions.' Then write some non-unit fractions: $\frac{2}{3}, \frac{2}{4}, \frac{3}{4}, \frac{2}{5}, \frac{3}{5}, \frac{4}{5}$ with a label 'non-unit fractions'.

 Introductory activity

Draw a number line from 0 to 1 and ask students to choose fractions with denominators of 5 or less. As each student chooses a fraction, they come out and mark its approximate position on the number line.

Ask other students whether they agree with where each fraction has been placed. If not, can they explain where it should be and why?

Direct students to look at the Think back on page 116 of the Student Book. Use this as a reminder of what a **unit fraction** and **non-unit fraction** are. You could ask them to add these definitions to the class display poster, as mentioned in the language support section. Ask students to work with a partner to write down as many non-unit fractions as they can with denominator 6. *Which ones can you find equivalent fractions for?*

Draw a number line marked in sixths (with 0 at one end, 1 at the other end and five equally spaced divisions in between) and mark the fractions and their equivalents on the line.

Repeat this for fractions with denominator 8, drawing a new number line marked in eighths.

Students then complete the activities on page 116 of the Student Book. Encourage students to write any equivalent fractions on the number lines in the Student Book for the five fractions that they choose on each line.

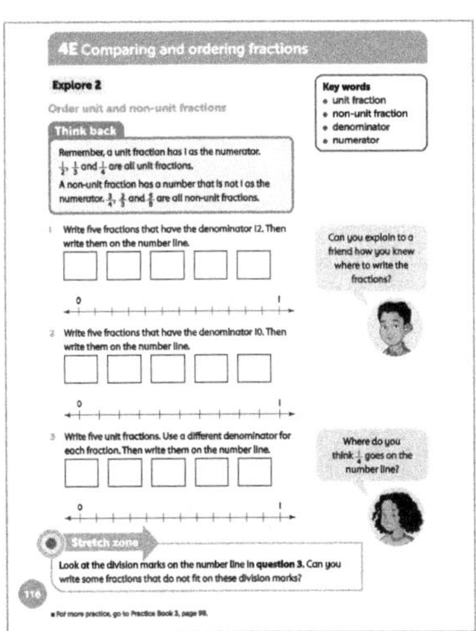

Differentiation

Supporting: Help students to find equivalent fractions using a fraction wall before they write them on the number line.

Consolidating: Ask students to order non-unit fractions by placing them on a number line. They can also find all the equivalent fractions and write these in the correct place on the number line.

Extending: Ask students to explore non-unit fractions where the numerator and denominator differ by 1 (e.g. $\frac{2}{3}, \frac{3}{4}, \frac{4}{5}$).

Stretch zone: *Look at the division marks on the number line in question 3. Can you write some fractions that do not fit on these division marks?*

Check that students understand that the marks on a number line will not be suitable for all denominators. For

example, the number line in question 3 is not marked for fifths or tenths. However, they could still mark these fractions in their approximate locations.

 Reflection time

Ask students to explain their strategy for placing their sets of fractions on the number lines. *How did they decide the **order** of the fractions? Is it harder to place unit fractions or non-unit fractions on the number line? Can you think of any unit fractions that are greater than a non-unit fraction? For example $\frac{1}{2}$ is greater than $\frac{2}{5}$.*

Practice Book: Students complete Practice Book page 98. They can do this directly after the main activity, as homework, or as the focus of a separate mathematics session to help students consolidate their learning and build fluency.

Students complete some more inequality statements, this time using three fractions with different denominators. Remind students what the < and > symbols mean. *Look carefully at which symbol is used in each question so that you are clear about is being asked of you.*

Differentiated outcomes	
All students	should recognise unit and non-unit fractions, with the help of a fraction wall, and write them on a number line.
Most students	will find all the non-unit fractions for a given denominator and some or all of their equivalents.
Some students	may find equivalent fractions for any unit and non-unit fraction.

Answers

Student Book page 116

1 Students can choose from $\frac{1}{12}, \frac{2}{12}, \frac{3}{12}, \frac{4}{12}, \frac{5}{12}, \frac{6}{12}, \frac{7}{12}, \frac{8}{12},$ $\frac{9}{12}, \frac{10}{12}, \frac{11}{12}, \frac{12}{12}$ and some of their equivalents: $\frac{2}{6}, \frac{1}{4}, \frac{1}{3}, \frac{1}{2},$ $\frac{2}{3}, \frac{3}{4}, \frac{5}{6}$. Check that they have placed them correctly on the number line.

2 Students can choose from $\frac{1}{10}, \frac{2}{10}, \frac{3}{10}, \frac{4}{10}, \frac{5}{10}, \frac{6}{10}, \frac{7}{10}, \frac{8}{10},$ $\frac{9}{10}, \frac{10}{10}$ and some of their equivalents: $\frac{1}{5}, \frac{2}{5}, \frac{1}{2}, \frac{3}{5}, \frac{4}{5}$. Check that they have placed them correctly on the number line.

3 Check that students have correctly placed the unit fractions they chose.

Practice Book page 98

There are many possible answers to these questions. Check that students have correctly identified fractions greater than or less than the ones given.

Stretch zone: Answers will vary. Check that students have correctly ordered five fractions, using the correct < and > symbols, for example $\frac{1}{12} < \frac{1}{10} < \frac{1}{8} < \frac{1}{6} < \frac{1}{5}$.

4 Fractions

Connect Student Book page 117

Big idea

- We can find fraction and decimal equivalents for tenths. We can find fractions of amounts. We can order and compare fractions. We can add and subtract fractions.

Global skills

- **Real-world skills:** presenting information
- **Interpersonal skills:** communication

Key vocabulary

- fraction, equivalent fraction

Resources

- fraction wall

Language support

Help students with the language associated with 'twenty-fourths' in the context of a day, and the various equivalents that they might come across, for example twelfths, eighths. Support them by modelling sentences that describe how an activity takes up a particular fraction of a day. For example, we might sleep for one third of the day (8 hours).

 Introductory activity

Tell students that they have worked hard at understanding how to represent fractions, and finding equivalent fractions. Explain that now they are going to use the 24 hours in a day to find out what fraction of each day they spend doing different activities: being at school, sleeping and so on. Discuss some possible things that take up part of your day, for example as follows.

| Sleeping (8 hours) | Eating (2 hours) | Travelling to work (2 hours) | Reading (3 hours) |
| Teaching (5 hours) | Marking books (2 hours) | Exercising (1 hours) | Watching TV (1 hour) |

Help students to work out the fraction of the day that each activity takes, using 24 as the denominator. For example, sleeping takes $\frac{8}{24}$ of the day. Remind students how we can divide the numerator and denominator to find an equivalent fraction. So, sleeping takes $\frac{1}{3}$ of your day.

Help students get good estimates for some activities. *How long in total do you spend having meals? How much of your day do you think you spend at school?* Their estimates should contain whole numbers only.

 Main activity

Students work individually on the activity in the Student Book on page 117. Allow them plenty of time to work out all the possible activities and then check that the time given to each activity totals 24 hours. Encourage students to look for equivalent fractions for their activities. You could refer them to the example in the second speech bubble to show how $\frac{6}{24}$ can be written as $\frac{1}{4}$.

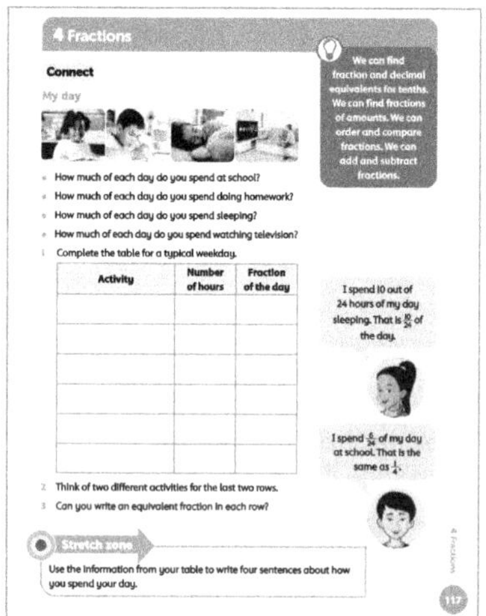

Differentiation

Supporting: Support students with turning all their activities into a fraction with denominator 24.

Consolidating: Encourage students to stick to whole hours only so that the fractions of a day are more straightforward.

Extending: Ask students to work out the fraction for things that don't take a whole number of hours. For example, mealtimes might total $2\frac{1}{2}$ hours out of 24. Remind them that they could multiply the numerator and denominator by the same number to find an equivalent fraction, so $\frac{2.5}{24}$ can be written as $\frac{5}{48}$.

Stretch zone: *Use the information from your table to write four sentences about how you spend your day.*

Check that students have written sentences that correctly describe parts of their day using fractions.

 Reflection time

Discuss with students what they found from their results. Refer back to the estimates they made in the introductory activity too and ask how good their estimates were. *Did any of the actual results surprise you?*

Did every activity have a fraction for it? Did all of the fractions have equivalents? Were there fractions with denominator 24 that did not have an equivalent? What can you say about the numerators of those fractions?

Differentiated outcomes	
All students	should find 24 hours' worth of activities.
Most students	will represent all their daily activities as a fraction and simplify some of their fractions.
Some students	may use fractions for activities that are not a whole number of hours.

4 Fractions

Review   Student Book page 118 · Practice Book page 99

Global skills

- **Self-development skills**: reflecting on learning

Student Book

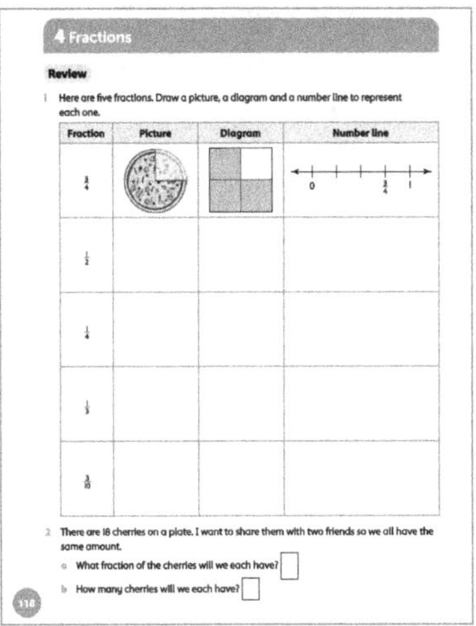

With young children, assessment activities are most effective when carried out as an everyday classroom activity. Students should be able to explore fractions in different contexts and use previous knowledge of calculating to add and subtract fractions. They should develop strategies for finding fractions of amounts and be able to use their knowledge of equivalent fractions to order and compare fractions, sometimes with the help of fraction walls or number lines. As they work on the Student Book activity on page 118, they should be able to represent each fraction as a picture, a diagram and on a number line. Point out to students that fractions often come in different representations, so it is important that they can recognise them in diagrams, in numbers, in everyday objects and as decimals. This will prepare them for Stage 4 when they will explore decimal numbers further. Point out also how fractions link to other units in mathematics: adding and subtracting, multiplying and dividing and so on.

Answers

Student Book page 118

1   Various answers are possible, so check that students have used pictures and diagrams that accurately

represent each fraction and see that it is located correctly on the number line.

2   **a** $\frac{1}{2}$

　　b   9 cherries

Practice Book

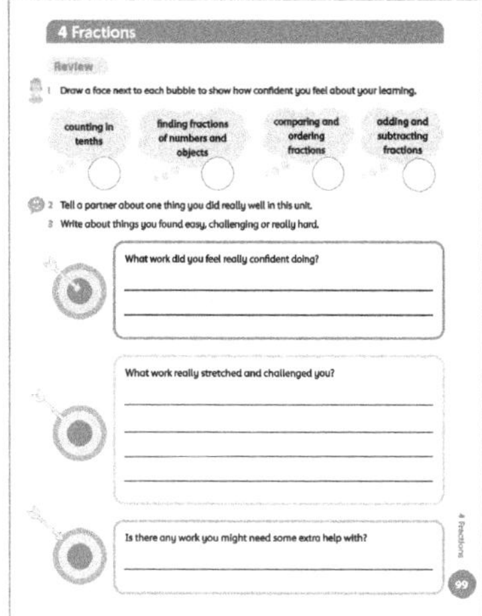

It is appropriate to complete this Practice Book review as a whole-class discussion. You may choose to keep a record of the class discussion or a copy of the review page for your own records. The review provides an opportunity for students to reflect on their learning from the unit, to discuss any areas of mathematics that they feel went particularly well, and any areas that they feel less confident about. Ensure that all students have a copy of the Student Book as a reminder of the areas of mathematics that they have worked on in this unit.

Allow students plenty of time for discussion before asking them to complete the Practice Book page individually, and then, if appropriate, to share their responses with the rest of the class. If students complete this self-assessment at home, encourage them to discuss this with adults. Make a note of areas that students still feel unsure about. As fractions are found in many everyday contexts, revisit these ideas regularly. Build fraction work into other aspects of everyday practice, for example when sharing things or finding a part of an amount or quantity.

Additional material

There are additional end-of-unit assessments available on the *Oxford Owl for School* website.

5 Length, mass and capacity

Overview

Big idea

One of the big ideas in this unit is for students to discover that there are a range of standard units that we use for measuring things. When making measurements, we make decisions about which units are appropriate for that measurement.

The other big idea is that we can convert between different standard units of measurement. These conversions are often based around factors of 100 and 1000. We can use the prefixes 'milli' and 'kilo' to help us understand the conversion factor: 1 kilogram is 1000 grams and 1 metre is 1000 millimetres.

Students use their ever-developing knowledge of measures to write and solve problems involving length, mass and capacity.

Look out for

- **Students who are lacking the fine motor control to measure carefully in smaller units (e.g. millimetres or grams).** Model for them how to measure in the relevant units and let them work with more competent peers as they practise reading scales using the smaller units.

- **Students who do not have a sense of how long one metre is, or how heavy a kilogram is, so are unable to use these as 'benchmarks'.** Provide opportunities for them to experience these standard measures regularly in and around the classroom with, for example, metre rules, 1 kg packets of food.

Possible misconceptions.

- **Students measure length from the end of rulers instead of the where 0 cm line is.** Model this for students, then let them practise while you watch, or arrange for them to work with peers who can do this.

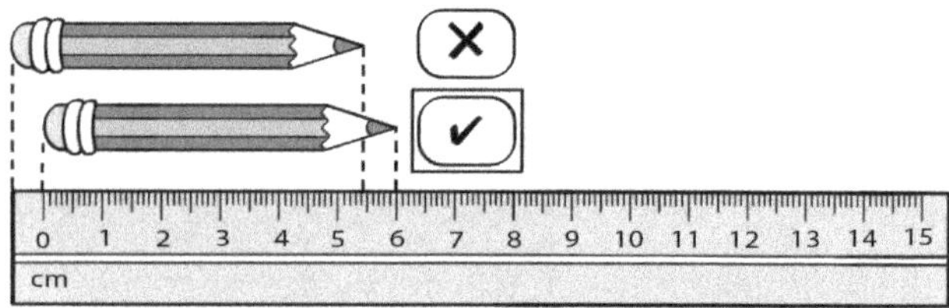

- **Students think that two measurements are different because the numbers are different, when they actually represent the same measurement (e.g. 1000 ml and 1 litre).** The best way to overcome this is to use two different forms for every measurement: for example, a line can be both 1 m 25 cm long and 125 cm.

- **Students think that any unit of length is appropriate to measure any length, or that any unit of mass is appropriate to measure mass, rather than choosing a suitable unit to measure. For example, they think that they can measure the distance from home to school in cm.** Discuss the most suitable units each time something is to be measured and have standard units available to use as benchmarks.

Key vocabulary

- measure, size, compare, estimate; scale, division
- approximately, too much, too little, too many, too few
- length, width, height, depth
- long, short, high, low, wide, narrow, thick, thin, longest, shortest, highest, lowest
- kilometre, metre, centimetre
- weigh, heavy, light, heavier, lighter, heaviest, lightest, mass
- kilogram, gram
- capacity, contains, empty, full; litre, millilitre, perimeter

Coverage in lessons

Learning objective	E	5A	5B	5C	5D	C	R
Measure, compare, add and subtract: lengths (m/cm/ mm); mass (kg/g); volume/capacity (l/ml).	✓	✓	✓	✓		✓	✓
Measure the perimeter of simple 2D shapes.					✓		

5 Length, mass and capacity

Big question

- How can I measure different things?

Global skills

- **Creative skills**: investigating
- **Interpersonal skills**: communication

Key vocabulary

- kilometre, metre, centimetre, kilogram, gram, litre, millilitre

Resources

- large sheets of paper, felt pens

Language support

- Read the standard units aloud with students. Place particular emphasis on the pronunciation of the units.

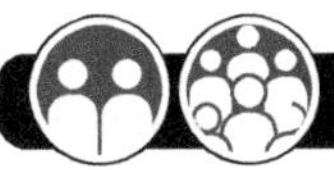 **Introductory activity**

Write 'distance in km' on the board. Give pairs two minutes to write down anything they can think of that they would measure in **kilometres** (km). One member of the pair should talk, while the other records. Repeat for the following standard units:

- length in **metres**
- length in **centimetres**
- mass in **kilograms**

- mass in **grams**
- capacity in **litres**
- capacity in **millilitres**.

Students should swap roles so that they take it in turns to both give answers, and record answers. Take feedback each time and discuss the appropriateness of the units.

 **Main activity**

Look together at page 119 of the Student Book. If you have access to an IWB you could use this. Discuss the images of measuring instruments and things being measured on the Student Book page. Use the speech bubble question and statements on the page as discussion prompts. Organise students into seven groups. Give each group a large sheet of paper. Each sheet should be labelled with one of the standard units mentioned in the introduction. Give the group five minutes to write down some measurements they know that use this unit. For example, they may write 'I can measure my height in metres and centimetres. I am 1 m 25 cm' or 'A spoonful of medicine is 5 millilitres'.

After five minutes, move the sheets of paper around the groups. Repeat this until each group has worked with each poster. Each group should generate new ideas by looking at the ideas that are already on the poster and adding new ones. Encourage discussion within the group, as this often produces new ideas.

Differentiation

Supporting: Point to objects around the room and ask students to estimate how long they are or how much they weigh. Students can then check by measuring.

Consolidating: Encourage students to be more independent when generating ideas for estimating and measuring.

Extending: Ask students to think of examples using all the units of measure: length, mass and capacity.

 **Reflection time**

Each group should tell you of one measurement that they knew from their completed poster and one that they learned from reading the poster. *How did you know that unit? When do you use it to measure?* Use one of the measurements that you have written on the board. Ask pairs of students to create a word problem with this measurement as the answer. An example using kilograms might be: I weigh 58 kg and my two friends weigh 44 kg and 49 kg respectively. A fairground ride has a weight limit of 150 kg per ride. Can my friends and I go on the ride together? Pairs exchange their word problems to solve.

5A Choosing appropriate units and converting units

Discover
Student Book page 120 • Practice Book page 100

Specific learning focus

- Choose and use appropriate units and equipment to estimate, measure and record measurements.
- Know the relationship between kilometres and metres, metres and centimetres, kilograms and grams, litres and millilitres.

Global skills

- **Creative skills**: problem solving
- **Interpersonal skills**: teamwork

Key vocabulary

- measure, metre (m), kilogram (kg), litre (l)

Resources

- a selection of measuring equipment including metre rules, tape measures, rulers
- kitchen scales, measuring cylinders

Language support

Ask students to describe the process they have used in the activity to encourage them to use the language of measurement with confidence. Support them by using the vocabulary of measurement as you ask, for example: *Why do you think that is approximately one metre? Why do you think that will be approximately one litre? What was your estimate?* Prepare a display of key words in measurement that can be added to during the unit.

Introductory activity

Take the class into a large space or outside. Ask students to estimate the distance between two objects such as a tree and a bench. Take a range of responses, asking students what knowledge they are using to help them estimate. Then use the measuring equipment to find the correct distance. Repeat this activity asking for an estimate of the height of the school. Discuss how you may calculate this. (You could **measure** the height of five bricks and use this information, or perhaps measure the height of a window.)

Main activity

Arrange the classroom so that there are three distinct areas. One area should contain the metre rules and other rules for measuring length; another area should contain the equipment for measuring capacity; and the last area should contain equipment for measuring mass.

Before you ask students to work through the activities in the Student Book on page 120, refer them to the Think back at the top of the page. You can also refer them to the first speech bubble, which will help them to estimate how long 1 metre is. Students should move between the different areas in small groups, and work on the tasks in the Student Book as they visit each area.

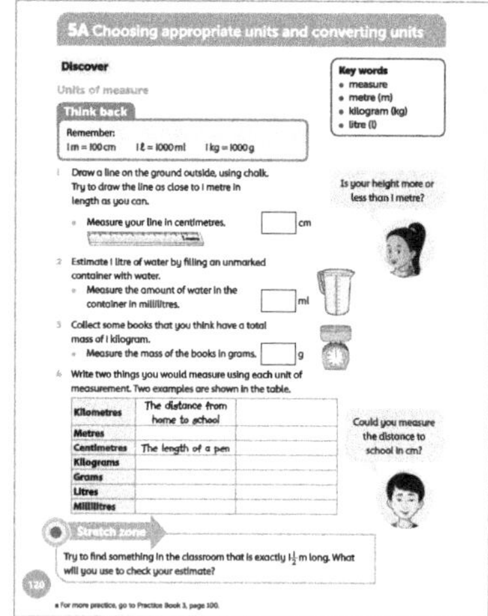

Once the groups have visited each area, they should sit in pairs to complete question 4. They should work to support each other by sharing information they know. Refer them to the question in the second speech bubble. This could help to prompt the sort of discussions they need to have about each unit of measurement.

Differentiation

Supporting: Support students by showing them how to read scales carefully.

Consolidating: Encourage students to discuss estimates as a group and to support each other with their reasoning.

Extending: Challenge students to find things in the classroom that are, for example, approximately 90 cm long, that weigh approximately 500 g or that hold approximately 750 ml.

Stretch zone: *Try to find something in the classroom that you estimate is exactly $1\frac{1}{2}$ m long. What will you use to check your estimate?*

Check that students are estimating and then measuring appropriately. *How can you use what you knew about the length of 1 metre to help you with this?*

 Reflection time

Ask each pair to write down something that you could measure in either metres or centimetres. This should be something that is only one or two metres long. Repeat this exercise for litres and millilitres, and kilograms and grams. Share answers with the whole class. Use this opportunity to correct any misconceptions that arise about units of measurement.

Practice Book: Students complete Practice Book page 100. They can do this directly after the main activity, as homework, or as the focus of a separate mathematics session to help students consolidate their learning and build fluency.

Students think of appropriate things to measure with each unit of measure. They then need to think of an object that matches a given measurement. They may need support from an adult if working on this activity at home.

Differentiated outcomes	
All students	should measure accurately with support.
Most students	will estimate and measure benchmark measurements accurately.
Some students	may estimate and measure different objects accurately using benchmarks to estimate other measurements.

5A Choosing appropriate units and converting units

Explore Student Book page 121 • Practice Book page 101

Specific learning focus

- Choose and use appropriate units and equipment to estimate, measure and record measurements.
- Know the relationship between kilometres and metres, metres and centimetres, kilograms and grams, litres and millilitres.

Global skills

- **Creative skills**: exploring
- **Real-world skills**: research
- **Interpersonal skills**: communication

Key vocabulary

- kilometre (km), centimetre (cm), gram (g), millilitre (ml)

Resources

- a selection of measuring equipment including metre rules, tape measures, rulers, scales and measuring cylinders
- map of the local area (professionally published or one that you have produced yourself) that has a scale
- string

Answers

Student Book page 120

Observe students while they are estimate and measure objects, and check that they are reading the scales on the measuring equipment correctly.

Practice Book page 100

Check that students' chosen objects would be measured in those units.

Language support

Ask students to repeat each of the statements. This gives them practice with the important vocabulary of 'longer', 'shorter', 'heavier', 'lighter' and so on.

 ### Introductory activity

Display the map on the board, or place it on the floor with students sitting in a circle around the map. Ask students to locate the school on the map and highlight it. Then ask them to name a place that they think is more than 1 km from the school. Use the scale to cut the string so that it represents 1 kilometre, and check the answer. Then draw a circle using the string to show students that everything inside the circle is closer than 1 km from the school, and everything outside the circle is further than 1 km. Students then use this information to complete question 1.

 ### Main activity

Refer students to the Think back to remind them of the conversions between units. Students work in pairs on the questions in the Student Book on page 121 and take it in turns to think of an answer, which they should both write down (if they agree that it is correct). *How can you check your answers for each question?* Discuss answers, ensuring that students mention the correct equipment for measuring each object.

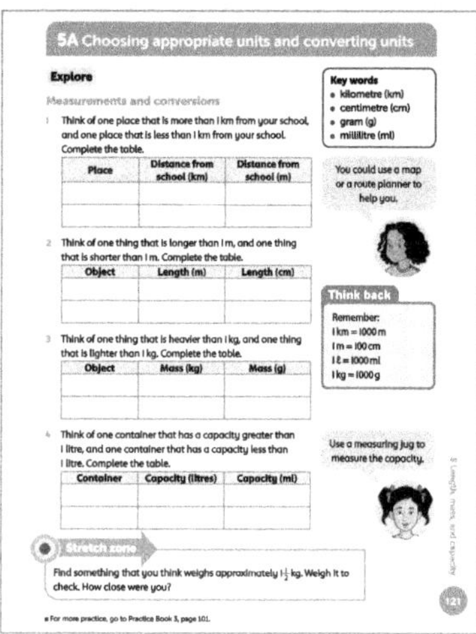

Differentiation

Supporting: Ask students which answers they are sure of and less sure of and support them to correct any incorrect information.

Consolidating: Encourage students to find answers that are close to the measure given, but above or below it. For example, for things longer and shorter than 1 m, they should look for things around 80 cm and 120 cm rather than 5 cm and 10 m.

Extending: Challenge students to estimate to a suggested accuracy, for example to the nearest cm.

Stretch zone: *Find something that you think weighs approximately* $1\frac{1}{2}$ *kg. Weigh it to check. How close were you?*

Check that students are estimating and then measuring appropriately. *How can you use what you knew about the mass of 1 kg to help you with this?*

 Reflection time

Revisit the opening activity. This time, students should look at the map and use the information to find places that are, for example:

- approximately 2 km from school
- approximately 500 m from school.

Choose one of the places you have discussed. Ask pairs to make up a story about walking to that place and the distances to the places that they stop at on the way.

Practice Book: Students complete Practice Book page 101. They can do this directly after the main activity, as homework, or as the focus of a separate mathematics session to help students consolidate their learning and build fluency.

Students convert units of metric measurements. There is a reminder of the key conversions to help them. Ask questions such as: *In question 1, will there be more metres than centimetres or more centimetres than metres? In question 6, will there be more kilograms or more grams?*

Differentiated outcomes	
All students	should know some objects and their related measurements to the nearest unit.
Most students	will find objects close to a given measurement.
Some students	may begin to estimate and measure with greater accuracy.

Answers

Student Book page 121

Check that students' chosen places and objects match the distances, lengths, masses and volumes given.

Practice Book page 101

1	5 m	**9**	1.5 kg
2	3.5 m	**10**	700 g
3	7000 m	**11**	4500 ml
4	150 cm	**12**	5.5 l
5	6.5 km	**13**	7 l
6	4500 g	**14**	0.5 l
7	500 g	**15**	1.4 l
8	8.5 kg		

Stretch zone: a thousandth, a thousand, a hundredth

5B Solving word problems involving measures

Discover Student Book page 122 • Practice Book page 102

Specific learning focus

- Choose and use appropriate units to solve word problems involving measures.

Global skills

- **Creative skills**: problem solving
- **Interpersonal skills**: communication

Key vocabulary

- length, mass, capacity

Resources

- lollipop sticks in a in a jar - each lollipop stick has a different student's name on it
- large sheets of paper

Language support

Ask students to say the questions aloud. You can help write the word problems and the answers for those students who find it difficult to write the problems down.

 **Introductory activity**

Write on the board 'How far is that in metres?' Ask students to look at the first picture on page 122 of the Student Book. Each pair should make up a word problem that finishes with the phrase 'How far is that in metres?' Direct students to the example in the speech bubble to help them. They should write their problem on a large sheet of paper. Collect the problems and select one at random to read aloud. Select a student by taking a lollipop stick from the jar. This student should try to solve the problem with help from the group who made it up. Repeat this three or four times and agree with the class a set of criteria for making up a 'good' word problem.

Main activity

Look together at each of the problems on page 122 and discuss the context for each to help students prepare to write word problems. For example, ask: *How many are there? Which units will you use?* Students can then work in pairs or small groups and discuss the contexts together before writing their word problems. While they are working, ask them to tell you the answer to one of their problems and the best method for working it out.

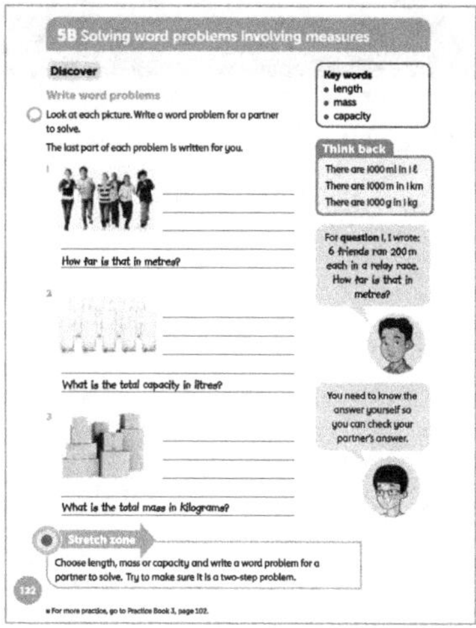

Differentiation

Supporting: Give students simple answers such as 1 metre or 1 litre and ask them to make up a problem with that answer.

Consolidating: Ask students to make up the hardest problem they can that they can solve themselves.

Extending: Ask students to create two-step problems.

Stretch zone: *Choose **length**, **mass** or **capacity** and write a word problem for a partner to solve. Try to make sure it is a two-step problem.*

Check students' problems for accuracy and suitability.

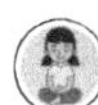 **Reflection time**

Students should swap their questions with a partner and solve them. When they have solved the questions, they can give each other feedback using the 'Two stars and a wish' technique: students tell one another two things they really liked about the problems and one thing that they would change.

Practice Book: Students complete Practice Book page 102. They can do this directly after the main activity, as homework, or as the focus of a separate mathematics session to help students consolidate their learning and build fluency.

Students solve measures word problems. They must give their answer in two ways, using different units of measurement. Remind them of the RUCSAC approach to solving problems (see page 154 in this Teacher's Guide).

Differentiated outcomes	
All students	should create simple word problems.
Most students	will create more complex word problems.
Some students	may create two-step word problems.

Student Book page 122

Check that students' word problems relate to the pictures.

Practice Book page 102

1 2000 m, 2 km

2 4000 g, 4 kg

3 1 ℓ, 1000 ml

Check that students' word problems use the facts provided and that their answers are correct.

5B Solving word problems involving measures

Explore Student Book page 123 · Practice Book page 103

Specific learning focus

- Solve word problems involving length, mass and capacity.

Global skills

- **Creative skills**: problem solving
- **Real-world skills**: interpreting information
- **Interpersonal skills**: communication

Key vocabulary

- longer, shorter, heavier, lighter

Resources

- none required

Language support

Continue using the RUCSAC acronym (see the introductory activity). Ask them to read the question aloud to you. You can help with this approach by asking questions about the key information.

What is the question asking you to do? What operation will you need to use? Is that a reasonable answer?

Introductory activity

Use one of the word problems written by students in the previous lesson and write it on the board. Write the following acronym on the board.

R Read the question carefully.

U Understand the question.

C Choose the operation.

S Solve the problem.

A Find the answer.

C Check the answer.

Solve the problem as follows. Ask students to discuss each step in pairs. Look at question 1 in the Student Book on page 123. Ask students to underline or highlight the key information. *What is part a asking? What calculation do you need to use? What unit of measurement do you need to include in your answer?* Repeat for parts b and c. Students may even find it helpful to draw the items. Tell them the drawings do not have to be accurate; just sketches will do.

Main activity

Students should work individually on the rest of the word problems on page 123 of the Student Book. Encourage them to talk to a partner to help make sense of each problem and to check their answers. As you work with students, ask questions such as: *Does that answer seem reasonable? How can you check your answer? What units of measurement do you need to include in your answer?* Encourage more-confident students to create their own problems using the same contexts as the problems in the Student Book.

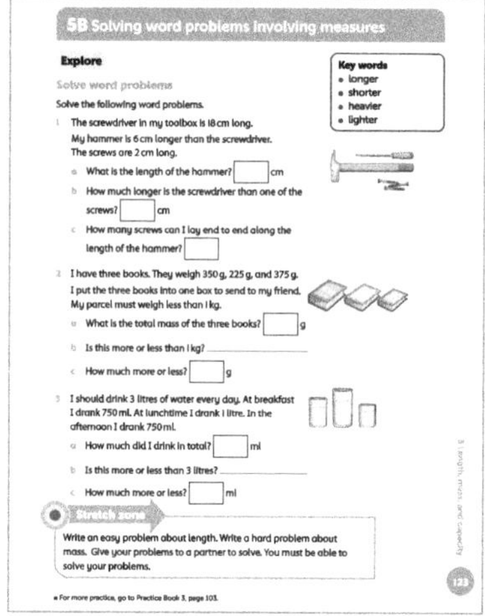

Differentiation

Supporting: Highlight key words for students. Encourage them to draw sketches to help solve the problems.

Consolidating: Ask students to check that the answer is reasonable in relation to the context of the problem.

Extending: Ask students to create their own problems using the same contexts as the problems in the Student Book.

Stretch zone: *Write an easy problem about length. Write a hard problem about mass. Give your problems to a partner to solve. You must be able to solve your problems.*

Check that students have written suitable problems and provided solutions. *What makes that an easy problem? What makes that a hard problem?*

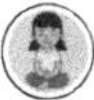 **Reflection time**

Solve some of the problems that have been created by your more-confident students in the stretch zone. To help all students, ask, for example: *What do you know about this problem? What units will you be using? Which operation will you be using? How do you know that answer is reasonable? How can you check the answer?*

Practice Book: Students complete Practice Book page 103. They can do this directly after the main activity, as homework, or as the focus of a separate mathematics session to help students consolidate their learning and build fluency.

Students write more of their own word problems. They are given a context and some ideas to include in their problem. Encourage them use vocabulary that tells the reader what calculation they need to do. *If the problem is an addition problem, what words might tell us that? What about subtraction?*

Differentiated outcomes	
All students	should highlight key words and solve problems with support.
Most students	will highlight key words and solve problems independently.
Some students	may write their own more complex word problems.

Answers

Student Book page 123

1   **a**   24 cm   **b**   16 cm   **c**   12

2   **a**   950 g   **b**   less   **c**   50 g

3   **a**   2500 ml   **b**   less   **c**   500 ml

Practice Book page 103

Check that students' word problems use the facts provided and that their answers are correct.

5C Reading scales

Discover   Student Book page 124 • Practice Book page 104

Specific learning focus

- Read scales to the nearest division or half division, using scales that are numbered or partially numbered.

Global skills

- **Creative skills**: exploring
- **Real-world skills**: research
- **Self-development skills**: reflecting on learning
- **Interpersonal skills**: communication

Key vocabulary

- litre, millilitres, scale

Resources

- a range of different containers that can hold liquid (If you prefer you could use pulses or grains)
- rulers

Language support

Ask students to explain the decisions they are making. Focus on the vocabulary of 'litres' and 'millilitres'. You should also model language such as 'halfway between'.

 **Introductory activity**

Hold up a metre stick but hold it back to front so there are no markings on it (or cover up the markings if it is double sided). Ask students to estimate where, for example, 50 cm, 75 cm, 10 cm are.

 **Main activity**

Use a measuring jug or cylinder to measure 1 litre of liquid. Select one of the containers and ask students to estimate how far up the container the liquid will come when you pour it in. Ask individuals to mark the container at the appropriate height. Ask where 500 millilitres would be. Encourage students to measure the height of 1 litre and use this information to make an estimate of 500 millilitres. Use the measuring cylinder to test the accuracy of their **scales**.

Give each small group three containers to work with. Groups should work on the activity in the Student Book on page 124. Students can each record their answers in their own book. As they work, ask them questions to guide their thinking, such as for question 2: *What will be the smallest and largest numbers on your scale?* For question 3 ask: *How will you find out how much each container holds?* You could suggest that students use a measuring jug and pour water from the container into the jug so they can read off the capacity from the scale on the jug.

Observe students while they are working. Note who can read the scale on the measuring jug confidently and draw their own scale, and who needs further practice.

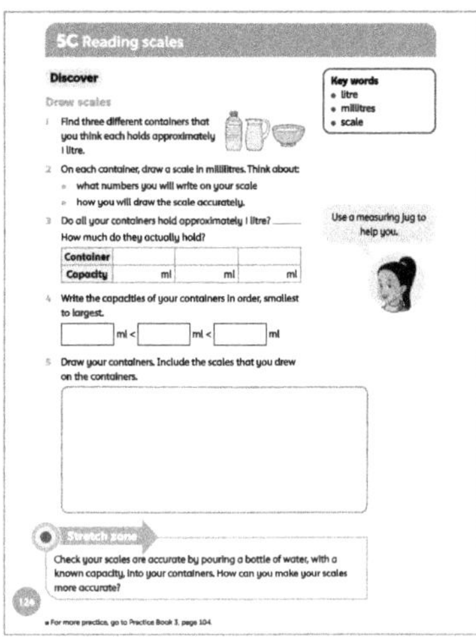

Differentiation

Supporting: Pair up less-confident students with more-confident students for support.

Consolidating: Ask students to measure exactly for each container.

Extending: Challenge students to use the measures they know to estimate capacities of other containers.

Stretch zone: *Check your scales are accurate by pouring a bottle of water, with a known capacity, into your containers. How can you make your scales more accurate?*

Check that students understand how to check accuracy of containers, using a measuring jug or cylinder with a scale marked in ml.

 **Reflection time**

Test one container from each group for accuracy using the measuring cylinder. The group that has made the most accurate scales should explain to the others in the class how they worked on the activity.

Practice Book: Students complete Practice Book page 104. They can do this directly after the main activity, as homework, or as the focus of a separate mathematics session to help students consolidate their learning and build fluency.

If this activity is being done at home, ensure that students have the correct equipment (measuring jug, tape measure and kitchen scales). If this is not possible, it will be better to complete the activity in the classroom, as a separate mathematics lesson.

Differentiated outcomes	
All students	should contribute to creating accurate measuring cylinders.
Most students	will use their cylinders to measure accurately to the nearest 100 millilitres.
Some students	may be able to read a scale to the nearest ml.

Answers

Student Book page 124

Answers will vary according to which containers are used. Check students' answers for accuracy based on their particular containers.

Practice Book page 104

Check that students' totals are accurate and are close to the requested amounts.

5C Reading scales

Explore 1   Student Book page 125 • Practice Book page 105

Specific learning focus

- Use a ruler to measure to the nearest centimetre.

Global skills

- **Creative skills**: exploring
- **Interpersonal skills**: communication

Key vocabulary

- centimetre, millimetre

Resources

- mini whiteboards and markers
- rulers
- set squares for drawing right angles

Language support

Model the language of measurement by asking appropriate questions.

- *How long is that?*
- *How many centimetres is that?*
- *Can you find something that is shorter?*
- *Can you find something that is longer?*

 **Introductory activity**

Draw six different lines on the board with lengths between 17 cm and 52 cm, but do not draw them in order of length, just in a random order. Write the length of one of the lines underneath that line. Label the lines with the letters A–F. Students should work in pairs to estimate the lengths of the lines. They write down their estimates on their whiteboards. They then write the lengths of the lines in order, shortest to longest. Finally, select pairs of students to come to the board and use rulers to measure the lines to check their answers. *How did the one length that I gave you help you estimate the other lengths?*

 **Main activity**

Students complete the activity in the Student Book on page 125. They should work in pairs to identify the objects they want to draw. One of the pair should measure the object and tell their partner the dimensions so that they can draw it. Remind students, when using a ruler, to place the '0' mark, not the edge of the ruler, at one end of the object.

Refer students to the speech bubble in the Student Book on page 125 and remind them that their drawings need to be accurate.

Students in each pair should then swap roles and repeat the activity.

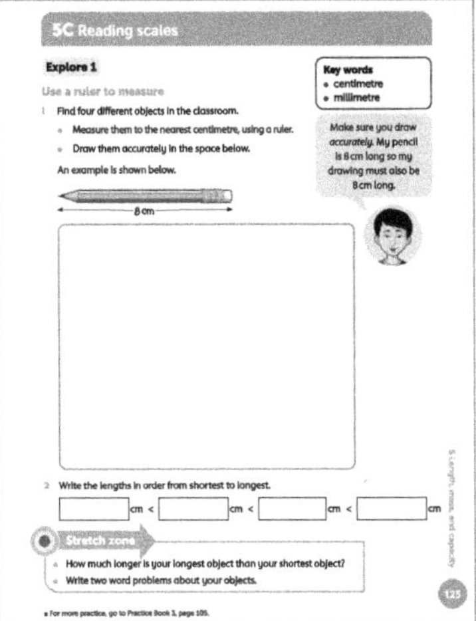

Differentiation

Supporting: Observe students measuring and model how to measure accurately if necessary.

Consolidating: Ask students to check each other's drawings to make sure that their partner has accurately drawn each object.

Extending: Show students how to measure using millimetres and then ask them to measure each object more accurately.

Stretch zone: *How much longer is your longest object than your shortest object?*

Write two word problems about your objects.

Check that students have calculated the difference correctly and have written suitable word problems.

 **Reflection time**

Play 'Guess my object'. Hide an object under a cloth and ask students to try to guess what it is by asking you questions to which you can only answer 'Yes' or 'No'. Encourage them to ask about the object's dimensions, and to use comparative language such as 'bigger than' and 'smaller than'.

Practice Book: Students complete Practice Book page 105. They can do this directly after the main activity, as homework, or as the focus of a separate mathematics session to help students consolidate their learning and build fluency.

If this activity is being done at home, ensure that students have the correct equipment (measuring jug, tape measure and kitchen scales). If this is not possible, it will be better to complete the activity in the classroom, as a separate mathematics lesson.

<table>
<tr><td colspan="2">Differentiated outcomes</td></tr>
<tr><td>All students</td><td>should measure accurately with support to draw objects.</td></tr>
<tr><td>Most students</td><td>will measure accurately independently to draw objects.</td></tr>
<tr><td>Some students</td><td>may be able to measure using millimetres.</td></tr>
</table>

Answers

Student Book page 125

Check how accurate students' measuring is. Is it close to the length of the objects they have drawn or do they need more practice?

Practice Book page 105

Check whether students' results seem reasonable. Ask them to explain how they measured the amount of water each person held and how much the water weighed.

Stretch zone: Students' explanations should make a connection between the height of a person and how much water they can hold.

5C Reading scales

Explore 2 Student Book page 126 · Practice Book page 106

Specific learning focus

- Read scales to the nearest division or half division, using scales that are numbered or partially numbered.

Global skills

- **Creative skills:** problem solving
- **Interpersonal skills:** communication

Key vocabulary

- mass, grams, kilograms, estimate

Resources

- scales

Language support

Model the language of measurement by asking by asking appropriate questions.

- *How heavy is that?*
- *How many grams?*
- *Can you find something that is heavier?*
- *Can you find something that is lighter?*

 **Introductory activity**

Show students a set of three objects of different **mass**. Ask a student to come to the front and put the objects in order of mass from lightest to heaviest by comparing them, two at a time, by picking them up.

Ask another student to come to the front and give them a 500 g mass to hold. Then ask them to pick up each of

the three objects in turn and say whether they think it is more or less than 500 g in mass. They can then pick up each object to compare it with the 500 g mass.

Ask a third student to come to the front and weigh each object on the scales to check the results of the ordering and whether each was more or less than 500 g.

Main activity

Students should work in pairs on the activity in the Student Book on page 126. They first select the objects they want to weigh. Each student in the pair should **estimate** the mass of each object and then use the scales to check and record the mass of the object, in grams.

Refer students to the speech bubble on the Student Book page and remind them for the need to include a unit of measurement each time. *Why is the unit of measurement important? If I said my metal water bottle weighs 200, what unit do you think I need to include? What unit do I need to include if I say that my bag (and its contents) weighs 1?*

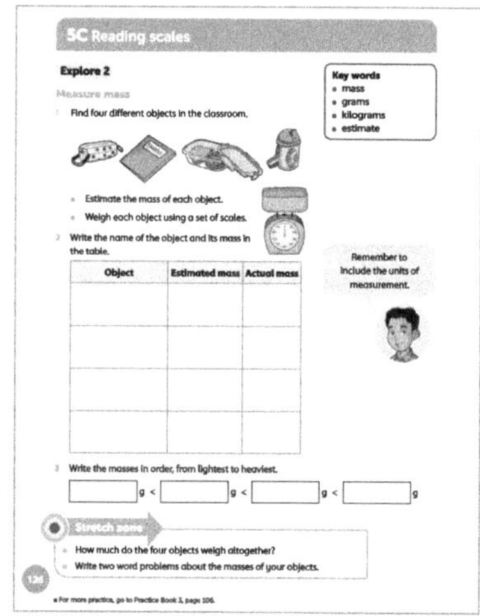

Differentiation

Supporting: Observe students estimating and weighing and model how to weigh accurately if necessary.

Consolidating: Ask students to check each other's measurements on the scales.

Extending: Challenge students to measure things weighing as little as 10 g, 20 g and 50 g.

Stretch zone: *How much do the four objects weigh altogether? Write two word problems about the masses of your objects.*

Check that students have written the correct answer and that they have written appropriate word problems.

 ## Reflection time

Play 'Guess the mass'. Hold up an object. Students guess its mass by asking you questions to which you can only answer 'Yes' or 'No'. Encourage them to use comparative language such as 'heavier than' and 'lighter than.'

Practice Book: Students complete Practice Book page 106. They can do this directly after the main activity, as homework, or as the focus of a separate mathematics session to help students consolidate their learning and build fluency.

Students mark various lengths and volumes on scales. *What two measurements is 75 cm between? Where will you mark it? What does each mark on the scales represent?*

Differentiated outcomes	
All students	should weigh objects accurately with support.
Most students	will weigh objects accurately to the nearest 100 g.
Some students	may measure lighter objects (with masses of less than 50 g) accurately.

Answers

Student Book page 126

As the objects chosen will have various masses, check that students have made reasonable estimates, read the scales correctly and then ordered the objects correctly by mass.

Practice Book page 106

Check that students have accurately labelled the measures on the rulers and measuring jugs.

Stretch zone: Students' explanations may suggest using a tape measure or a piece of string and then using a ruler to measure the length of string.

5D Perimeter

Discover · Student Book page 127 · Practice Book page 107

Specific learning focus

- Calculate the perimeter of shapes made from squares.

Global skills

- **Creative skills:** problem solving

Key vocabulary

- perimeter, length, width

Resources

- squared paper
- square tiles

Language support

Provide clear examples to demonstrate that a perimeter is a measure of the length around the outside of a shape. Label a diagram with 'length', 'width' and 'perimeter' and reinforce the notion that perimeter is measured in the same units as lengths (e.g. millimetres, centimetres, metres).

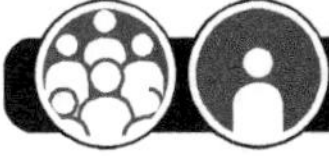 ## Introductory activity

Draw a rectangle on the board and label the sides as 5 cm by 4 cm. Ask students to draw the rectangle on squared paper. *How long is each side of the rectangle?* Students should be able to say that two of the sides are 5 cm and the other two are 4 cm.

Now ask students to add up the **lengths** of all four sides: 5 cm + 5 cm + 4 cm + 4 cm = 18 cm. *The total of all the side lengths is called the **perimeter**. It is the length all the way around the outside of the rectangle.*

 ## Main activity

Look together at page 127 of the Student Book. If you have access to an IWB you could use this. Refer students to the worked examples. Point out that each shape is made up of 24 squares. (You could mention that the area of each shape is therefore the same.) Talk through each example, the first being a rectangle 6 m by 4 m, and the second being an irregular shape made from squares. Ask a student to explain why the first shape has a perimeter of 20 m, encouraging them to describe the perimeter in terms of the sum of the separate sides.

Then ask another student to explain why the perimeter of the second shape is 22 m. Help the class to see that each of the straight edges has been added in turn to get 22 m.

Students can now complete the questions in the Student Book on page 127, using 24 square tiles to make shapes that they then draw onto squared paper. Explain that tiles must be joined along complete edges. Students investigate the different perimeters that these same 24 squares can make.

Refer students to the second speech bubble. *Did you think that all the shapes would have the same perimeter?* (You may want to point out that shapes that have the same area can all have different perimeters. There is no link between the size of the area and the perimeter.)

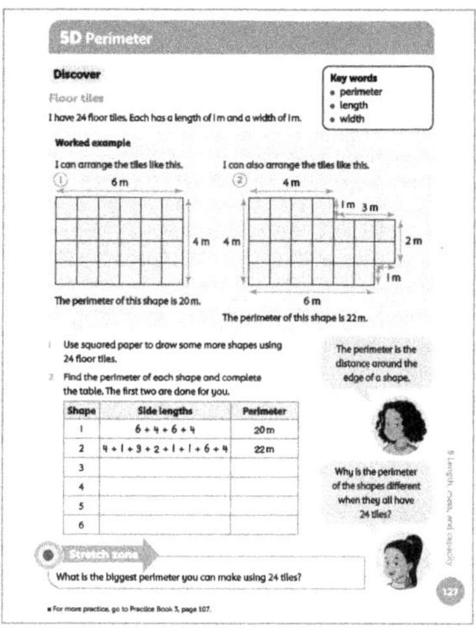

Differentiation

Supporting: Support students in counting the square edges around their shapes.

Consolidating: Ask students to count the square edges on the outside of the shapes before totalling them to find the perimeter.

Extending: Ask students to make complex shapes using their 24 square tiles and find the perimeter.

Stretch zone: *What is the biggest perimeter you can make using 24 tiles?*

Check that students have calculated the perimeter correctly and found the largest possible, which is 50, made from 24 tiles in a single line.

 **Reflection time**

Ask students to describe what they found from measuring perimeters of their shapes. What was the largest perimeter they found? What was the smallest? What did they notice about shapes with smaller perimeters?

Practice Book: Students complete Practice Book page 107. They can do this directly after the main activity, as homework, or as the focus of a separate mathematics session to help students consolidate their learning and build fluency.

Students draw shapes made from six squares onto a grid. They compare the perimeters of each of the shapes. *Can you estimate what the largest perimeter will be before you draw your shapes?*

Differentiated outcomes	
All students	should work out the perimeter of rectangles.
Most students	will work out the perimeter of non-rectangular shapes made from square tiles.
Some students	may understand that shapes with the same area can have different perimeters.

Answers

Student Book page 127

Students may have a variety of answers, so check that each perimeter is correct for the shapes they made.

Practice Book page 107

Students may have a variety of answers, so check that each perimeter is correct for the shapes they made, and that they have been drawn accurately.

Stretch zone: 14 cm

5D Perimeter

Explore Student Book page 128 · Practice Book page 108

Specific learning focus

- Calculate the perimeter of objects in the classroom.

Global skills

- **Creative skills:** investigating
- **Interpersonal skills:** team work

Key vocabulary

- perimeter, length, width

Resources

- rulers, tape measures

Language support

Provide clear examples for students that demonstrate that a perimeter is a measure of the length around the outside of a shape. Label a diagram with 'length', 'width' and 'perimeter' and reinforce the notion that perimeter is measured in the same units as lengths, namely millimetres, centimetres, metres and so on.

 **Introductory activity**

Tell students that you are going to measure the perimeter of the board. Ask students to tell you what shape it is. (rectangle) Ask what they need to measure to find the perimeter. (the total length of all its sides).

Ask students to estimate the perimeter first, in centimetres, by estimating the length and height of the board. If the top of the board is within reach for students, ask two students to come out and measure the length of the board and two others to measure the height. Write down the measurements. For example, they could be 150 cm and 100 cm.

Now ask students to add up the lengths of the sides of the board: 150 cm + 150 cm + 100 cm + 100 cm = 500 cm. *So the perimeter of the board is 500 cm.*

Main activity

Tell students they are going to find the perimeter of some rectangular objects in and around the classroom or school. For each shape they choose to measure, they must be sure that it is a rectangle. They will be measuring the length and **width** (or height) and recording these in the table on page 128 of the Student Book. Direct students to look at the speech bubbles and use these as prompts for discussion. Point out how the first row of the table has been completed.

Once students have measured the length and width (height) of four rectangular objects, they should then use the measurements to calculate the perimeter of each shape.

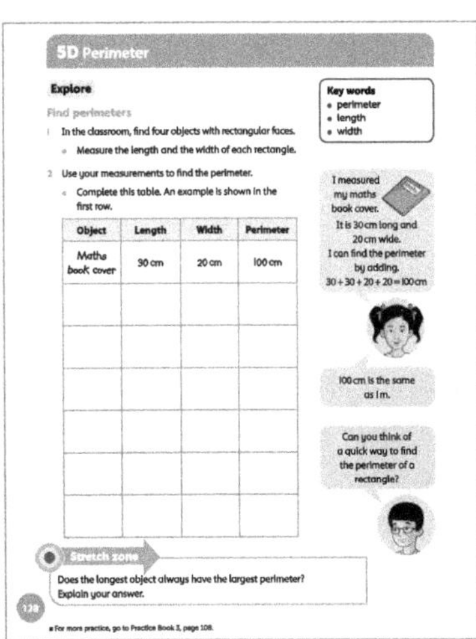

Differentiation

Supporting: Support students in measuring the sides of their shapes.

Consolidating: Ask students to estimate the side lengths before totalling them to find the perimeter.

Extending: Ask students to work out a quick way of working out the perimeter of a rectangle.

Stretch zone: *Does the longest object always have the largest perimeter? Explain your answer.*

Check that students understand that the perimeter depends on the length and the width.

 **Reflection time**

Ask students to describe what they found from measuring perimeters of their objects. What was the largest perimeter they found? What was the smallest?

Refer students to the third speech bubble on page 128 of the Student Book. Ask students whether they found an easy way to find the perimeter. Some students may have reasoned that they only need to measure one length and one width (height) to get the perimeter as these can just be doubled if the shape is a rectangle.

Practice Book: Students complete Practice Book page 108. They can do this directly after the main activity, as homework, or as the focus of a separate mathematics session to help students consolidate their learning and build fluency.

It would work well if students did this activity at home because they would be able to find different objects from those used in the main activity. Remind them that the objects need to be rectangles or have rectangular surfaces. They will need a tape measure and a ruler to measure the perimeters.

Differentiated outcomes	
All students	should measure the lengths of all the sides of rectangular objects with support.
Most students	will measure the lengths of all the sides of rectangular objects.
Some students	may find a quick or easy way to find the perimeter of a rectangle by finding the length and width and doubling it.

Answers

Student Book page 128

Students may have a variety of answers, so check that each perimeter is correct for the objects they used.

Practice Book page 108

Students may have a variety of answers, so check that each perimeter is correct for the objects they used.

Stretch zone: No, we do not always need to measure every side. For example, if the shape is a rectangle, we know that each pair of opposite sides is the same length, so we only need to measure one of the sides in each pair and then double it.

5 Length, mass and capacity

Connect Student Book page 129

Big idea
- I know the different units and the equipment I need to use to measure things accurately.

Global skills
- **Creative skills**: investigating
- **Real-world skills**: presenting information
- **Self-development skills**: reflecting on learning

Key vocabulary
- metres, centimetres

Resources
- thick card, or planks of wood and sets of books to make ramps
- toy cars, lorries and other toy vehicles of different sizes
- rulers, tape measures
- large sheets of paper

Language support
The language support in this activity comes from working together. Organise groups that match the more-confident English speakers with those who are less confident.

 **Introductory activity**

Make two ramps using some books and two planks of wood or thick card. Use a different number of books for each ramp so that the ramps are different heights. Select two different-sized vehicles, for example a small car and a lorry. Put one of the vehicles at the top of each ramp. *If I let go of these two vehicles at the same time, which one do you think will travel furthest?* Students write their predictions. *Can you explain your answer?*

Discuss whether this experiment is fair. *What could we do to make it fair?*

 **Main activity**

Students should work through the investigation in the Student Book on page 129. They should work in small mixed-attainment groups. Encourage them to make notes in draft form first and to only complete the questions in the Student Book when they have finished the investigation and talked to you about their findings.

As you work with the groups, talk to them about fair testing (that is, that only one variable should be changed at a time). In the first experiment, this variable is the height of the ramp. In the second experiment, it is the mass of the car.

When each group has finished the activity, they should plan a poster for reflection time. *What was the most important thing you found out?* They can make the posters on large sheets of paper and you can take photographs to paste into the Student Books to show the results and the most important findings of the groups' investigations.

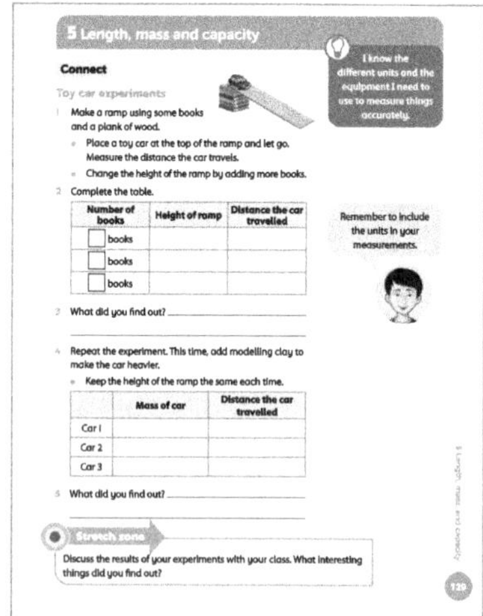

Differentiation
Supporting: Encourage students to carry out the measuring.

Consolidating: Encourage students to check the measurements and to create a recording system.

Extending: Encourage students to make predictions about how far the car will go.

Stretch zone: *Discuss the results of your experiments with your class. What interesting things did you find out?*

 **Reflection time**

Each group should present their findings. Allow them one presentation slide, or one poster-sized piece of paper, to illustrate their most important finding.

Differentiated outcomes	
All students	should contribute to the work of the group, measuring and recording.
Most students	will measure accurately and develop appropriate recording methods.
Some students	may make accurate predictions.

5 Length, mass and capacity

Global skills

- Self-development skills: reflecting on learning

Student Book

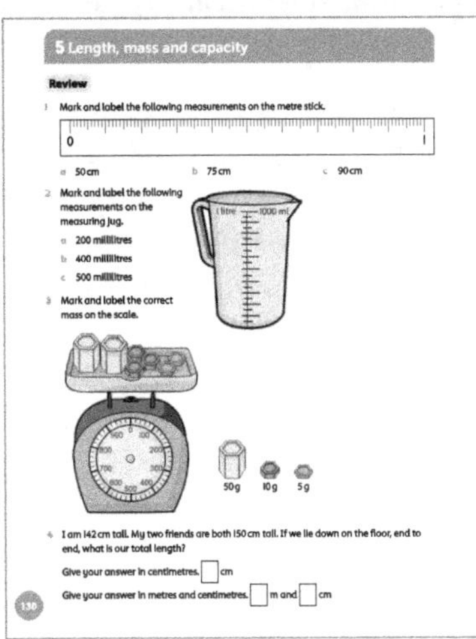

With young children, assessment activities are most effective when carried out as an everyday classroom activity. Students should have used rulers, tape measures, containers of different capacities and a range of scales to support them through the unit.

When students have marked the measurements on the ruler for question 1, ask them to find something in the classroom to match each length. Repeat for the measuring jug in question 2 – can they find containers with those capacities? Do the same for mass in question 3 – can students find something that weighs the same as the scales are showing?

As students are working, reinforce the importance of estimating before measuring, as they would when doing calculations to get a good sense of what a reasonable answer would be.

Watch as students complete the activities that require them to estimate, measure and record using units of length, mass and capacity. Check that students understand the language of measuring.

Answers

Student Book page 130

1 and **2** Check that the answers are accurate in each case.

3   The weights sum to 140 g

4   442 cm, 4 m and 42 cm

Practice Book

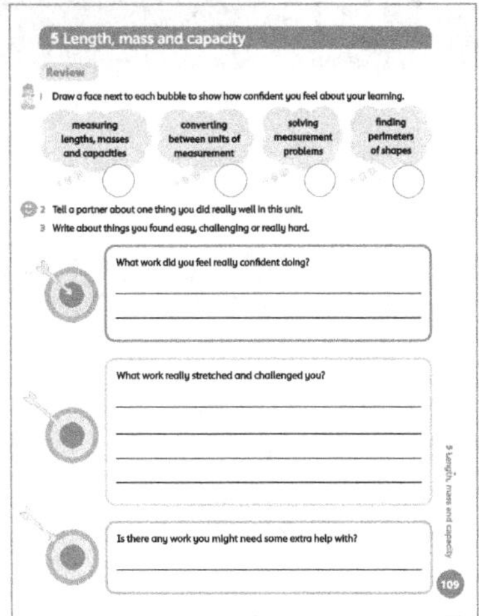

It is appropriate to complete this Practice Book review as a whole-class discussion. You may choose to keep a record of the class discussion or a copy of the review page for your own records. The review provides an opportunity for students to reflect on their learning from the unit, to discuss any areas of mathematics that they feel went particularly well, and any areas that they feel less confident about. Ensure that all students have a copy of the Student Book as a reminder of the areas of mathematics that they have worked on in this unit.

Allow students plenty of time for discussion before asking them to complete the Practice Book page individually, and then, if appropriate, to share their responses with the rest of the class. If students complete this self-assessment at home, encourage them to discuss this with adults. Make a note of areas that students still feel unsure about. For example, how do students feel about the different measures they have learned about? Do they feel confident when estimating and measuring lengths, masses and capacities? Are they confident at converting between units of measurement? Do they understand why using the correct unit of measurement, particularly in their answer, is important?

Additional material

There are additional end-of-unit assessments available on the *Oxford Owl for School* website.

Overview

Big idea

The idea of money as a measure of the exchange value of goods is a very advanced idea. Students only know that a price is attached to goods and they know what notes or coins to offer. This means that we must treat money differently from other measures such as length, mass and so on although, in this unit, students will begin to calculate with amounts of money using addition and subtraction.

Increased use of bank cards is reducing students' opportunities to observe money being handled, so it is crucially important that teachers introduce students to 'real' money in the classroom so that students can learn the currency.

The three key ideas needed for students to learn about and understand money are coin recognition, equivalence and practical situations. Students may or may not be familiar, from real-life contexts, with amounts of money written in decimal form. You should choose whether to use the standard decimal form for these amounts, or to read and write them with, for example, separate dollars and cents. So, $1.50 would be written (and certainly read) as $1 dollar and 50 cents.

Look out for

- **Students who may not have much experience of handling notes and coins, and consequently lack awareness of the values and the exchange process when using money.** Provide support by providing real or 'fake' notes and coins to manipulate and work with. When adding or subtracting amounts, students will need to use their understanding of place value.

- **Students who may not be able to exchange one coin for two coins that have the same total value (e.g. 10c = 5c + 5c).** Model this process for students, using precise language and always asking the value of each coin that is used.

Possible misconceptions

- **Students think, when paying for goods, that the customer only gives the shopkeeper money and should never receive any money back.** Explain to students that 'giving change' is something that happens when a customer gives too much money for goods. For example, if they give a $10 note for something that costs $5.50, they will receive $4.50 back.

- **Students misunderstand the place value of prices and so have difficulty adding and subtracting with money.** Help students to add in pence (or cents) to begin with before moving on to adding pounds and then pence (or dollars and cents).

Key vocabulary

- money, notes, coins, cost, buy
- add, subtract, change, total
- currency, pounds sterling (£), pence (p)

Coverage in lessons

Learning objective	E	6A	6B	C	R
Add and subtract amounts of money, using both £ and p, in practical contexts (or using your own currency).	✓	✓		✓	✓
Give change, using both £ and p, in practical contexts (or using your own currency).	✓		✓	✓	✓

6 Money

Engage Student Book page 131

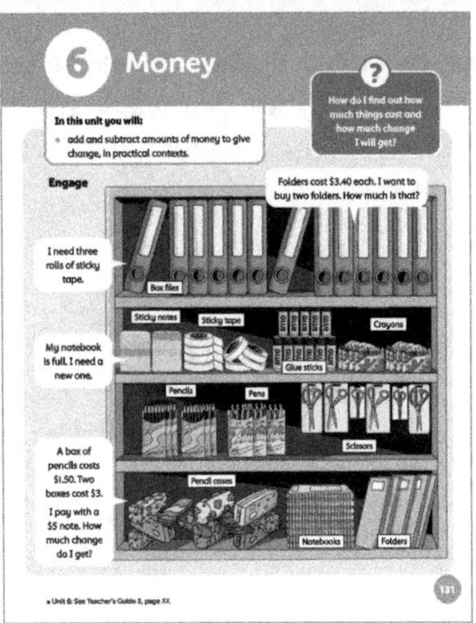

Big question

- How do I find out how much things cost and how much change I will get?

Global skills

- **Creative skills:** problem solving
- **Real-world skills:** financial literacy
- **Interpersonal skills:** communication

Key vocabulary

- money, notes, coins, cost, buy, add, subtract, change

Resources

- pricelist or price labels

Language support

Set up a money poster in the local currency. Display images of notes and coins with sentences such as: 'We use notes and coins to buy things in shops.' Include labels for the notes and coins showing their values. For example, for the UK they need the words 'pounds' and 'pence', while for the USA they need 'dollars', 'cents', 'nickels', 'dimes', 'quarters'.

 **Introductory activity**

Ask students what they can tell you about *money. Do you know what the **notes** and **coins** in our county are called?* Write the names of the different denominations and have examples of each to pass round, as well as screen images to show what the coins and notes look like.

Ask students whether they can explain what happens when they **buy** something and give a note or coin that is more than the amount they are paying.

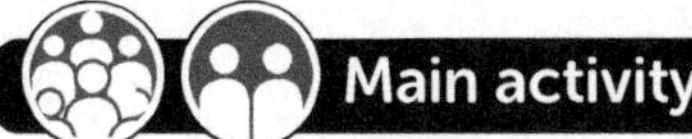 **Main activity**

Look together at page 131 of the Student Book. If you have access to an IWB you could use this. Ask students to look at the picture. It shows a selection of items of stationery and some prices in dollars and cents. Remind them of the RUCSAC approach to solving word problems, helping them to identify the key words and numbers in the questions.

The question about pencils already states that two boxes will **cost** $3 and that a $5 note is used to pay for them. *How can we work out how much change will be needed?* They should be able to say that you need to **subtract** the cost ($3) from the note value ($5). $5 − $3 = $2, so there would be $2 **change**.

Ask students to work in pairs and make up some prices for the other stationery items they can see. You can continue to work in dollars, or in your local currency. In pairs, students take a turn at choosing two items (the same or different), working out the total cost and choosing a note or coin greater than the total to 'pay' with. Their partner should then subtract the cost of the items from the note or coin to work out how much change they would give back.

 **Reflection time**

Ask students to describe how they worked out the amounts of change. Do they know whether this is how it is done in shops? Talk to students about the way shopkeepers give change, starting from the total price of the goods and counting on to the value of the note or coin given.

For example, goods cost $3.50 and a $10 note is used to pay. The shopkeeper will start from $3.50 and count on 50¢ to $4, then count on $6 to get to $10, so they would give $6.50 in change.

Ask students for other examples like this.

6A Adding and subtracting with money

Discover
Student Book page 132 • Practice Book page 110

Specific learning focus

- Use addition and subtraction to find the total cost of items.

Global skills

- **Creative skills:** problem solving
- **Real-world skills:** research/financial literacy
- **Interpersonal skills:** communication

Key vocabulary

- currency, notes, coins

Resources

- price lists
- notes and coins (plastic or real) or place-value counters

Language support

Reinforce the vocabulary of 'addition', 'altogether' and 'total', and 'subtraction', 'more than', 'less than', 'difference', 'how much more?', and include here the language of 'exchange' and the local currency names, such as dollars and cents.

 **Introductory activity**

Look at the items in the table on page 132 of the Student Book. Ask, for example: *Where can we get these items from? How can we find out how much each item costs?* If appropriate, visit the shop, or pre-record a video of you visiting a shop, focusing on the items in the shop to show to students. If this is not possible, students can research the prices using the internet by visiting, for example, a large supermarket's online shop.

Then make a class price list for students to refer to.

 **Main activity**

Ask one student to choose two items from the list. Write them on the board and then find their costs and write these beside the items. Model addition using money or place-value counters to represent money, either by adding the dollars and then the cents, or converting the costs into cents only.

Then also model subtraction to find the 'difference in cost' between the two items.

Ask students to work in pairs. One chooses either two or three items. These can be different items or more than one of the same item. They then both write down the prices and work out the total or difference, using a suitable method.

Students then complete the activity in the Student Book on page 132 where they will be making up addition and subtraction word problems about the total or difference of items from the list.

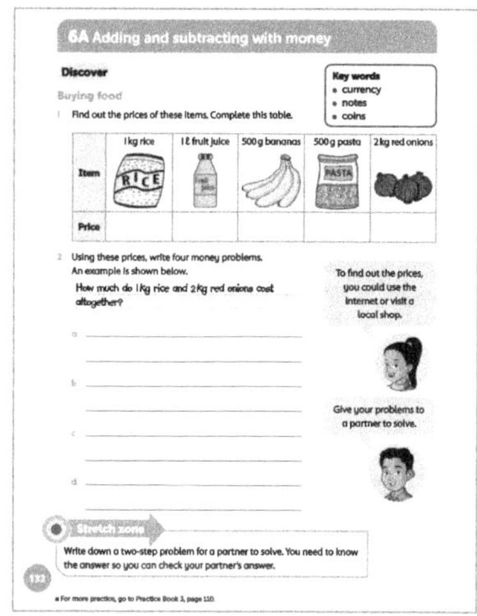

Differentiation

Supporting: Ask students to make simple totals and help them with the addition and subtraction. Model how to use place-value counters or real coins to do exchanges from cents to dollars, for example.

Consolidating: Ask students to check the total or difference by using an inverse calculation.

Extending: Ask students to find the total or difference using a mental method.

Stretch zone: *Write a two-step problem for a partner to solve. You need to know the answer so you can check your partner's answer.*

Check whether students have written and solved a suitable two-step problem. Students' problems shouldn't include giving change (as that is covered in the next lessons) so an example might be: *How much more is a bag of rice and a bunch of bananas than a bottle of orange juice?*

 **Reflection time**

Ask some students to share the items and prices they chose and explain how they added or subtracted the amounts to get the total. Did they look for easy ways to do this mentally? Did they start with the most expensive item first?

Set a challenge for students either to find two items that add up exactly to a given amount, or to find two items as close to that amount as possible, but less than the amount. For example, they might have $5 to spend, so can they find two items that cost $5 or just less than this?

Practice Book: Students complete Practice Book page 110. They can do this directly after the main activity, as homework, or as the focus of a separate mathematics session to help students consolidate their learning and build fluency.

In this activity, the cost of the items is given in pounds and pence, so ensure that students are happy working with this currency before they attempt the questions. Alternatively, the prices could be amended to reflect the local currency.

Differentiated outcomes	
All students	should add two prices to get a total or subtract to find the difference, with support if needed.
Most students	will correctly add two or more prices or find the difference between two prices.
Some students	may find the total or difference in a range of ways, including mentally.

6A Adding and subtracting with money

Explore Student Book page 133 • Practice Book page 111

Specific learning focus

- Use addition and subtraction to work out costs of items and amount of change.

Global skills

- **Creative skills:** problem solving
- **Real-world skills:** financial literacy
- **Interpersonal skills:** communication

Key vocabulary

- pounds sterling (£), pence (p), or local currency equivalents

Resources

- list of items from a shop, with prices
- coins (plastic or real), place-value counters

Language support

Reinforce the vocabulary of 'addition': 'altogether' and 'total', and 'subtraction': 'how much more', 'difference'.

Practice Book

Answers

Student Book page 132

Students' answers will vary. Check the wording of their problems to see that they have made up a suitable question.

Practice Book page 110

1 £10.20

2 £8.20

3 Answers may vary here so check that the lists total less than £15.

4 Jian spends £7.15, Marco spends £7.55. Marco spends 40p more.

 **Introductory activity**

Show students a list of items with their prices, perhaps food items from a shop. Ask students to say how much each item costs, for example saying $2.50 as '2 dollars and 50 cents'. *Which item is the cheapest? How do you know? Which item is the most expensive? How do you know?*

Ask them to work in pairs to take turns choosing two items from the list and working out how much more one costs than the other. Students can use coins or place-value counters to help them work out the difference.

 **Main activity**

Set up a class role-play with the items listed in the introductory activity. Choose a student to be the shopkeeper, and another as the customer. The customer goes into the shop and chooses two items to buy. You tell them how much money they have. The other students say the total cost of the items, say whether they have enough money to buy them and give reasons.

Students then work on the activity on page 133 of the Student Book. They answer questions about the cost of a selection of items from a sports shop and whether they can afford to buy the items with a limited amount of money. The prices are all given in pounds and pence, as the sports shop is based in London.

For questions 3 and 4, ask students to explain to a partner how they worked out what to buy and why they chose those items. They will use these explanations later in reflection time.

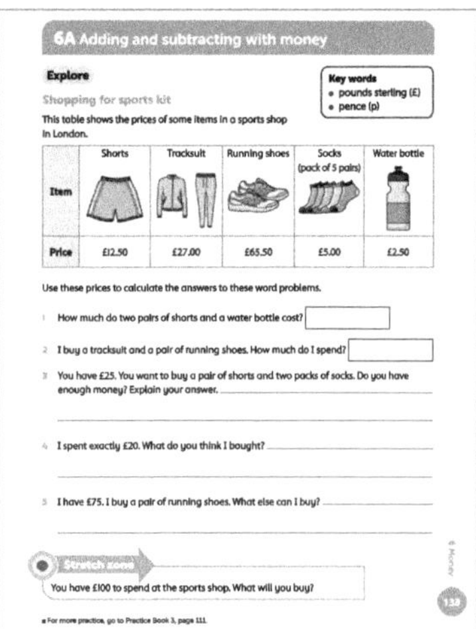

Differentiation

Supporting: Help students with the calculations. Model how to use place-value counters or coins to do exchanges from pence to pounds, for example.

Consolidating: Ask students to check their calculations using inverse operations.

Extending: Challenge students to calculate the amounts mentally.

Stretch zone: *You have £100 to spend at the sports shop. What will you buy?*

Check that students have calculated accurately and not spent more than $100.

 **Reflection time**

Ask students to share how they spent their money in question 4. *How did you know whether you had spent too much? Did you have enough money to buy another item? How did you calculate the final total? What else would you buy if you had an extra £5?*

Practice Book: Students complete Practice Book page 111. They can do this directly after the main activity, as homework, or as the focus of a separate mathematics session to help students consolidate their learning and build fluency.

Students write their own word problems based on real-life prices of food items. They will need to find out the prices of the items first, for which they may need adult support.

Differentiated outcomes	
All students	should add 2 prices below $25.
Most students	will spend up to $25 and calculate the change.
Some students	may work out how to spend an amount exactly with no change.

Answers

Student Book page 133

1 £27.50

2 £92.50

3 Yes, I have enough, the total is £22.50.

4 Various answers possible, for example four packs of socks.

5 Socks and water bottle, or three water bottles.

Practice Book page 111

Answers will vary. Check that students have written suitable problems and calculated the correct answer.

6B Giving change

Discover Student Book page 134 • Practice Book page 112

Specific learning focus

- Work out the amount of change needed after buying items.

Global skills

- **Creative skills:** problem solving
- **Real-world skills:** financial literacy

Key vocabulary

- total, notes, coins, change

Resources

- coins and notes of local currency, real or otherwise

Language support

Display some items with a price label and the change if the item was paid for with $1 or $5, as appropriate. Explain the meaning of the word 'change' in this context. Price labels can be written as, for example, both $3 and 50¢ and $3.50.

 **Introductory activity**

Say to students that you are going to buy an item for $14 and 70¢. You will give the shopkeeper a $20 note. *How much change should I get?* Model for students how to work out how much change, using place-value counters, notes and coins and then drawing a number line. For example, subtract $14 and 70¢ from $20 to leave $5 and 30¢:

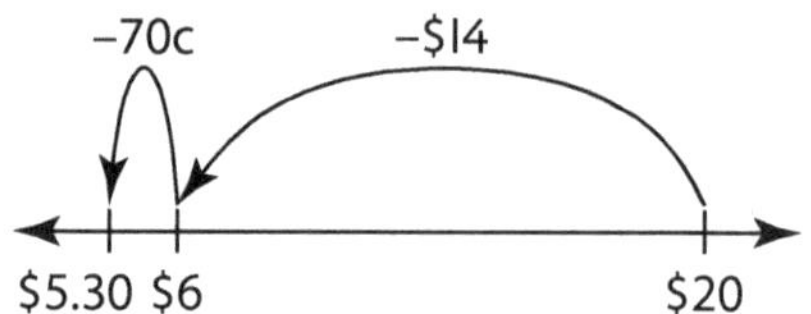

or, using the 'shopkeeper method' count on from $14 and 70¢ to $15, then $5 to $20 to get to $5 and 30¢ change altogether.

Which method do you find easier?

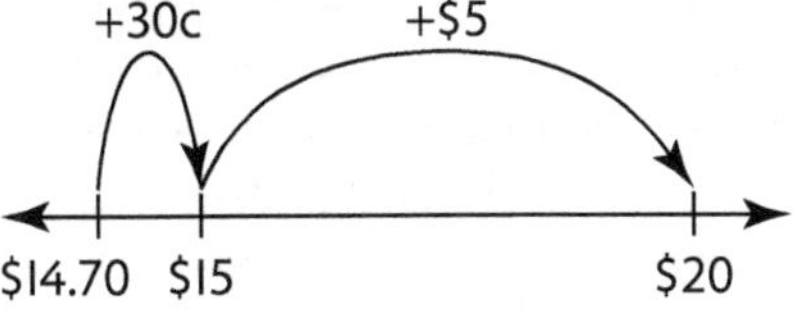

Discuss, as a whole class, how to find out the prices of the items in the table on page 134 of the Student Book. You could either visit a shop or use the internet to find out the prices and then write them in the table.

Use class role-play with the items listed in the table. Choose two students to be the shopkeeper and customer. The customer goes into the shop and chooses two items to buy. Tell them the amount they have to pay with, for example $10. The other students say the **total** cost of the items and how much change they will get, and give reasons.

Tell students that they are going to use a list of items to write their own problems that will involve adding prices and then subtracting from a note of larger value to find out how much change there will be. They can then complete the activity on page 134 of the Student Book.

They may follow the example given, which gives two items, or they may decide to include multiple items, either some the same or all different.

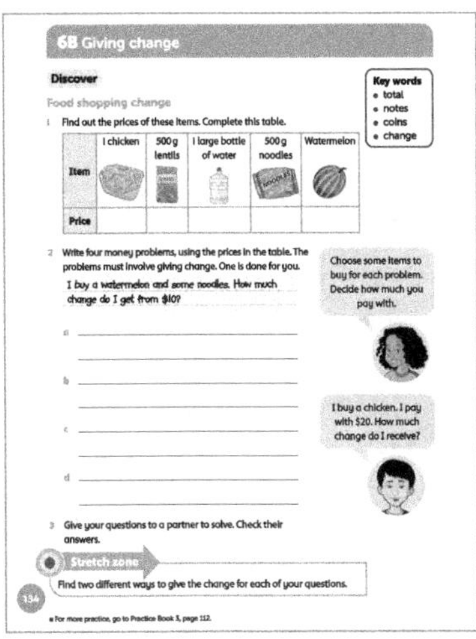

Differentiation

Supporting: Allow students to use notes and coins to work out the change and show them how to count on from the item cost, perhaps using a number line.

Consolidating: Students can use number lines or jottings to work out the change.

Extending: Ask students to find different ways of using notes and coins to make the right amount of change.

Stretch zone: *Find two different ways to give the change for each of your questions.*

Check that students can correctly show different ways to make the change using the available currency. *Can you find the way with the fewest number of coins?*

Reflection time

Ask students to describe how they worked out the change for each question in the activity on page 134 of the Student Book. *Which way used the fewest number of notes or coins? Why might shopkeepers give change in a different way?* There are many possible reasons for this. For example: they want to give some of their smaller denomination coins so there are not as many coins left in their till, they might only have a few of some coins so they want to save as many of those as possible, they want to use the fewest possible coins as it is easier to count fewer coins than lots of coins, or the customer might ask for specific change for a reason such as needing it for parking.

Practice Book: Students complete Practice Book page 112. They can do this directly after the main activity, as homework, or as the focus of a separate mathematics session to help students consolidate their learning and build fluency.

Students first have to add two or more items together and then work out the change from $50. The prices are all given in decimal notation so you may want to go through the meaning of this again, before students attempt the questions.

Differentiated outcomes	
All students	should understand when change is needed.
Most students	will work out how much change and what coins or notes can be given.
Some students	may find more than one way to make the change amount.

Answers

Student Book page 134

Answers will vary. Check that students have written suitable problems and calculated the correct answer.

Practice Book page 112

1 $18.70

2 85¢

3 $3.50

4 $18.20

5 $6.85

Stretch zone: Check students' answers. For example, you could buy a tracksuit, running shoes and a pack of socks and you will have $3 left. Or you could buy 39 water bottles and have $2.15 left.

Explore
Student Book page 135 • Practice Book page 113

Specific learning focus

- Work out the amount of change from spending.

Global skills

- **Creative skills:** problem solving
- **Real-world skills:** financial literacy
- **Interpersonal skills:** communication

Key vocabulary

- total, notes, coins, change

Resources

- coins and notes of local currency, real or otherwise

Language support

Display some items with a price label and the change if the item was paid for with $10 or $50, as appropriate. The labels can show prices as, for example, $2.75 or $2 and 75 cents.

 **Introductory activity**

Show students page 135 of the Student Book, which contains a list of items from an electrical goods shop, with their prices. Ask students to say how much each item costs in full, for example '10 dollars and 75 cents'. *Which item is the cheapest? How do you know? Which item is the most expensive? How do you know?*

Ask students to work in pairs to take turns choosing two items from the list and working out the total cost of the items, or finding the difference between the two items.

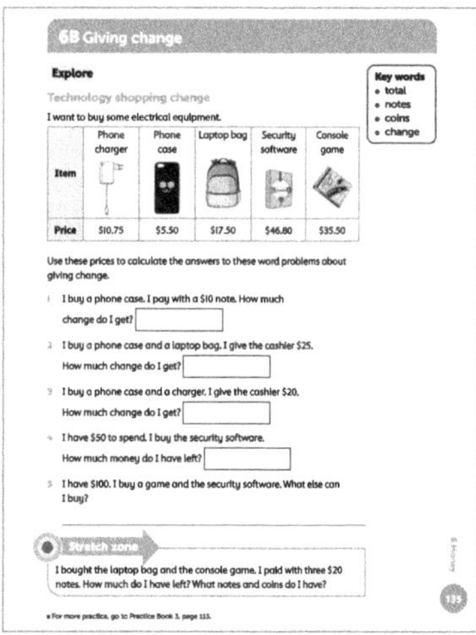

 **Main activity**

Say to students that you are going to choose some of items of electrical goods from the introductory activity, which they have already added together to find a total price. Remind students how they should find the change needed for the items when they pay with, for example, a $50 note or a $100 note. They may work out the change from a given note in different ways. For example, they may subtract the price from the note value or they may count on from the price to the note value in the way shopkeepers would.

Students should work through the questions on page 135 of the Student Book to consolidate their ability to solve problems that involve adding prices and then subtracting from a note of larger value to find how much change there will be.

Differentiation

Supporting: Allow students to use notes and coins to work out the change and show them how to count on from the item cost, perhaps using a number line.

Consolidating: Students can use number lines to work out the change.

Extending: Ask students to find different ways of using different notes and coins to make the right amount of change.

Stretch zone: *I bought the laptop bag and the console game. I paid with three $20 notes. How much do I have left? What notes and coins do I have?*

Check that students can correctly show different ways to make the change using the available currency. *Which way uses the fewest coins?*

 **Reflection time**

Ask students to describe how they worked out the change for each question on page 135 of the Student Book. *Give me one way of making that amount of change. Can you give me another? Which way used the fewest number of notes or coins?*

Practice Book: Students complete Practice Book page 113. They can do this directly after the main activity, as homework, or as the focus of a separate mathematics session to help students consolidate their learning and build fluency.

Students practise giving the correct amount of change in three different ways, consolidating their understanding of the different notes and coins. The activity uses dollars and cents but they could also do this activity using their own currency if that is more appropriate.

Differentiated outcomes	
All students	should understand when change is needed.
Most students	will work out how much change and what coins or notes can be given.
Some students	may find more than one way to make the change amount.

Answers

Student Book page 135

1 $4.50

2 $2

3 $3.75

4 $3.20

5 Phone charger and case, or laptop bag

Practice Book page 113

Answers will vary but may include:

1 $1.60: $1, 50¢, 10¢ $1, 25¢, 25¢, 10¢
$1, 50¢, 5¢, 5¢

2 40¢: 25¢, 10¢, 5¢   10¢, 10¢, 10¢, 10¢
25¢, 5¢, 5¢, 5¢

3 $2.25: $1, $1, 25¢ $1, $1, 10¢, 10¢, 5¢
$1, 50¢, 50¢, 25¢

4 85¢: 25¢, 25¢, 25¢, 10¢   25¢, 25¢, 25¢, 5¢, 5¢
25¢, 25¢, 10¢, 10¢, 10¢, 5¢

Stretch zone: The cost was $7.65, I paid with a $10 note; the cost was $12.65, I paid with a $10 and a $5 note; the cost was $17.65, I paid with a $20 note.

6 Money

Connect   Student Book page 136

Big idea

Price labels tell us how much items cost. When we pay for an item, we can work out our change by counting up from the cost of the item to the amount of money we pay with.

Global skills

- **Creative skills:** problem solving
- **Real-world skills:** financial literacy
- **Interpersonal skills:** teamwork

Key vocabulary

- cost, notes, coins, change

Resources

- price list of possible snack items

Language support

Produce labels for the different areas of the shop, for example: 'counter', 'till', 'basket', and labels for the people involved, which they can wear: 'shopkeeper', 'customer'. Refer students to the display they made at the beginning of the unit to remind them of the different coins and notes.

 ## Introductory activity

Look together at page 136 of the Student Book. If you have access to an IWB you could use this. Explain that if students were to have a healthy snack shop, they could use some of the things they have explored in this unit.

Explain that they will need to decide what they want to stock in their shop. Suggest that they go to a supermarket (if appropriate) to find out the cost of items, or research them using a shop's website. Tell students what they are going to do to prepare the shop, stock the items (or pictures of them) with price labels and get ready for selling the items.

 ## Main activity

Talk with students about how they decided on the prices for the items in their shop, and how they might consider prices that would be easier to manage and to give change for.

Ideally, set up the shop with some students and have a snack day at the end of this unit. You could choose to sell other things that you know students would prefer.

Students should work out how much it would cost to stock the snack shop and how much money they would make from sales.

They should then make up some word problems about the items in the snack shop.

Differentiation

Students should work in mixed-attainment groups so that less-confident students hear and see appropriate vocabulary and strategies being used.

Stretch zone: *If you sell all your items, can you work out how much profit you will make?*

Check that students have calculated correctly.

---

6 Money

Review Student Book page 137 • Practice Book page 114

Global skills

- **Self-development skills:** reflecting on learning

Student Book

With young children, assessment activities are most effective when carried out as an everyday classroom activity. Students should have notes and coins available to support them. If necessary, you can rewrite the prices on the menu to make them more accessible to everyone. These could either by using whole dollar and 50¢

 Reflection time

Bring together the results from the various groups to find out what the healthy snack shop would sell and how much it would cost to set up. Look together at some of the word problems about the shop items and ask how each could be solved. Finally, if any students worked on the Stretch zone question, you could discuss the profit made.

Differentiated outcomes	
All students	should contribute to the group discussion.
Most students	will listen to each other's calculations, correcting if necessary.
Some students	may explain their calculations clearly for others to understand.

Answers

Student Book page 136

Check that the calculations needed for the healthy snack shop are correct.

---

amounts only, or writing them as, for example, $4 and 75¢, rather than using decimal notation.

Watch as students work out the process needed and the change required and decide on the best way to make up different amounts using the coins available.

Encourage students to use the coins. Extend by asking a range of similar questions but with more items to buy or asking students to make their own money problems, using the items in the menu.

Answers

Student Book page 137

1 a $6.25   b $3.75
   c $1, $1, $1, 25¢, 25¢, 25¢ or $1, $1, $1, 25¢, 25¢, 10¢, 10¢, 5¢

2 No, the pizzas will cost $11.

3 Various answers that could include pizzas and/or side orders. Check that the items chosen cost less than $20 in total.

4 a $9.75   b $5.25   c $5, 25¢ or $5, 10¢, 10¢, 5¢

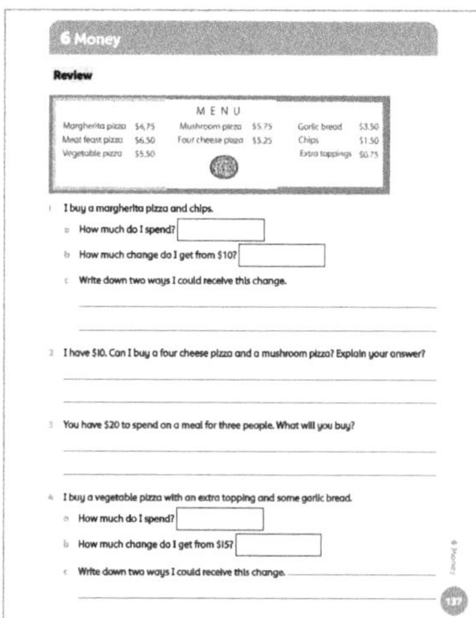

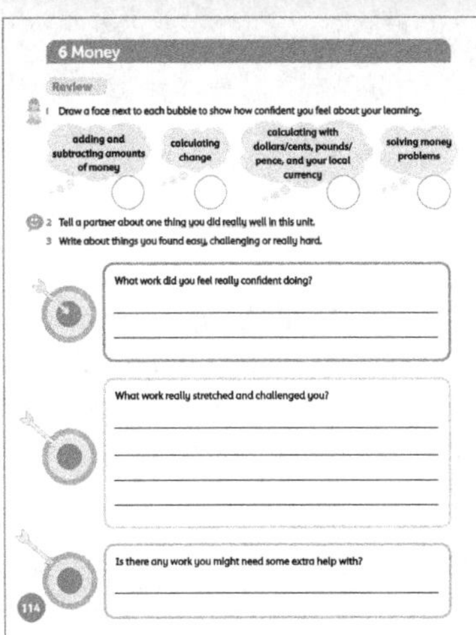

It is appropriate to complete this Practice Book review as a whole-class discussion. You may choose to keep a record of the class discussion or a copy of the review page for your own records. The review provides an opportunity for students to reflect on their learning from the unit, to discuss any areas of mathematics that they feel went particularly well, and any areas that they feel less confident about. Ensure that all students have a copy of the Student Book as a reminder of the areas of mathematics that they have worked on in this unit.

Allow students plenty of time for discussion before asking them to complete the Practice Book page individually, and then, if appropriate, to share their responses with the rest of the class. If students complete this self-assessment at home, encourage them to discuss this with adults. Make a note of areas that students still feel unsure about. They will become more familiar with decimal notation as they move into Stage 4, through their work in fractions and decimals, but also through continued work on money, and calculations using money in decimal form.

Additional material

There are additional end-of-unit assessments available on the *Oxford Owl for School* website.

7 Time

Overview

Big idea

One of the big ideas in this unit is that time is used as a measure of 'duration', that is, how long something takes. So, for example, a TV programme may be 1.5 hours long. The .5 of an hour represents half an hour, or 30 minutes. We can write this duration as 1 hour 30 minutes. When we describe the time of day, we display the minutes after the hours, rather than using a decimal unit of an hour. For example, 01:30 is half-past one in the morning. So, time can be both a 'time of day' and the 'time something takes'.

Another big idea is that 'time' is linked to 'base 12'. That is, we work in units of 12 rather than units based on the decimal system (which is base 10). There are 12 hours around the face of a clock and 2 × 12 hours in a 24-hour day. Similarly, there are 60 seconds in a minute (5 × 12) and 60 minutes in one hour (5 × 12).

Students will find it useful to use a number line as an image of the day. This can be labelled 'Midnight to Midnight' and should use 24 hour clock notation as well as a.m. and p.m. when necessary.

Roman numerals are encountered in this unit and students will most likely see them on clock faces, so will need to tell and write the time from an analogue clock showing Roman numerals from I to XII.

Look out for

- **Students who are more accustomed to using digital technology than they are to using analogue technology. For example, they will only read the time '12:35' as 'twelve thirty-five', rather than also saying 'twenty-five to one'.** These students may attain well in other areas of mathematics but find telling the time difficult. Support students by providing various opportunities to see and read analogue clocks.

- **Students who struggle with times *to* the hour.** Use both analogue and digital clocks to read the time beyond half-past to see that they can refer to how many minutes to the next hour.

- **Students who confuse the minute and hour hands.** When reading the time on analogue clocks, point to the hands and name them as you ask questions, for example: *Where does the hour hand point at 4 o'clock? Where does the minute hand point at quarter to 8?*

Possible misconceptions

- **Students think that o'clock time only occurs once in a day, instead of twice in one day, in the morning and the evening.** Encourage students to think of things they might be doing at, for example, 7 a.m. and 7 p.m., such as having breakfast and having a bath before bedtime.

- **Students think the hour hand points to the hour at half past or quarter past times. For example, for half past 4, they draw the hour hand pointing to 4 rather than halfway between 4 and 5.** Use geared analogue clock faces to show how the hour hand moves during an hour, and point out the position of the hour hand on a class clock when it is quarter past, half past or quarter to the hour.

Key vocabulary

- days of the week: Monday, Tuesday, Wednesday, Thursday, Friday, Saturday, Sunday
- months of the year: January, February, March, April, May, June, July, August, September, October, November, December
- seasons: autumn, winter, spring, summer
- day, week, fortnight, month, year
- calendar, date; morning, afternoon, evening, night, midnight, a.m., p.m., today, yesterday, tomorrow
- How long ago? How long before? hour, minute, second
- o'clock, half past, quarter to, quarter past
- clock, watch, hands, stopwatch, Roman numeral
- timetable, duration, days of the week

Learning objective	E	7A	7B	7C	7D	C	R
Tell and write the time from an analogue clock, including using Roman numerals from I to XII, and 12-hour and 24-hour clocks.		✓					✓
Estimate and read time with increasing accuracy to the nearest minute; record and compare time in terms of seconds, minutes and hours; use vocabulary such as o'clock, a.m./p.m., morning, afternoon, noon and midnight.				✓			✓
Know the number of seconds in a minute and the number of days in each month, year and leap year.	✓		✓	✓		✓	✓
Compare durations of events [for example to calculate the time taken by particular events or tasks].	✓		✓		✓	✓	✓

7 Time

Engage Student Book page 138

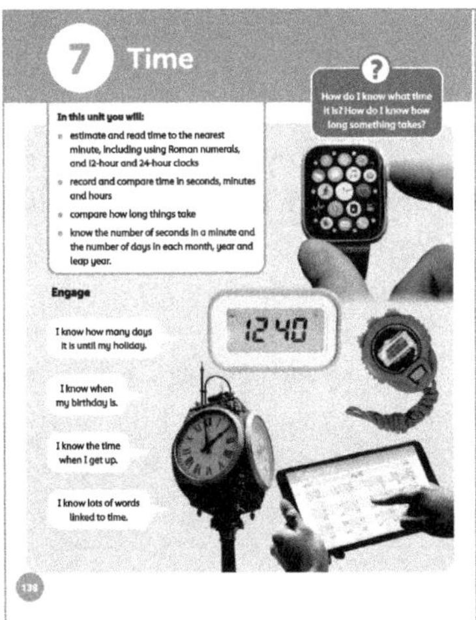

Big question

- How do I know what time it is? How do I know how long something takes?

Global skills

- **Creative skills:** problem solving
- **Interpersonal skills:** communication

Key vocabulary

- second, minute, hour, day, week, month, year, how long ago? how long before?, calendar, date, morning, afternoon, evening, night, midnight, a.m., p.m., today, yesterday, tomorrow, o'clock, half past, quarter to, quarter past, clock, watch, hands, Roman numerals

Resources

- magazines and newspapers, a calendar
- an analogue clock and a digital clock displayed in the classroom

Language support

Group students who are less-confident English speakers with others who are more confident.

Work with pairs or groups and read the vocabulary from the posters aloud. Where possible, ask students to use each word in a sentence.

Introductory activity

Look together at page 138 of the Student Book. If you have access to an IWB you could use this. Point out the speech bubbles and ask individual students whether they can tell you, for example, how many days there are until their birthday, or what time they get up.

Ask students to talk in pairs about all the different things that they think of when you say 'time'. They should use page 138 of the Student Book as a stimulus and write down as many words and phrases as they can relating to time. Take feedback from each pair until you have written a class list on the board of all the words or phrases they have identified.

Main activity

When you have completed the list of words and phrases, ask students to move into groups of four or five. These groups should be mixed-attainment groups, so those less confident in English hear the vocabulary modelled by others. Give each group a selection of magazines and newspapers. Groups should create posters using images and text from the magazines and newspapers, alongside the key vocabulary of time. Groups should classify the words on their poster in any way they want to. If the class have not already identified the words and phrases listed in the key vocabulary section above, add these to the list on the board, along with common words that they may already know (for example days of the week, months of the year, seasons).

Differentiation

Supporting: Ask students to use the words that they know in sentences. Introduce them to some of the new vocabulary to model the pronunciation.

Consolidating: Ask students to use new vocabulary in sentences.

Extending: Ask students for any more vocabulary that they know that is linked to time.

 **Reflection time**

Groups should share their posters and explain why they decided to classify their words as they did. The posters can be displayed for the whole unit. Add to them during the unit with any new words or phrases that are introduced.

7A Telling the time

Discover 1 Student Book page 139 • Practice Book page 115

Specific learning focus

- Estimating and telling the time to the nearest minute on analogue clocks.

Global skills

- **Creative skills:** problem solving
- **Real-world skills:** interpreting information
- **Interpersonal skills:** communication

Key vocabulary

- a.m., p.m., 24-hour clock, nearest minute, digital

Resources

- geared analogue clock and digital clock displayed in the classroom

Language support

Make a poster of illustrations with key words for typical activities done during the day: eat breakfast, go to school, eat lunch, do homework, play outside, watch TV, eat dinner, go to bed.

 **Introductory activity**

Ask students to discuss in pairs what they do at different times of the day. Ask pairs to the front of the class: one student uses an analogue clock to show the time they do an activity and says if it is **a.m.** or **p.m.**, then the other student writes the time on the board using 24-hour digital notation.

Main activity

Model for students how to draw the hands correctly on a blank analogue clock face, making sure that the hour hand is, for example, halfway between the two hours if it is a 'half past' time.

Students then complete the activity in the Student Book on page 139. Refer students to the speech bubbles for some hints about writing **24-hour clock** times and also to remind them to write the times to the **nearest minute**. Using the examples in the Student Book, students should think about the times they do each of the listed activities. They should try to use the most accurate time they can for each event and record the time to the nearest minute on both the analogue and **digital** clocks on the page.

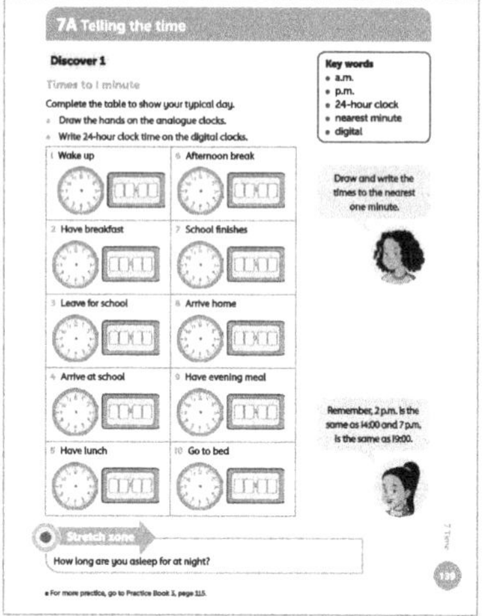

Differentiation

Supporting: Ask students to use the time words that they know when discussing the time of each event. Help them match the spoken time with the clock time to the nearest 15 minutes or the nearest 5 minutes, if they are able to.

Consolidating: Ask students to use vocabulary related to times to the nearest five minutes on both analogue and digital clocks.

Extending: Ask students to identify the finishing time for each event and ask them to show it on a clock to the nearest minute.

Stretch zone: *How long are you asleep for at night?*

Check that students have a good idea about sleep start and finish times and can work out the duration that they sleep. *Do you sleep for longer at the weekend? If so, how much longer?*

 **Reflection time**

Discuss with students the times they used on their clocks. *Who had the earliest breakfast? Who went to bed the latest? How did you represent times that were a.m. or p.m. on each clock? Which type of clock do you find easiest to use? Why?*

Practice Book: Students complete Practice Book page 115. They can do this directly after the main activity, as homework, or as the focus of a separate mathematics session to help students consolidate their learning and build fluency.

Times are given to the nearest minute for students to write on a digital clock and to draw the hands on an analogue clock. They must write the digital times using 24-hour clock notation. *Is it always useful to give a time to the nearest minute? When do you think an approximate time is better and when should we use a precise time?*

Differentiated outcomes	
All students	should identify times of doing daily events to the nearest 15 minutes.
Most students	will record the time on an analogue and digital clock to the nearest five minutes.
Some students	will record the time accurately on both analogue and digital clocks to the nearest minute.

Answers

Student Book page 139

Check that students' times match on both clocks and are reasonably accurate.

Practice Book page 115

Check that students have drawn the times correctly on the analogue and digital clocks.

7A Telling the time

Discover 2
Student Book page 140 · Practice Book page 116

Specific learning focus
- Read the time in Roman numerals.

Global skills
- **Creative skills**: problem solving
- **Real-world skills**: interpreting information

Key vocabulary
- a.m., p.m., Roman numerals, nearest minute, o'clock

Resources
- clockfaces with Roman numerals
- blank clock face (Resource sheet 7.1)

Language support

Display a list of the Roman numerals to 12 alongside their Hindu-Arabic equivalents (I = 1, II = 2, VI = 6 and so on). Point out to students that on a clock, 4 is sometimes shown as IIII even though the Roman for 4 is IV.

 Introductory activity

Display a large clock face on the board with **Roman numerals**. Explain to students that I means 1, V means 5 and X means 10. Ask students to model some of the numerals using their hands. For example, they can model I by holding up one finger, II by holding up two fingers. V can be made by holding up 5 fingers and creating a V shape between your thumb and index finger, and X can be made by holding up 10 fingers and crossing your two thumbs.

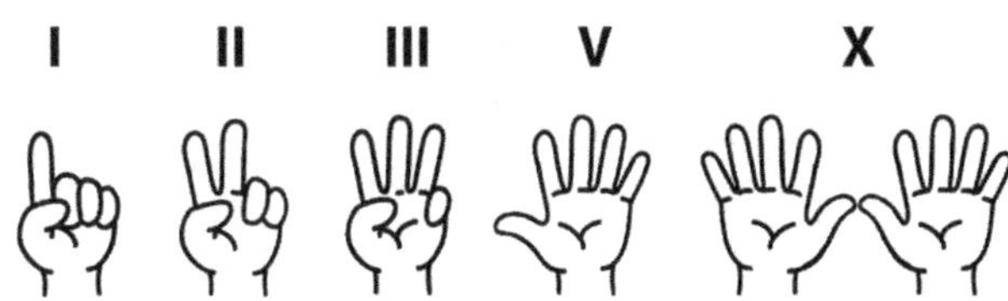

Explain that Roman numerals can be combined to make other numbers, putting letters before or after each other to represent different numbers. For example: V = 5, then IV is 'one before 5' (4) and VI is 'one after 5' (6).

Go around the clock stopping at each number and asking students to say the number and how it is made up. For example, they say '9 is shown as IX' (one before 10).

 Main activity

Set the hands on the Roman numeral clockface to read different 'o'clock' times. Students can use their knowledge of the Roman numerals to say the time. Then move on to show some 'half-past' times, and finally place the hands at random times for the students to read and say.

Ask students to look at page 140 in the Student Book. The first example shows a clockface time as well as the approximate time in words and exact time in digits. *How do you know the times are correct? How is that time an approximate time?* Talk about rounding up or down to the nearest five minutes. *Is it always useful to give a time to the nearest minute? When do you think an approximate time is better and when should we use a precise time?*

Students should complete the remaining times, and write an approximate time for each clock, on page 140 of the Student Book. Refer students to the second speech bubble. As with a normal analogue clock, they do not know, just by looking at these clocks, whether the time is a.m. or p.m. However, the second column of the table tells them this information, so they can write the correct 24-hour time in the final column.

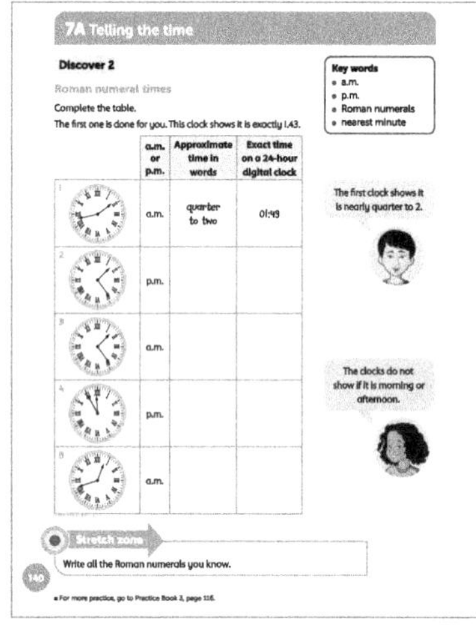

Differentiation

Supporting: During the first part of the main activity, ask students to estimate to the nearest five minutes.

Consolidating: During the first part of the main activity, ask students to estimate to the nearest minute when reading the analogue clock.

Extending: Ask students to record the exact time on each clockface. They can write their own times and mark the position of the hands on the Roman numeral clockface.

Stretch zone: *Write all the Roman numerals you know.*

Students can use a blank clock face (Resource sheet 7.1) to write the numbers to 12 but encourage them to work out numbers beyond 12. They may also know C for 100 and M for 1000 as they may have seen these written for dates.

 Reflection time

Show students some more times on the Roman numeral clockface and ask them questions about each time. For example, ask: *What time will it be 20 minutes later? or What was the time 25 minutes earlier?* Students can respond orally, or draw the time on a Roman numeral clockface.

Practice Book: Students complete Practice Book page 116. They can do this directly after the main activity, as homework, or as the focus of a separate mathematics session to help students consolidate their learning and build fluency.

Students read the time from Roman numeral clockfaces and then write the time in words and in 24-hour digital notation. If they are not confident writing the times to the nearest minute, they could write an approximate time instead.

Differentiated outcomes	
All students	should estimate the time to the nearest five minutes.
Most students	will estimate to the nearest minute and read the time accurately.
Some students	may estimate and read accurately any time on the clockface.

Answers

Student Book page 140

1 a.m. quarter to two 01:43
2 p.m. twenty-five past one 13:24
3 a.m. twenty-five past one 01:24
4 p.m. five to twelve 23:55
5 a.m. twenty to one 00:42

Practice Book page 116

1 a.m. 17 minutes past 8 in the morning 08:17
2 p.m. 32 minutes past 5 in the afternoon 17:32
3 a.m. 25 minutes to 1 in the morning 00:35
4 a.m. 12 minutes to 11 in the morning 10:48
5 p.m. 9 minutes to 8 in the evening 19:51

Stretch zone: Check students' answers, for example: age 7 is VII, age 8 is VIII.

7A Telling the time

Explore 1   Student Book page 141 · Practice Book page 117

Specific learning focus

- Read and tell the time to the nearest minute on 12-hour digital and analogue clocks.
- Use a.m., p.m. and 12-hour digital clock notation.

Global skills

- **Creative skills:** problem solving
- **Real-world skills:** interpreting information
- **Interpersonal skills:** communication

Key vocabulary

- 12-hour clock, digital, nearest minute

Resources

- geared analogue clock
- mini whiteboards and markers

Language support

Continue to ask students to tell the time throughout the day, week and term until you feel that they are confident. You could set a phone alarm at different times during the day. When it sounds, ask students: *What time is it?*

 **Introductory activity**

Revise the meaning of 'noon'. Ask students to explain 'a.m.' and 'p.m.'. Write these digital times on the board:

15:40   05:55   21:38   16:45   11:59

Ask students to say the time to a partner as 'minutes to', for example: 3:40 is 'twenty minutes to four'. Show students the 20 minutes on an analogue clock if necessary.

Ask students to work in pairs on similar questions. One should write a time down on their whiteboard and the other should say how many minutes to the next hour it is.

 **Main activity**

Explain to students that they are going to write the same times, using 'past' and 'to' the hour format. Ask, for example, *How would you write in words the time when it shows 4:35?* Students should know that this can be either 'four thirty-five' or 'twenty-five to five'. Choose other examples such as 1:40 or 6:45. Now work through the first example in the Student Book on page 141 together. Ask a student to read the item on the analogue clock. Refer students to the speech bubble to see how this time is spoken.

Look at questions 1 and 2 together to compare them. *How many minutes past 5 is this? How many minutes to 6 is this? Which time is easier to say?*

Ask students to work individually on the remaining questions. In question 5, the two hands are on top of one another. Ask students to explain to each other why only the minute hand can be seen. Even though they are working individually, pairs should check each other's answers and agree a correct answer if they differ. They should then find the mistake together.

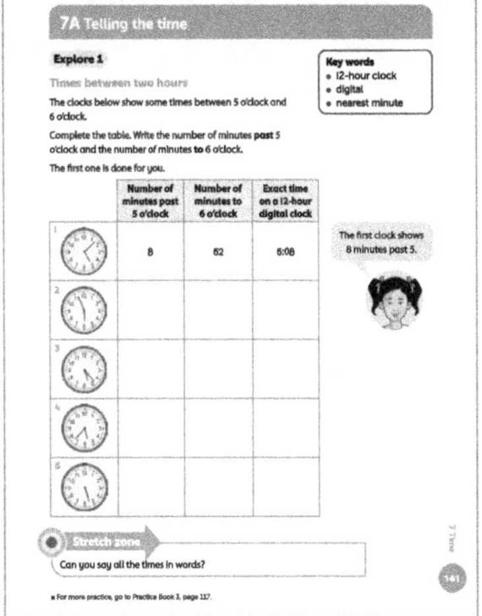

Differentiation

Supporting: Support students by using a geared analogue clock.

Consolidating: Encourage students to check each other's answers.

Extending: Challenge students to write times that are not multiples of 5 minutes using the 'past' and 'to' forms, such as 4:37.

Stretch zone: *Can you say all the times in words?*

Check that students can write the time on each clock in words, in the same format as the example in the speech bubble. For example, they write '4 minutes to 6', and '56 minutes past 5'.

 **Reflection time**

Look at students' answers together and discuss. Select students who are confident to share the strategies that they used. Ask, for example: *How did you decide which hour to refer to when saying the time 2:50? Did you prefer 'two fifty' or 'ten to three'? Did you count back the minutes from o'clock, or count on to the hour from the time given?*

Practice Book: Students complete Practice Book page 117. They can do this directly after the main activity, as homework, or as the focus of a separate mathematics session to help students consolidate their learning and build fluency.

Students read time on a digital clock and then draw the same time on an analogue clock and write it in words, using the format of the example given in question 1.

Differentiated outcomes	
All students	should estimate the time on each clock using 'past' and 'to' the hour.
Most students	will tell the time accurately on each clock using 'past' or 'to' to the nearest five minutes.
Some students	may quickly and accurately tell the time on each clock using 'past' or 'to' to the nearest minute.

Answers

Student Book page 141

1	8	52	5:08
2	56	4	5:56

3	23	37	5:23
4	37	23	5:37
5	27	33	5:27

Practice Book page 117

Check that students have shown the times correctly on the analogue clocks.

1 twenty-five to 7

2 ten to seven

3 twenty-eight past 7

4 twenty-three to eight

5 twenty-five past 8

6 quarter to 2

Stretch zone: Accept any answer between 1.40 and 1.50 a.m. or p.m.

7A Telling the time

Explore 2 Student Book page 142 · Practice Book page 118

Specific learning focus

- Read and tell the time to the nearest minute on 12-hour digital and analogue clocks.
- Use a.m., p.m. and 12-hour digital clock notation.

Global skills

- **Creative skills:** problem solving
- **Real-world skills:** interpreting information
- **Interpersonal skills:** communication

Key vocabulary

- a.m., p.m., 24-hour clock, nearest minute, half past, quarter past, quarter to

Resources

- clock faces, a geared analogue clock
- mini whiteboards and markers

Language support

Work with students when they are using the clocks. Focus on the times of day and revise the names for 'o'clock', **'half past'**, **'quarter past'**, **'quarter to'**. You can also support students when they are converting from digital times such as 4:50 to the more usual analogue description of 'ten to 5'.

 **Introductory activity**

Ask students the difference between a.m. and p.m. *What time is it when we change from a.m. to p.m.?* Explain that the word 'noon' is a very old word that meant the ninth hour after sunrise. Over time, it was changed to 'midday' meaning 12 'o'clock, when the sun is at its highest in the sky. However, 'noon' is still used and this is where we get the word 'afternoon', meaning after 12 o'clock.

Ask students to give you times before and after noon which match given clues, for example: *Give me time in the morning that includes 42 minutes. Give me a time in the afternoon that includes 'past'. Give me a time in the afternoon that includes 'to'.* Students say the specific times and write the time on their whiteboards using 24-hour digital notation.

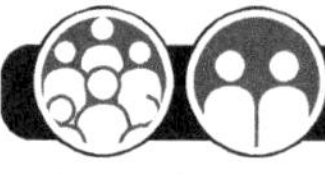 **Main activity**

Ask students to think about how long it takes them to travel to school. They might estimate it to be 10 minutes, 20 minutes, half an hour or more. Give them some typical times for your own journey to school, to the nearest minute. For example, you leave home at 06.53 and arrive at school at 07.38. *How long is my journey to school?* Students can use analogue clock faces to help them, if necessary, to work out the journey time as 45 minutes.

Repeat with other similar questions, for example: *I finish school at 15.37 and arrive home from school at 16.12. How long does it take me to walk home from school?*

Direct students' attention to the speech bubble on page 142 of the Student Book. *How do you know if a time is morning, afternoon or evening? Does the activity suggest which it is? Does the use of a.m. and p.m. help?*

Students then work in pairs to the activity on page 142 of the Student Book. They should make sure that the minute hand is accurately on the correct number of minutes and the hour hand is approximately correct between the hours.

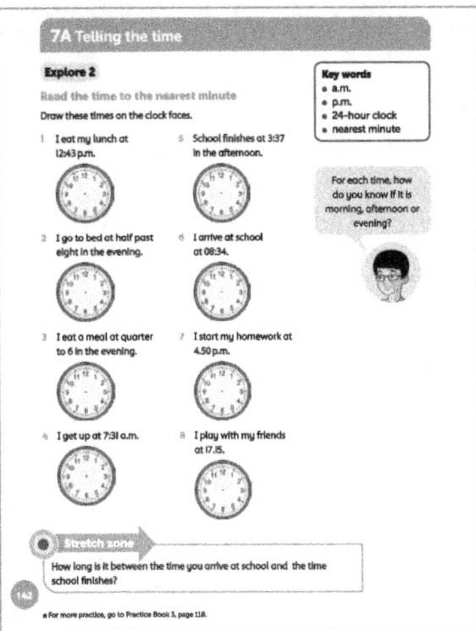

Differentiation

Supporting: Help students work with a clock to see how the hands move and to work out how long something takes.

Consolidating: Ask students to check in pairs to see whether they have placed the hands correctly.

Extending: Challenge students to place the hands based only on seeing the 24-hour digital time, and use this to help them calculate how long things take.

Stretch zone: *How long is it between the time you arrive at school and the time school finishes?*

Check that students can count carefully to get the minutes accurately. *How will you work this out? How can you use a number to help you to calculate the duration?* You could repeat with other durations, if you have time.

 **Reflection time**

Discuss how the hands turn in relation to each other. *How far does the minute hand turn when the hour hand has moved halfway between two numbers? How much has the hour hand turned in 20 minutes?*

Ask questions about durations of time, for example: *How long is it between starting homework and eating an evening meal? How long is it between eating lunch and school finishing?*

Practice Book: Students complete Practice Book page 118. They can do this directly after the main activity, as homework, or as the focus of a separate mathematics session to help students consolidate their learning and build fluency.

Differentiated outcomes	
All students	should position the hands correct to the nearest five minutes.
Most students	will position the minute hand correctly and the hour hand approximately.
Some students	may position both hands accurately and use the clock to work out how long something takes.

Answers

Student Book page 142

Check students' estimates for the positions of the hour hands and the accuracy of the minute hands for each time.

Practice Book page 118

Check students' estimates for the positions of the hour hands and the accuracy of the minute hands for each time.

Stretch zone: 00:01 means the time is one minute past midnight, or one minute past 12 a.m. or one minute past 12 in the morning.

7B Reading a calendar

Discover Student Book page 143 • Practice Book page 119

Specific learning focus

- Read and mark dates on a calendar.

Global skills

- **Creative skills:** exploring
- **Real-world skills:** interpreting information
- **Interpersonal skills:** communication

Key vocabulary

- months of the year, calendar, date

Resources

- calendars for the current year

Language support

Allow all students access to a calendar throughout this unit or have a large calendar on display. Talk about today's date with the students every day. This could be at the beginning of each day or at the start of a session, or both.

 **Introductory activity**

Show students a **calendar** for a year and discuss the features on it. Ask students to notice the order of the **months of the year**, which months have 31 or 30 days and that February has fewer than 30 days and that it has a different number in a leap year than a non-leap year.

Ask students to work in pairs and find any months for that year's calendar that have five Mondays, or only four Thursdays and so on.

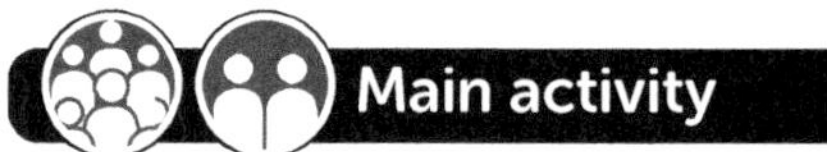 **Main activity**

Invite students to find any special **dates** they know on the calendar, for example their birthday, relevant festivals or public holidays. Choose two of these special dates and ask students to tell you which of the special dates their birthday is closer to.

Explain that, in the in activity on page 143 of the Student Book, they are going to mark class birthdays on the calendar. Provide each pair with a list of the class birthdays.

Pairs should work together to complete the activity in the Student Book. They start by marking each birthday on the calendar by circling the date. When they have completed as far as question 7, they should ask each other further questions using the birthdays marked on the calendar. Ask them to use questions such as: Which month has the most birthdays? Which month has the fewest birthdays? How many birthdays are there in the shortest month? Whose birthday is closest to 1st January?

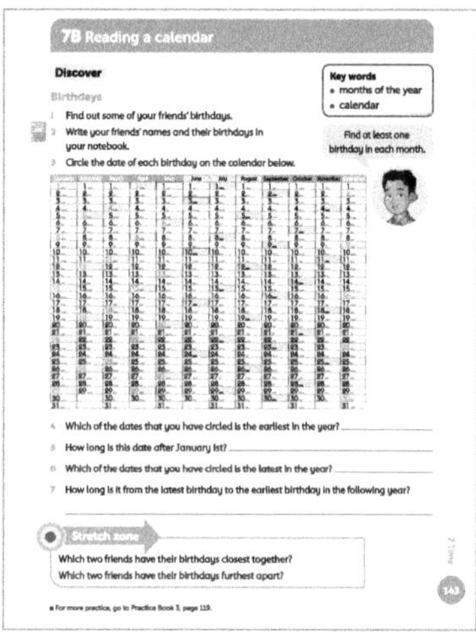

Differentiation

Supporting: Ask students to point out dates that they know are special and say why.

Consolidating: Ask students to describe dates on the calendar that have interesting mathematical features, such as January 1st (01/01) and November 11th (11/11).

Extending: Ask students to measure how many days it is between given dates, perhaps two birthdays or two festival dates.

Stretch zone: *Which two friends have their birthdays closest together?*

Which two friends have their birthdays furthest apart?

Check that students have calculated correctly from the dates used.

 **Reflection time**

Ask students to share with the class any interesting facts or figures about the calendar. *Do any two (or more) of you share the same birthday? Do any of you share a birthday with another member of your family? Does anyone have a birthday on a festival or public holiday? When are the school holidays? Who has a birthday during one of the school holidays?*

Practice Book: Students complete Practice Book page 119. They can do this directly after the main activity, as homework, or as the focus of a separate mathematics session to help students consolidate their learning and build fluency.

Students write the dates of five special events that happen during the first half of the year. They write a word problem using the dates they have written.

Differentiated outcomes	
All students	should locate any date on the calendar finding the month first, then the number.
Most students	will mark significant dates on the calendar and describe their position in relation to other dates.
Some students	may use the calendar to find how many days are between given dates.

Answers

Student Book page 143

Answers will vary according to the birthdays used. Check that the questions have been answered correctly for the dates used.

Practice Book page 119

Check that the questions have been answered correctly for the dates used.

Explore Student Book page 144 • Practice Book page 120

Specific learning focus

- Find out birth dates and use them to calculate ages.

Global skills

- **Creative skills:** investigating
- **Real-world skills:** research
- **Interpersonal skills:** communication

Key vocabulary

- months of the year, date of birth/birth date

Resources

- calendars

Language support

Revisit language from previous lessons on months of the year and how to write and say dates of birth. Students may need help with reading the names of famous people.

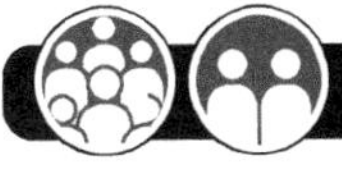

Introductory activity

Write on the board the name of a celebrity or famous person who is still alive and their **date of birth**. For example, give details for footballer Cristiano Ronaldo, born 5 February 1985. Ask students to work in pairs to find out Ronaldo's age in years, months and days up to today's date.

Help students by prompting them to count on in years from 1985 to this year, then counting on in months from February to this month and then from the 5th to today's date. If you are working on this lesson before 5 February, first count the whole years, then the months, then the days.

Check the accuracy of the calculation by comparing the answer from different pairs of students.

Main activity

Ask students to think of some famous people they know and to research their dates of birth. Discuss with them what it means to be 'famous' – someone is only famous if you and many others have heard of them and what they do. This could include, for example, sports people, musicians or TV personalities.

Students should then complete the tables on page 144 of the Student Book.

Remind students how to find the age difference between people, using the same method you used to find out Cristiano Ronaldo's age, that is, first counting the whole years, then the months, then the days. Tell students that they can use jottings to help them but should work out the total mentally.

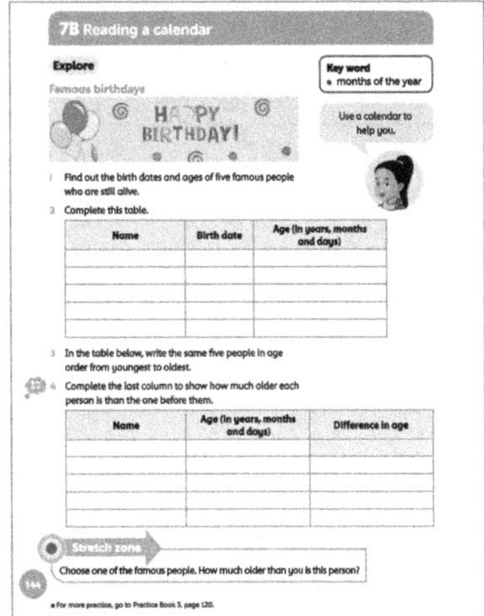

Differentiation

Supporting: Help students find the dates of birth of the famous people they choose.

Consolidating: Students can use their knowledge of days in each month to help them work out the ages. Allow them to use calendars as well.

Extending: Ask students to choose five classical composers and work out who lived the longest.

Stretch zone: *Choose one of the famous people. How much older than you is this person?*

Check that students have calculated correctly for the person they chose. *Which unit of time did you work out first? And then which?*

 Reflection time

Ask different students to describe the people they chose and how they worked out their ages. *What strategy did they use if the birthday fell in a month that hasn't been reached yet this year?*

Who chose the oldest famous person? Who chose the youngest famous person?

Practice Book: Students complete Practice Book page 120. They can do this directly after the main activity, as homework, or as the focus of a separate methematics session to help students consolidate their learning and build fluency.

Students list eight people and work out exactly how old they are in years, months and days. They are reminded of the fact there is a leap year every four years so they need to factor this in to their calculations.

Differentiated outcomes	
All students	should find dates of birth of famous people.
Most students	will work out ages from birth dates.
Some students	will mentally calculate difference in ages between any two people.

Student Book page 144

Answers will vary according to which famous people were chosen. Check that students have calculated their ages correctly.

Practice Book page 120

Answers will vary according to which people were chosen. Check that their ages have been calculated correctly.

7C Units of time

Discover Student Book page 145 • Practice Book page 121

Specific learning focus

- Suggest and use suitable units to measure time and know the relationship between different units of time.

Global skills

- **Creative skills:** investigating
- **Interpersonal skills:** teamwork

Key vocabulary

- second, minute, hour, day, stopwatch

Resources

- calendar and clocks for display
- stopwatches
- small calendar for each group

Language support

Work with students when they use the calendar. Focus on the names of the days and the months. Ask students what they notice about the number of days in a week and in each month. You can then also revise the other units of time: **seconds, minutes, hours** and **days.**

 **Introductory activity**

Tell students to close their eyes. Say that you want them to estimate how long one minute is. *When you think that one minute is up, put up your hand and open your eyes.*

On the board, draw the following table:

Time	Less than 40 seconds	40–59 seconds	50–59 seconds	Exactly right	61–70 seconds	Over 70 seconds
Number of students						

As students put up their hands, complete the table. If anyone gets it exactly right, ask them how they did it. Talk about ways you can estimate one minute, such as counting and saying a particular word between each number (one–elephant–two–elephant–, and so on.).

 **Main activity**

Now repeat the activity and encourage students to use one of the methods the class have just discussed. The student who was most accurate last time can do the recording this time.

Students should work in small groups on the activities in the Student Book on page 145. They should take it in turns to use the **stopwatch**. They will also need to use a calendar to support them in question 3.

Refer students to the speech bubbles, which give some suggestions of the sort of activities they could select to complete their tables.

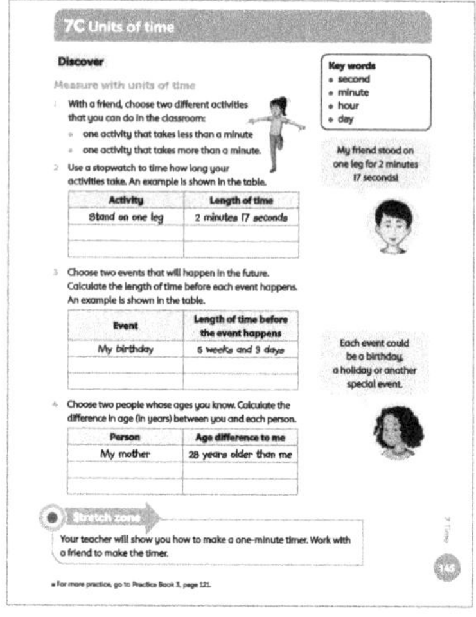

Differentiation

Supporting: Ask students to work in a pair while they time the events so they learn how to use the stopwatch.

Consolidating: Ask students to work in pairs and check each other's timings.

Extending: Ask students to predict how long events will take.

Stretch zone: *Your teacher will show you how to make a one-minute timer. Work with a friend to make the timer.*

Students make a timer that measures one minute accurately. They can search for instructions for this using the internet. The timer may be a sand timer, a water timer, or it could be a piece of apparatus that takes exactly one minute to roll a ball through.

 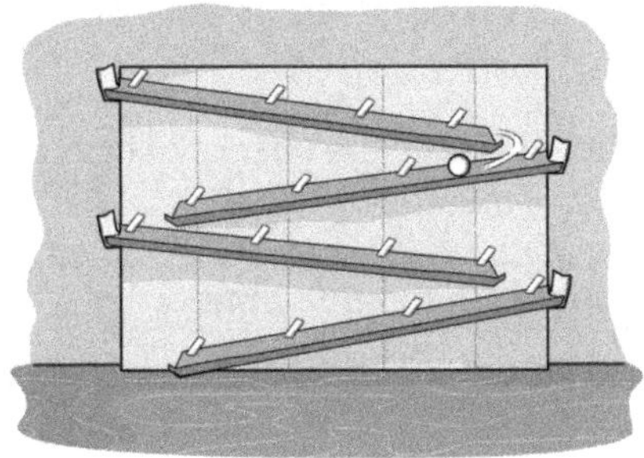

 Reflection time

Ask groups to report back on their activities. They should be encouraged to describe the activities they chose to time, the future events they selected and the people whose ages they worked with. Ask, for example: *Did you estimate the time before doing each activity? How did you calculate how long it was until your special events? How did you calculate the difference between the ages?*

Practice Book: Students complete Practice Book page 121. They can do this directly after the main activity, as homework, or as the focus of a separate mathematics session to help students consolidate their learning and build fluency.

Students answer questions relating to durations of events on a calendar. Ensure that they remember how to read a calendar before they start the activity. *Why are some of the rectangles shaded on the calendar? What day does the month start on?*

Differentiated outcomes	
All students	should time events accurately and complete activities with support.
Most students	will complete activities independently.
Some students	may be able to support others in completing the activities.

Answers

Student Book page 145

Answers will vary according to the activities, events and people that students have chosen. Check that the answers seem reasonable.

Practice Book page 121

1 Check that the events have been circled correctly on the calendar.

2 7

3 30

4 5

5 9

6 23

7 8

7C Units of time

Explore 1   Student Book page 146 • Practice Book page 122

Specific learning focus

- Convert between units of time.

Global skills

- **Creative skills:** problem solving

Key vocabulary

- second, minute, hour, day, week, month, year

Resources

- display of time units and their equivalents

Language support

Display a list of all the different time units and their relationships, for example: second, minute, hour, day, week, month, year, and the various connections such as 60 seconds = 1 minute.

Introductory activity

Teach the children how to remember which **months** have 31 days. Tell them to place their left hand down with fingers spread. Start from the left. Fingers will be months with 31 days, spaces between fingers have 30 (or fewer, in the case of February).

So, little finger is January = 31 days, space = February (28), ring finger = March (31), space = April (30), middle finger = May (31), space = June (30), first finger = July (31).

Now work back the other way starting with first finger = August (31), space = September (30), middle finger = October (31), space = November (30) and ring finger = December (31).

Main activity

Remind students about how to solve word problems using RUCSAC. Then ask: *In 5 weeks time, I will be going on holiday. How many days is that?* Ask students what they need to know to solve this problem. (how many days are in 5 **weeks**) *What is the calculation that we need to do to find this out?* They should be able to say that each week has 7 days, so they need to work out $5 \times 7 = 35$.

Now ask them how many seconds it takes to have a shower if it takes 10 minutes. Using the fact that 1 minute = 60 seconds, students should work out $10 \times 60 = 600$ seconds.

Students should then complete the questions on page 146 of the Student Book. They may use the display of time units to help them. Refer them to the two speech bubbles. The first tells them to use their answers to question 1 to help them answer the problems in question 2. The second speech bubble asks them about leap years. Ensure that all students know what a leap year is and how it differs from other years.

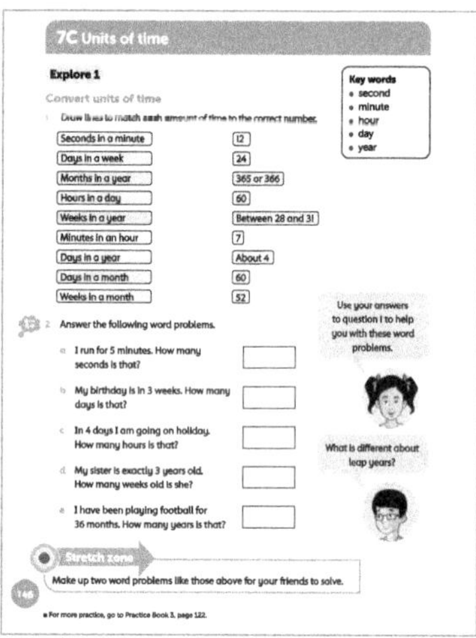

Differentiation

Supporting: Work with students to model solving the questions using the display of time units.

Consolidating: Ask students to carry out calculations using calendars, for example, with jottings.

Extending: Ask students to create additional word problems for each other.

Stretch zone: *Make up two word problems like those above for your friends to solve.*

Check that students have devised suitable problems and that their friends are able to solve them correctly.

 ### Reflection time

Check through each of the questions on page 146 of the Student Book to make sure that all students have the correct answers. Did they use the RUCSAC approach to solve each of the problems? *Can you explain the method you used? How did you know which calculation to use?*

Practice Book: Students complete Practice Book page 122. They can do this directly after the main activity, as homework, or as the focus of a separate mathematics session to help students consolidate their learning and build fluency.

Students complete a series of facts about converting units of measure.

Differentiated outcomes	
All students	should be able to find the correct time equivalences from the display.
Most students	will answer all questions including the calculations.
Some students	may create and solve one- or two-step word problems about time units.

Answers

Student Book page 146

1 Seconds in a minute = 60

Days in a week = 7

Months in a year = 12

Hours in a day = 24

Weeks in a year = 52

Minutes in an hour = 60

Days in a year = 365 or 366

Days in a month = between 28 and 31

Weeks in a month = about 4

2 a 300
   b 21
   c 96
   d 156
   e 3

Practice Book page 122

1	60	12	31
2	60	13	31
3	24	14	30
4	7	15	31
5	52	16	30
6	365	17	31
7	366	18	31
8	600	19	30
9	300	20	31
10	168	21	30
11	104	22	31

7C Units of time

Explore 2
Student Book page 147 · Practice Book page 123

Specific learning focus
- Convert from one unit of time to another.

Global skills
- **Creative skills:** problem solving
- **Self-development skills:** reflecting on learning

Key vocabulary
- second, minute, hour, day, week, fortnight

Resources
- display of time units and their equivalents, calculators

Language support
Display a list of all the different time units and their relationships, for example: second, minute, hour, day, week, month, year, and the various connections such as 60 seconds = 1 minute.

 **Introductory activity**

Write on the board: 'How many hours are there in 5 days?' *What is the calculation that we need to do to find this out?'* Students should be able to say that there are 24 hours in each day, so 5 days will be 24 × 5. *How will you work this out?* Accept any method. If no one suggests the following, you could explain that 24 × 5 is the same as 12 × 10 (halve the first number and double the second) or 24 is the same as 2 × 12, so we could also write 2 × 12 × 5, which could be reorganised into 2 × 5 × 12.

Now write on the board: 'How many minutes in May?' Students should be able to tell you that May has 31 days, and each day has 24 hours and each hour is 60 minutes. So, the calculation is 31 × 24 × 60. It is acceptable for students to use a calculator to find that this is 44 640.

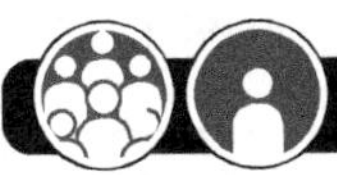 **Main activity**

Refer to the display of time units and explain that students can use this to complete the activities on page 147 of the Student book. Remind them always to work up by one unit at a time. So, for example, to turn seconds into hours, they need to multiply by 60 to get to minutes, then by 60 again to get to hours.

Students can use any calculation method of their choice, but they can use a calculator if needed.

Refer students to the second speech bubble to ensure that they know what a **fortnight** is.

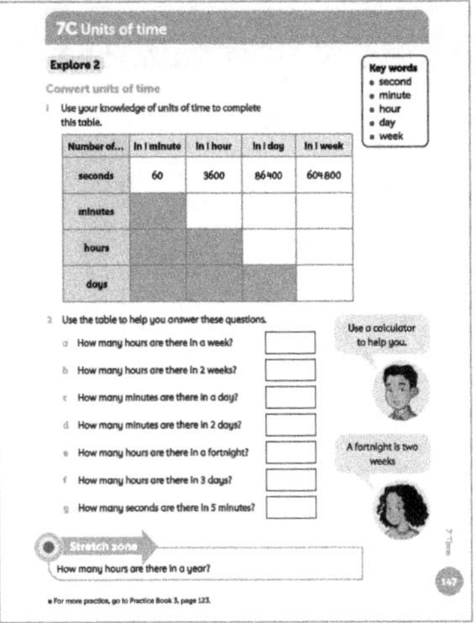

Differentiation

Supporting: Work with students to calculate the conversions with a calculator.

Consolidating: Ask students to tell you the calculation and to estimate before using the calculator.

Extending: Ask students to make up similar questions for converting time units.

Stretch zone: *How many hours are there in a year?*

Check that students have calculated correctly: 24 × 365 = 8760. *How did you work it out?* Students may be able to use the grid method to work this out if you explain how to set out the partitioned numbers.

 **Reflection time**

Ask several students to share their calculations between time units. *How did you decide what calculation to do? What methods did you use for the calculation? How did you check your calculation?*

Practice Book: Students complete Practice Book page 123. They can do this directly after the main activity, as homework, or as the focus of a separate mathematics session to help students consolidate their learning and build fluency.

Students answer a variety of time problems involving units of time and telling the time.

Differentiated outcomes	
All students	should be able to convert one unit to the next larger one, for example minutes to hours.
Most students	will convert from one unit to another and explain their method.
Some students	may be able to convert forwards and backwards between time units.

Student Book page 147

1

Number of...	in 1 minute	in 1 hour	in 1 day	in 1 week
seconds	60	3600	86 400	604 800
minutes		60	1440	10 080
hours			24	168
days				7

2 a 168 **b** 336 **c** 1440 **d** 2880 **e** 336 **f** 72 **g** 300

Practice Book page 123

1 61 days

2 95 minutes

3 18 days

4 270 minutes

5 24 months

6 28 days

7 22:35 Check that hands are drawn correctly.

8 18:45 Check that hands are drawn correctly.

Stretch zone:

208 weeks (approximately)

1461 days (365 × 3 normal years plus 366 for the leap year)

35 064 hours (1461 × 24)

7D Calculating time intervals

Discover Student Book page 148 • Practice Book page 124

Specific learning focus

• Calculate how long different daily events last.

Global skills

• **Creative skills:** problem solving
• **Real-world skills:** interpreting information

Key vocabulary

• timetable, duration, 24-hour times

Resources

• none required

Language support

Support students with an explanation of timetables, with some visual aids, including bus and train timetables, school timetables and TV schedules, to show all the different types. Then focus on the features of a school timetable in preparation for this lesson.

 **Introductory activity**

Ask students to look at page 148 of the Student Book. Explain that they will be listing some activities and working out how long they spend on each one. Ask them, initially, just to make a list of all the activities they do in a typical day. Base it on a school day and ask them to include each lesson as a separate activity. Write the ideas on the board for students to refer to throughout the main activity.

 **Main activity**

Ask students to look at the first time in the first column of the **timetable** on page 148 of the Student Book: 06:00-07:00. *What is the first thing you do in the morning? What are you usually doing between 6 o'clock and 7 o'clock in the morning?* Most students will probably say sleeping but there may be some other suggestions. Model how they write their activity in the second column of the table and explain that they will need to list all the activities they do, in order, in this column.

Point out the third column in the table and the word '**duration**'. *What do you think we need to write in here?*

They might not always spend exactly one hour on each activity. For example, they might have their breakfast between 07:00 and 08:00 but it only lasts 30 minutes. They could also split the Activity column if, for example, they are asleep until 07:30 and then having breakfast between 07:30 and 08:00.

Show them how to populate the columns to show split activities.

Time	Activity	Duration
06:00–07:00	Sleep	1 hour
07:00–08:00	Sleep/Have breakfast	30 mins/30 mins
08:00–09:00	Get ready for school/ Travel to school	15 mins/45 mins
09:00–10:00	Registration, then Mathematics	15 mins/45 mins

Students should be able to carry out this activity individually as you have modelled the process for them. As you work with individuals, ask questions based on time intervals. For example: *How long do you spend in lessons? How long is it between waking up and leaving for school?* This will prepare students for the activity in the next lesson.

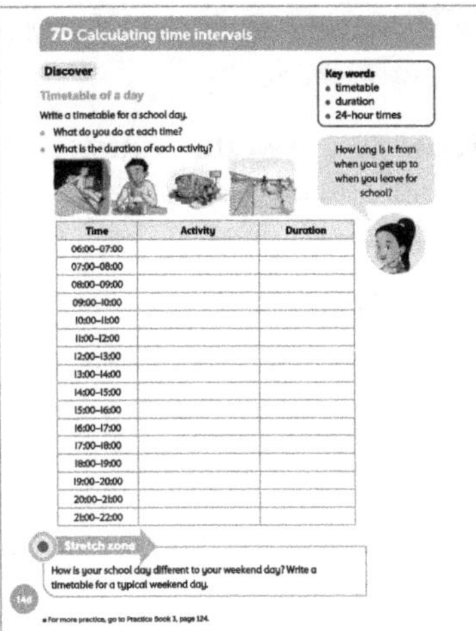

Differentiation

Supporting: Encourage students to write approximate times only so they stick to one-hour durations for each activity.

Consolidating: Ask students to describe the how long they spend on each activity during their school day.

Extending: Ask students to adapt the grid, as described in the main activity, to show the exact times of each event.

Stretch zone: *How is your school day different from your weekend day? Write a timetable for a typical weekend day.*

Check that students have completed the timetable reasonably.

 **Reflection time**

Students can work in pairs to compare the information in their timetables. *Who wakes up the earliest? By how much? Who has breakfast the latest? By how much?*

Students should make up their own questions to ask each other.

Practice Book: Students complete Practice Book page 124. They can do this directly after the main activity, as homework, or as the focus of a separate mathematics session to help students consolidate their learning and build fluency.

Students answer questions based on a timetable of a business trip. The durations are given in hours and minutes.

Differentiated outcomes	
All students	should complete the grid with durations lasting whole hours only.
Most students	will answer simple word problems based on the grid.
Some students	may adapt grid to show the time to the nearest 15 minutes.

Answers

Student Book page 148

Answers will vary according to the activities students include. Ask students to tell you how long different activities last, and to explain how they know or can use the timetable to work it out.

Practice Book page 124

1 1 hour 20 minutes

2 3 hours 20 minutes

3 2 hours 45 minutes

4 1 hour 30 minutes

5 11 hours 15 minutes

7D Calculating time intervals

Explore Student Book page 149 • Practice Book page 125

Specific learning focus

- Calculate how long different daily events last.

Global skills

- **Creative skills:** problem solving
- **Real-world skills:** interpreting information
- **Interpersonal skills:** communication

Key vocabulary

- 24-hour clock, how long before? how long between? how long after?

Resources

- TV schedule magazine or a TV schedule online
- strips of paper

Language support

Focus on the language of time intervals. For example, use these phrases.

- *How long before …?*
- *What time will that finish?*
- *When does that programme start?*

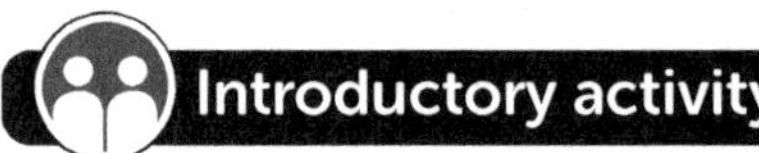
Introductory activity

Each pair should look at a page from the TV schedule magazine or a day from an online TV schedule. They discuss what they notice. You can prompt them asking questions such as: *What sort of programmes are on early in the morning? When are news programmes on?*

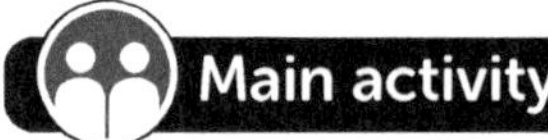
Main activity

Students work in pairs on this activity. This allows them to discuss the possible arrangements.

Look together at page 149 of the Student Book. If you have access to an IWB you could use this. Go through the activity in the Student Book. They complete the activity in their notebook but, first, encourage them to write the name of each programme, with its duration, on strips of paper so that they can move the programmes around on the timetable before they come to their final decision.

Once they have completed their TV schedule, they should answer the questions from question 2, directly into the Student Book, based on their individual TV schedules.

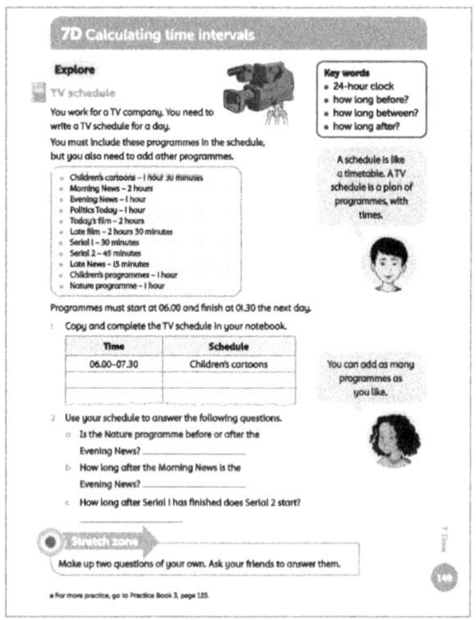

Differentiation

Supporting: Help students to construct their TV schedules by asking questions based on the programmes' start and finish times.

Consolidating: Ask students questions based on their schedules about duration of programmes and make sure that there are no gaps between the programmes.

Extending: Encourage students to adapt their schedules to the nearest five minutes.

Stretch zone: *Make up two questions of your own. Ask your friends to answer them.*

Check that students have written sensible questions based on their individual TV schedule.

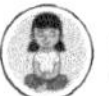
Reflection time

Ask students to come to the front to share their plans for the programme listings. They should offer a reason for their choices. Other students in the group should offer feedback based on the 'two stars and a wish' process (two things they like and one thing they would change).

Practice Book: Students complete Practice Book page 125. They can do this directly after the main activity, as homework, or as the focus of a separate mathematics session to help students consolidate their learning and build fluency.

Students answer questions based on a given TV schedule, with multiple channels. Ensure that students understand how to read this type of TV schedule by asking some questions before they complete the page, for example: *What four programmes could I be watching if I am watching TV between 08.30–09:00?*

Differentiated outcomes	
All students	should be able to make a TV schedule.
Most students	will talk about the TV schedule that they have made.
Some students	may construct a TV schedule to the nearest five minutes.

Answers

Student Book page 149

Check that students have filled up the schedule and that there are no gaps. Although answers will vary, check that their answers to the questions on page 149 of the Student Book match the information they have listed in the schedule.

Practice Book page 125

1   120 minutes

2   *The Quick Quiz* and *Dynamite Gang*

3   1 hour 30 minutes

4   *The Last Tiger* and *Rainbow Falls*

Stretch zone: Check that students' choices of programmes add up to 90 minutes, for example: *Football Round-Up,* or *Inside the Zoo, Dinosaur Kingdom* and *Pickle and Pop.*

7 Time

Connect Student Book page 150

Big idea

- I can tell the time using 12-hour and 24-hour clocks and I understand different units of time.

Global skills

- **Creative skills:** problem solving
- **Real-world skills:** interpreting information
- **Interpersonal skills:** team work

Key vocabulary

- days of the week; months of the year; second, minute, hour, day, week, month, year, how long ago? how long before?

Resources

- calculators

Language support

- You may choose to work with students with weaker language skills. Carry out the calculation for them but ask them to read out the number on the calculator. This will support them in the development of the language of place value and units of time.

 ### Introductory activity

Write on the board your date of birth, or that of a famous person if you prefer. Say that you are going to work out the person's age in years, weeks and days. For example, work out the age of Usain Bolt. His date of birth is 21 August 1986. (The calculation below works out his age at the end of 2020, but adjust for the current date by adding on the number of years, months and days since 31 December 2020.)

Model the calculation as follows.

How many days left in August 1986 after his birthday? (10)

How many days until the end of 1986?
(September + October + November + December
= 30 + 31 + 30 + 31 = 122)

The number of days Usain Bolt was alive in 1986 is 132, which is 18 weeks and 6 days.
How many years are there from the end of 1986 to the end of 2020? (34)

So at the end of 2020, Usain Bolt was 34 years, 18 weeks and 6 days old.

 ### Main activity

Ask students to estimate how long you have spent asleep in your life. Write their estimates on the board. Model the example calculation given in the Student Book on page 150, in the first row of the table.

- Write down how many hours you sleep a night, that is, in one 24-hour period.
- Calculate how long you sleep in a week by multiplying the figure above by 7.
- Multiply this by 52 to find out how long you sleep in a year.

Use a whole number of hours for your initial number to make calculations easier.

Students work through the activity in the Student Book on page 150, which follows the same method as the Student Book example. They should work in groups to support each other as this is quite a complex activity, but they should each work out the answer based on their individual age in years, months and days. Support them by comparing answers from other groups so that they can see whether they are getting sensible answers.

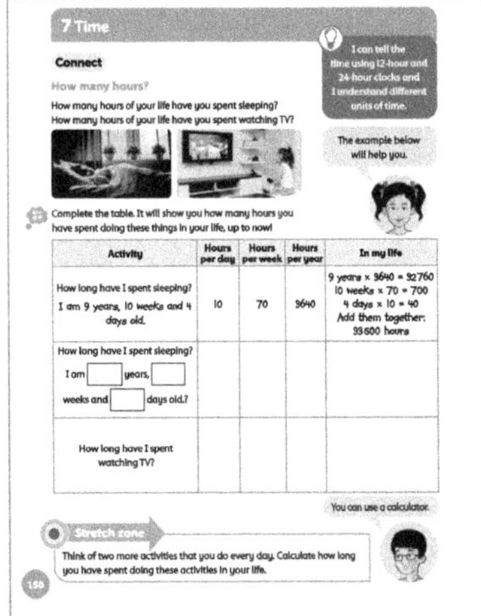

Differentiation

Supporting: Ask students to tell you how they are working the answers out, explaining each step of their method so you can check it.

Consolidating: Ask students to share their strategies with others.

Extending: Ask students to check answers for accuracy.

Stretch zone: *Think of two more activities that you do every day. Calculate how long you have spent doing these activities in your life.*

Check that students have completed some appropriate calculations.

 **Reflection time**

Ask the groups who completed the stretch zone to name the extra activities they identified. Before they give the times they estimated for these activities, other students should estimate what a sensible time might be. As this activity progresses, students should be able to use previous answers to develop the accuracy of their estimates.

Differentiated outcomes	
All students	should participate in the group work and understand the relationship between units of time.
Most students	will use the relationship between units of time to carry out calculations.
Some students	may be able to support others in carrying out calculations and check that answers are sensible.

7 Time

Review Student Book page 151 • Practice Book page 126

Global skills

- Self-development skills: reflecting on learning

Student Book

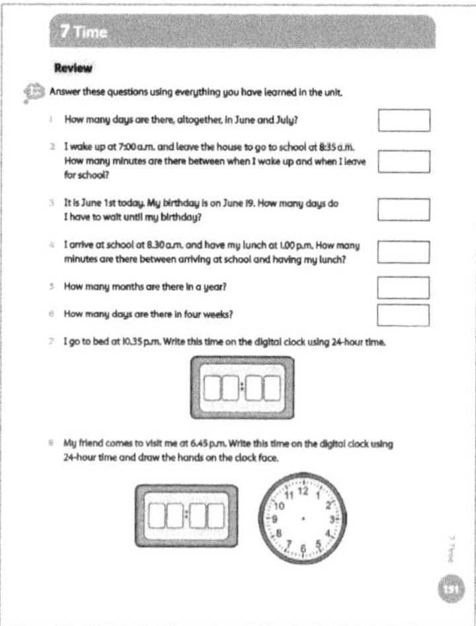

With young children, assessment activities are most effective when carried out as an everyday classroom activity. Students should have calendars and both analogue and digital clocks available to support them.

Watch as students calculate the number of months, days and minutes for the questions, and observe their strategies for working them out.

With the clock questions, notice how they place the missing digits and hands.

Answers

Student Book page 151

1	61 days	**5**	12 months
2	95 minutes	**6**	28 days
3	18 days	**7**	22:35
4	270 minutes	**8**	18:45

Practice Book

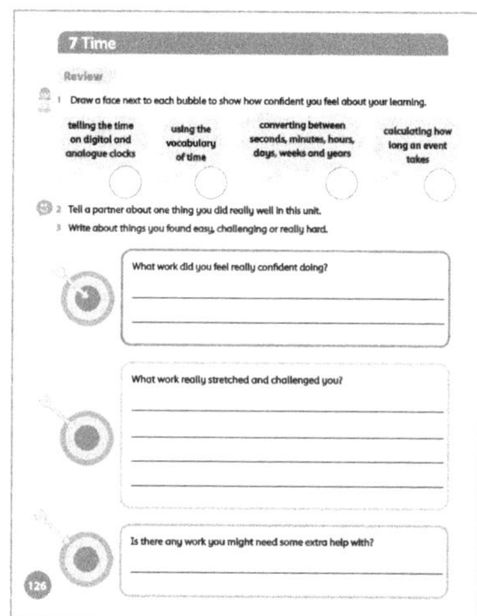

It is appropriate to complete this Practice Book review as a whole-class discussion. You may choose to keep a record of the class discussion or a copy of the review page for your own records. The review provides an opportunity for students to reflect on their learning from the unit, to discuss any areas of mathematics that they feel went particularly well, and any areas that they feel less confident about. Ensure that all students have a copy of the Student Book as a reminder of the areas of mathematics that they have worked on in this unit.

Allow students plenty of time for discussion before asking them to complete the Practice Book page individually, and then, if appropriate, to share their responses with the rest of the class. If students complete this self-assessment at home, encourage them to discuss this with adults. Make a note of areas that students still feel unsure about. These may include, for example, changing p.m. times into 24-hour times, or placing the hour hand correctly on a clock when the time is not 'o'clock'. Give students plenty of opportunities to experience how the hands move on analogue clocks and compare the time with digital clocks for both a.m. and p.m. times.

Additional material

There are additional end-of-unit assessments available on the *Oxford Owl for School* website.

8 Geometry – properties of shapes

Overview

Big idea

The big idea in this unit is that shapes can be described and classified using many different properties, for example, the number of sides, vertices, faces, edges, the size of the vertices, and whether the shape has symmetry or not.

Students will also notice more differences between two-dimensional (2D) and three-dimensional (3D) shapes. They will realise that images of 3D shapes (such as a drawing of a cube on paper) are 2D representations of those 3D shapes. They also continue to develop their skills of visualisation; a vital skill in all work linked to shapes and geometry. In this unit, students are encouraged to visualise through activities using 'feely bags' (a bag in which shapes are placed so that students can feel them but not see them), and by guessing a shape from its properties.

Students will become more aware of angles in shapes as they explore and describe the properties of 2D and 3D shapes and the relationships between them. An understanding of a right angle is essential because it is a property of so many shapes. Students will go on to use right angles to compare and describe other angles as smaller than a right angle or larger than a right angle.

Look out for

- **Students who only think of regular representations of shapes, for example, they may think that an equilateral triangle represents all triangles, or that a pentagon is represented by a regular pentagon.** There are lots of opportunities for students to see different representations of shapes in this unit, so they will see that, for example, any closed shape with five straight sides is a pentagon, not just those with same-length sides.

- **Students who confuse the names of 2D and 3D shape properties, for example sides, angles, faces, edges, vertices.** By having examples of 2D and 3D shapes around the classroom, students can have plenty of experience at handling and describing them, and labels can be added.

Possible misconceptions

- **Students only recognise squares and rectangles when 'sat' on one side, not at an angle.** You can support students by showing them shapes in a range of different orientations, so they realise that a shape's properties remain the same even if the orientation is changed.

- **Students may think that every surface on a 3D shape is a 'face', including those on spheres, cones and cylinders, for example.** Support students in understanding that where it is flat, it is called a face, otherwise it is a curved 'surface'.

Key vocabulary

- shape, 2D shapes, 3D shapes
- round, straight, curved
- face, side, edge, surface, corner
- angle, right-angled, vertex, vertices
- circle, semi-circle; square, rectangle, rectangular, quadrilateral; triangle, equilateral triangle, isosceles triangle, scalene triangle; pentagon, pentagonal, hexagon, polygon, hexagonal, octagon, octagonal; cube, cuboid, prism, pyramid, sphere, cone, cylinder
- bigger, larger, smaller; symmetrical, line of symmetry, fold, mirror line, mirror image, reflection, reflective symmetry, pattern, repeating pattern, straight line
- regular, irregular
- parallel, perpendicular

Learning objective	E	8A	8B	8C	8D	8E	C	R
Draw 2D shapes and make 3D shapes using modelling materials; recognise 3D shapes in different orientations and describe them.	✓	✓	✓				✓	✓
Recognise angles as a property of shape or a description of a turn.					✓			
Identify right angles, recognise that two right angles make a half-turn, three make three quarters of a turn and four a complete turn; identify whether angles are greater than or less than a right angle.		✓			✓		✓	
Identify horizontal and vertical lines and pairs of perpendicular and parallel lines.						✓		
Identify symmetry in shapes and pictures.				✓				

8 Geometry – properties of shapes

Engage Student Book page 152

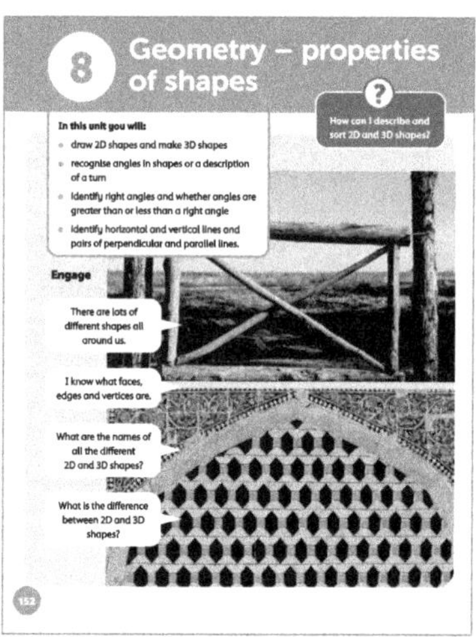

Big question

- How can I describe and sort 2D and 3D shapes?

Global skills

- **Creative skills**: exploring
- **Interpersonal skills**: communication

Key vocabulary

- 2D shape, 3D shape, straight, curved, face, side, edge, vertex (vertices), triangle, square, rectangle, quadrilateral, pentagon, hexagon, octagon, circle, semi-circle, cuboid, sphere, pyramid, prism

Resources

- magazines and newspapers
- large sheets of paper
- digital cameras (optional)
- access to the internet

Language support

Encourage students to present their posters and read out the names of the shapes.

 **Introductory activity**

Go for a walk with students around the school and the local environment. Point out the different **2D shapes** and **3D shapes** you see. If you have digital cameras, take photographs that you can print to use on students' posters. If going for a walk is not possible, find examples of 2D and 3D shapes in the classroom.

 **Main activity**

Look together at page 152 of the Student Book. If you have access to an IWB you could use this. Students work in small groups. Use the photographs in the Student Book (and any that you have taken) and give out magazines and newspapers to each group. On the board, write the following shape names: **triangle, square, rectangle, quadrilateral, pentagon, hexagon, octagon, circle, semi-circle, cube, cuboid, sphere, prism, pyramid.**

Ask groups to make posters of all the shapes they can see on the photographs and in the magazines and newspapers. They should label all the shapes with the correct names. These posters can be displayed for the rest of the unit.

Differentiation

Supporting: Point to shapes and ask students their names. Model the correct vocabulary for them for the some of the new shapes.

Consolidating: Ask students to find examples of all the shapes written on the board, including those shapes that are new to them.

Extending: Ask students to research to find the names of any shapes they do not know the names of. They should describe them by their properties.

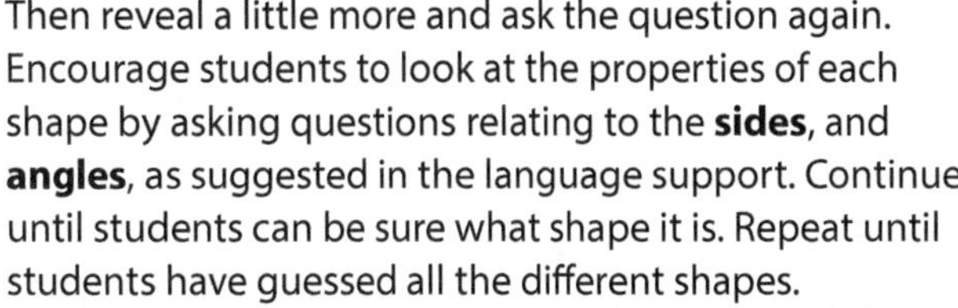

8A Identifying and classifying 2D shapes

Discover 1   Student Book page 153 • Practice Book page 127

Specific learning focus

- Identifying 2D shapes and recognising the angles in 2D shapes.

Global skills

- **Creative skills**: exploring
- **Real-world skills**: research

Key vocabulary

- polygon, pentagon, hexagon, octagon, side, angle, corner, straight, quadrilateral, vertices, regular, irregular

Resources

- feely bag
- shapes cut out of card: triangles, a square, a rectangle, quadrilaterals, a pentagon, a hexagon, an octagon, a circle, and a semi-circle
- mini whiteboards and markers

Language support

As you play the game in the introductory activity, model the use of the vocabulary 'sides', 'vertices' and 'right angles' ('square **corners**'), asking questions, for example: *How many sides can you see? Are there any right angles? How many vertices does this shape have?* Focus on the more complex words such as 'vertex' and 'vertices'.

Introductory activity

Place the card shapes in the feely bag. Pick a shape and gradually slide it out of the bag. Just reveal a small portion of the shape. *What do you think the shape is?*

Reflection time

Ask groups for examples of the different shapes. If any of the shape names do not have an example, ask students to look for these shapes in the environment or at home and to bring examples in. Start a display table for examples of 3D shapes. Make labels for 'Cube', 'Cuboid', 'Sphere', 'Prism', 'Pyramid', and find different examples to display during the unit.

Then reveal a little more and ask the question again. Encourage students to look at the properties of each shape by asking questions relating to the **sides**, and **angles**, as suggested in the language support. Continue until students can be sure what shape it is. Repeat until students have guessed all the different shapes.

Main activity

Explain the meaning of '**polygon**': a closed shape with **straight** sides. Ask students to say which of the shapes from the introduction were polygons and which were not, and why. Hold up a square and an oblong and explain that the square is a **regular quadrilateral** because it has four sides that are equal in length and four angles that are the same. The oblong is an **irregular quadrilateral** because its sides are not all the same length. Squares and oblongs are both rectangles because they both have four **right-angles**. Show students some examples of regular and irregular pentagons, hexagons and octagons and ask them to identify the regular shapes.

Ask students whether they can remember what a right angle is? *Can you point to a right angle in any of these shapes? Can you name any shapes that include right angles?* Students complete the activity in the Student Book on page 153 individually. As you work with students, continue to ask questions that model the correct language, for example:

- *What is the name of that shape?*
- *How many **vertices** does it have?*
- *Does it have any right angles?*
- *How many sides does it have?*

Refer students to the second speech bubble and encourage them to use the glossary at the back of the Student Book to help them if they are unsure of any of the names of the shapes.

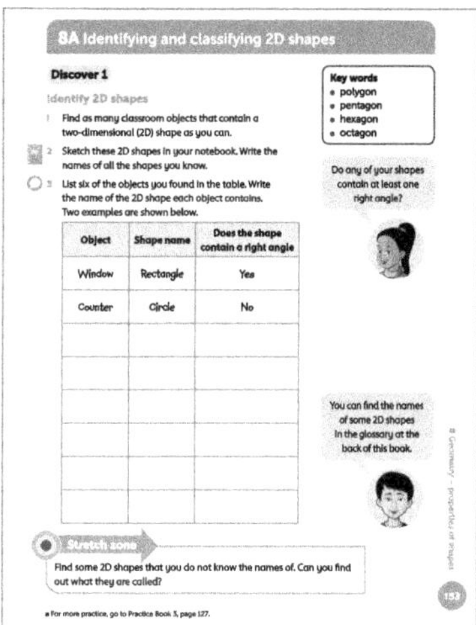

Students match 2D shapes with the correct shape names. *How do you know that shape is regular? What makes that shape irregular?*

Differentiated outcomes	
All students	should draw the shapes accurately with support.
Most students	will independently draw the shapes accurately and answer questions about their properties.
Some students	may draw the shapes accurately and describe their properties.

Answers

Student Book page 153

Check that students have drawn shapes that match what is required in each part of the table. In particular, check that students' regular shapes do have roughly equal sides and angles.

Practice Book page 127

1 irregular hexagon

2 regular pentagon

3 rectangle

4 equilateral triangle

5 regular octagon

6 irregular quadrilateral

7 irregular octagon

Stretch zone: In a regular shape, the side lengths and angles are all the same size. In an irregular shape, the side lengths and angles can be all different sizes.

Differentiation

Supporting: Ask questions to support students. Remind them of the meaning of 'polygon', 'hexagon' and so on.

Consolidating: Ask students to describe the properties of the shapes they are drawing.

Extending: Focus on the vocabulary 'regular' and 'irregular'. Ask students to define terms for you and to find other examples of the shapes around the classroom.

Stretch zone: *Find some 2D shapes that you do not know the names of. Can you find out what they are called?*

Look to see whether students have found the correct shape names.

 **Reflection time**

In pairs, students play 'Guess my shape'. One student draws a shape on their mini whiteboard. Their partner has to guess what they have drawn by asking questions to which their partner can only answer 'Yes' or 'No'.

Practice Book: Students complete Practice Book page 127. They can do this directly after the main activity, as homework, or as the focus of a separate mathematics session to help students consolidate their learning and build fluency.

8A Identifying and classifying 2D shapes

Discover 2 — Student Book page 154 • Practice Book page 128

Specific learning focus

- Identifying 2D shapes and recognising the angles in 2D shapes.

Global skills

- **Creative skills**: problem solving

Key vocabulary

- polygon, pentagon, hexagon, octagon, vertex

Resources

- isometric dotted paper
- selection of regular and irregular 2D shapes

Language support

As students draw and discuss the shapes they draw, support them with the correct names and pronunciation of the properties, for example 'regular' and 'irregular'.

 **Introductory activity**

Show students one shape at a time from a selection of 2D shapes. Ask questions about the shapes such as:

- *How many sides does this shape have?*
- *Are the sides the same length?*
- *What is this shape called?*

 **Main activity**

Students should choose a shape from those available and draw it using the isometric dotted paper. Encourage them to focus on the number of sides and the size of the angles of each shape. Ask students to name each shape correctly. *Does the isometric paper make it easier to draw the shapes?* Look together at page 154 of the Student Book. If you have access to an IWB you could use this. Refer students to the shapes drawn in the Student Book. These are some examples of shapes already drawn on isometric paper. Point at how sometimes the sides of the shapes follow the dots on the paper but sometimes the sides of the shapes are drawn through gaps in the dots. Each **vertex** of the shapes is, however, on one of the dots.

Students then complete the activity on page 154 of the Student Book. They are asked to draw a variety of quadrilaterals, pentagons, hexagons and octagons. Allow students to refer to the examples in the Student Book to help them.

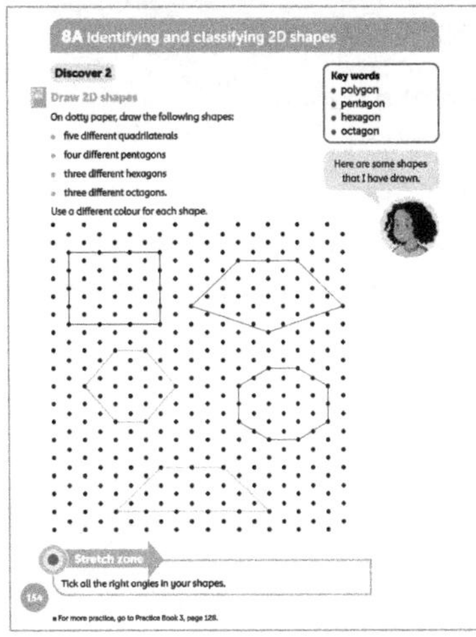

Differentiation

Supporting: Ask students to draw simple shapes such as squares, rectangles and triangles.

Consolidating: Ask students to draw irregular shapes with more sides such as hexagons and octagons.

Extending: Ask students to draw regular shapes and name them correctly.

Stretch zone: *Tick all the right angles in your shapes.*

Check that students have identified all the right angles correctly.

 **Reflection time**

Ask some students to share the shapes they drew and some information about them. For example, can they name their shape and describe its sides and angles?

Practice Book: Students complete Practice Book page 128. They can do this directly after the main activity, as homework, or as the focus of a separate mathematics session to help students consolidate their learning and build fluency.

Students find examples of 2D shapes in the classroom or at home. They draw them and describe some of the shape's properties. Remind students that a triangle that is not equilateral is irregular.

Differentiated outcomes	
All students	should draw simple shapes such as squares and triangles.
Most students	will draw and name irregular shapes.
Some students	may draw and name regular shapes.

Answers

Student Book page 154

Students should use the dots for the vertices of their shapes. Check that students have drawn:

1. five different quadrilaterals

2. four different pentagons

3 three different hexagons

4 three different octagons.

Practice Book page 128

Students answers will vary according to which shapes they find. Check that the drawing is accurate and that the properties are correct.

8A Identifying and classifying 2D shapes

Explore 1 Student Book page 155 • Practice Book page 129

Specific learning focus

- Identify, describe, and draw regular and irregular 2D shapes including pentagons, hexagons, octagons and semi-circles.
- Relate 2D and 3D shapes to drawings of them.

Global skills

- **Creative skills**: investigating
- **Interpersonal skills:** communication

Key vocabulary

- regular, irregular, right angle, polygon

Resources

- mini whiteboards and markers
- lollipop sticks in a jar – each lollipop stick has a different student's name on it

Language support

Continue to ask students questions, using the precise vocabulary of shape, for example:

- *What is the name of that shape?*
- *How many vertices does it have?*
- *Does it have any right angles?*
- *How many sides does it have?*

Ask them to repeat the answers to you.

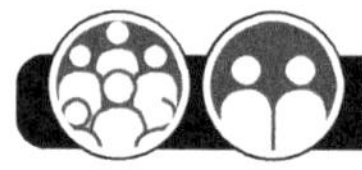

Introductory activity

Select one of the following shapes and draw it on the board: a triangle, a square, a rectangle, a quadrilateral, a pentagon, a hexagon, an octagon, a circle or a semi-circle. Describe the properties of the shape you have drawn so that students can make a copy of it on their own whiteboards. Make sure that your vocabulary includes vertices, sides and right angles. You will also need to

Stretch zone: An example might be the floor of their classroom, if it is an 'L' shape. It could be made up of two rectangles.

describe the relative lengths of sides. Repeat this several times checking that students draw the shapes accurately.

Main activity

Look together at some of the shapes that you drew on the board in the introductory activity and ask students to say whether each one is regular or irregular and explain why.

Refer them to the two speech bubbles on page 155 of the Student Book. *What is another name for a regular quadrilateral? How do you know whether a shape is regular?* Students should refer to the sides all being equal and the angles all being equal, so a square is a regular quadrilateral.

Students will need to use the posters from the Engage lesson and may need to refer to the display table with 3D shapes on it, and magazines for other 2D shapes. Students should work in pairs and take it in turns to find a shape that they can draw in one of the boxes on page 155 of the Student Book.

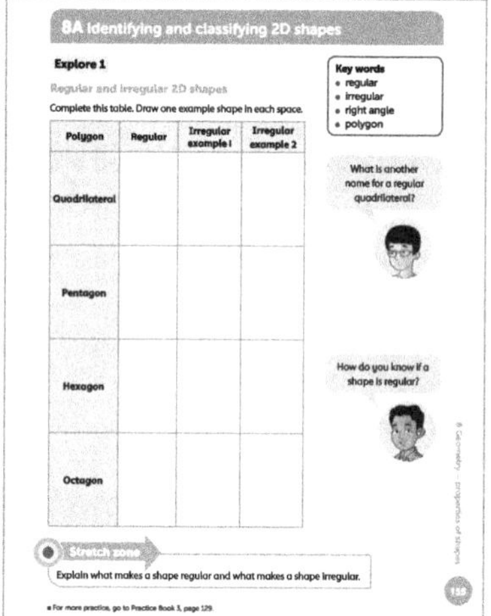

Differentiation

Supporting: Ask questions to support students. Remind them of the meaning of 'polygon' and 'hexagon' and so on.

Consolidating: Ask students to describe the properties of the shapes they are drawing.

Extending: Focus on the vocabulary 'regular' and 'irregular'. Ask students to define terms for you and to find other examples of the shapes around the classroom.

Stretch zone: *Explain what makes a shape regular and what makes a shape irregular.*

Check that students can correctly describe the difference between regular and irregular shapes and ask them to draw some examples.

 ## Reflection time

Take lollipop sticks from the jar to select individual students. Ask them to come to the front and sketch one of their examples on the board. They should describe their shape as they draw it, for example: 'This is an example of an irregular quadrilateral. It has 4 sides and 1 right angle'. Have sentence frames on the board to help them do this, for example: 'This shape has ___ sides and ___ angles. It is called a ______. It is regular/irregular because ______.'

Practice Book: Students complete Practice Book page 129. They can do this directly after the main activity, as homework, or as the focus of a separate mathematics session to help students consolidate their learning and build fluency.

Students read clues (properties of a shape) and they have to work out what the shape is, based on the clues given. They write the name of the shape. *If all the sides are different lengths, does this mean the shape is regular or irregular?*

Differentiated outcomes	
All students	should draw the shapes accurately with support.
Most students	will independently draw the shapes accurately and answer questions about their properties.
Some students	may draw the shapes accurately and describe their properties.

Stretch zone

Explain what makes a shape regular and what makes a shape irregular.

Check that students can correctly describe the difference between regular and irregular shapes.

Answers

Student Book page 155

Check that students have drawn shapes that match what is required in each part of the table. In particular, check that their regular shapes do have roughly equal sides and angles.

Practice Book page 129

1 rectangle

2 irregular hexagon

3 equilateral triangle

4 scalene triangle

5 irregular pentagon

6 circle

7 semi-circle

8 regular octagon

9 square

8A Identifying and classifying 2D shapes

Explore 2 Student Book page 156 · Practice Book page 130

Specific learning focus

- Identify, describe, and draw regular and irregular 2D shapes including pentagons, hexagons, octagons, and semi-circles.

Global skills

- **Creative skills**: investigating
- **Interpersonal skills:** communication

Key vocabulary

- regular, irregular, right angle, polygon, lines of symmetry

Resources

- lollipop sticks in a jar – each lollipop stick has a different student's name on it

Language support

Ask students about the angle properties of shapes, asking, for example:

- *Is that angle 90 degrees?*
- *Is it more or less than 90 degrees?*

Ask them to repeat the answers to you.

 **Introductory activity**

Ask students to describe a right angle. *How can we test whether an angle is a right angle?* Show students how to use the corner of a piece of paper, or the corner made when folding paper in half and half again. Ask students to check around the classroom for right angles. *Can you find any right angles on the window, the door, the whiteboard? Can you spot any angles that are smaller than*

or greater than a right angle? Sketch some shapes that have at least one right angle. Can you draw a shape with no right angles?

Main activity

Ask one student to sketch a triangle with one right angle. Ask another to sketch a different triangle with one right angle. Then ask whether it is possible to sketch a triangle with two right angles. Ask at least one student to try. Is it possible? Explain that if it is not possible to sketch a shape, they should write 'impossible' in that section of the table on page 156 of the Student Book.

Students will need to use the posters from the Engage lesson and magazines for other 2D shapes. Students should work in pairs and take it in turns to find a shape that they can draw in one of the boxes in the table in the Student Book.

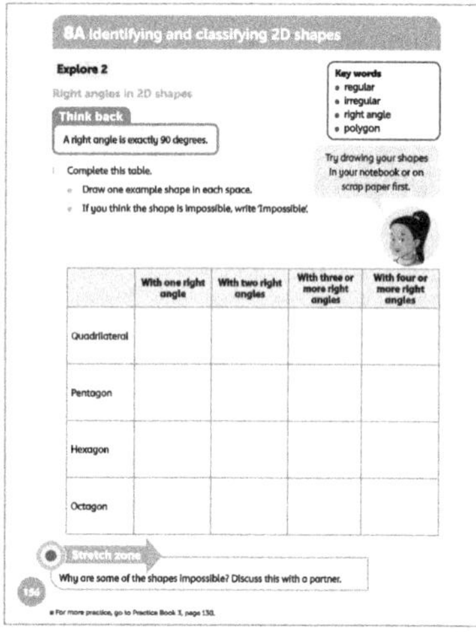

Differentiation

Supporting: Ask questions to support students. Remind them of the meaning of the names of the shapes. For example, remind them that a pentagon has five sides.

Consolidating: Ask students to describe all the properties of the shapes they are drawing.

Extending: Focus on the vocabulary 'right angle'. Ask students to define terms for you and to find other examples of shapes around the classroom.

Stretch zone: *Why are some of the shapes impossible? Discuss this with a partner.*

Check that students can explain why some shapes are impossible, using those shapes' properties.

Reflection time

Take lollipop sticks from the jar to select individual students. Ask them to come to the front and sketch one of their example shapes on the board. They should describe their shape as they draw it, for example: 'This is an example of a shape with 2 right angles. It has 5 sides, 2 right angles and 3 other vertices.' Have sentence frames on the board to help them do this.

Practice Book: Students complete Practice Book page 130. They can do this directly after the main activity, as homework, or as the focus of a separate mathematics session to help students consolidate their learning and build fluency.

Students sketch shapes with a given number of right angles and **lines of symmetry**. Before they complete the page, revise what a line of symmetry in a shape is. *How many lines of symmetry does a square have? How many does a rectangle have? What about a regular hexagon?*

Differentiated outcomes	
All students	should draw the shapes accurately with support.
Most students	will independently draw the shapes accurately and answer questions about their properties.
Some students	may draw the shapes accurately and describe their properties.

Answers

Student Book page 156

Check that students have drawn shapes that match what is required in each part of the table, with the correct number of right angles.

Stretch zone: A pentagon with four or more right angles is impossible.

Practice Book page 130

Check that students have drawn shapes that match what is required in each part of the table with the correct number of right angles and lines of symmetry.

A triangle with two right angles and a rectangle with no lines of symmetry are both impossible.

Stretch zone: Yes, a triangle can have one angle greater than a right angle, with two acute angles. All the angles will add to 180°. For example, an isosceles triangle could have one angle of 100° and two angles of 40°.

Explore 3 Student Book page 157 • Practice Book page 131

Specific learning focus

- Classify and sort 2D shapes using a Carroll diagram.

Global skills

- **Creative skills**: problem solving
- **Interpersonal skills**: communication
- **Self-development skills**: reflecting on learning

Key vocabulary

- regular, irregular, side, angle, vertex/vertices

Resources

- set of assorted 2D shapes

Language support

Help students understand the distinction between a shape having a property or *not* having that property to help them place shapes in Carroll diagrams. For example, a shape could have equal sides or not have equal sides, or it could have right angles or not have right angles.

 **Introductory activity**

Give students a set of various 2D shapes, regular and irregular, to sort. They can decide for themselves what the sorting criteria are, but remind them to sort based on shape properties, and not by colour or size. For example, they might decide to sort the shapes based on how many sides they have or whether they have right angles or whether they have line symmetry. As a reminder, ask students whether they can describe what line symmetry means and give an example of a shape with line symmetry. They should work in pairs and be able to describe to their partner why each shape belongs in one set or another.

Ask some students to share their sorting criteria and give an example of a shape in each set.

 **Main activity**

Direct students to look at the set of shapes on page 157 of the Student Book. Give them some time to discuss the shapes with a partner, then collectively work through the set of shapes and try to give them the correct name. Students should be using names such as triangle, quadrilateral, regular hexagon or irregular pentagon, and

refering to properties of being regular or not, or having symmetry or not, for example.

Now ask students to complete the Carroll diagrams in the Student Book on page 157. They should look carefully at the criteria for each diagram, then select shapes accordingly for each section. They write the number of the shape, rather than the name, in the correct section of the Carroll diagram.

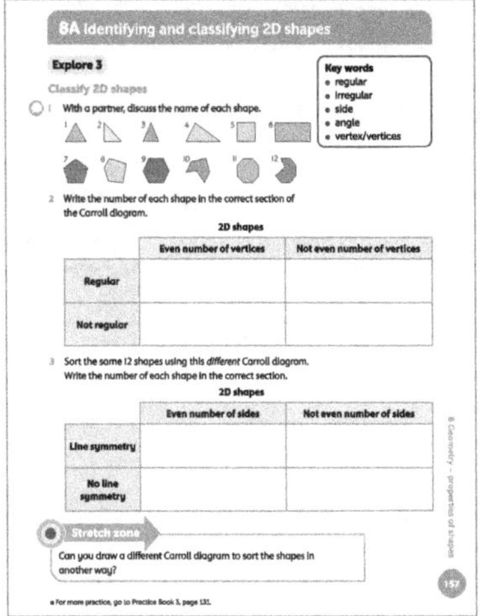

Differentiation

Supporting: Help students to count the vertices for the shapes.

Consolidating: Ask students to explain their strategies for placing the shapes in each section of the Carroll diagram.

Extending: Ask students to tell you another shape that would fit in each section.

Stretch zone: *Can you draw a different Carroll diagram to sort the shapes in another way?*

Check that students have found an alternative Carroll diagram and sorted the shapes correctly.

 **Reflection time**

Ask students to share their reasoning for placing each shape. Choose a shape and ask a student to say which section of the Carroll diagram they placed it in and why. Repeat several times, allowing a different student to share their reasoning each time.

Practice Book: Students can complete Practice Book page 131. They can do this directly after the main activity, as homework, or as the focus of a separate mathematics session to help students consolidate their learning and build fluency.

Students complete the Carroll diagrams with pre-selected sorting criteria. They are asked to draw eight shapes in each section. *You can draw the shapes, or you can write the names if you prefer. Include as many as possible in each section.*

Differentiated outcomes	
All students	should find at least one shape for each section of the Carroll diagram.
Most students	will place the set of shapes in the correct sections.
Some students	may find additional shapes to go in each section.

Answers

Student Book page 157

	Even number of vertices	Not even number of vertices
Regular	5, 9, 11	1, 7
Not regular	6, 10, 12	2, 3, 4, 8

	Even number of sides	Not even number of sides
Line symmetry	5, 6, 9, 11	1, 2, 3, 7
No line symmetry	10, 12	4, 8

Practice Book page 131

Check that the shapes drawn or written in each section of the Carroll diagrams are correct for the criteria given.

8B Properties of 3D shapes

Discover Student Book page 158 · Practice Book page 132

Specific learning focus

- Identify, describe, and make 3D shapes including prisms and pyramids.
- Identify 3D shapes in the environment.
- Classify 3D shapes according to the number and shape of faces, number of vertices and edges.

Global skills

- **Creative skills**: exploring

Key vocabulary

- 3D shape, edge, vertex/vertices, face

Resources

- feely bag
- set of assorted 3D shapes, including a sphere, a cube, other cuboids, a range of prisms and pyramids
- lollipop sticks in a jar – each lollipop stick has a different student's name on it
- card, scissors, sticky tape
- geometric construction shapes, or other 3D modelling equipment, or cardboard boxes and other recycling materials

Language support

As students are making the models, ask:
- *How many vertices does your shape have?*
- *What shape is each face of your 3D shape?*

Write the following on the board as a model sentence. Ask students to use it when they respond to your questions: 'My shape has ___ vertices. Each face is a __________ [or a __________]. My shape is a __________.' For example, they might say: 'My shape has 8 vertices. Each face is a square. It is a cube'.

 **Introductory activity**

Hide one 3D shape in a feely bag. Choose a student to come to the front and feel in the bag. They should describe the shape so that the other students can guess what it is. Encourage them to talk about vertices, **edges**, and the shape of each **face**. Repeat with different shapes and different students.

 **Main activity**

Students look around the classroom and, if possible, different areas of the school to collect examples of 3D shapes. Make sure that they find examples of prisms and pyramids. Then they should work through the activity on page 158 of the Student Book, using geometric construction shapes or cardboard boxes and other recycling materials to make models of the shapes they have found.

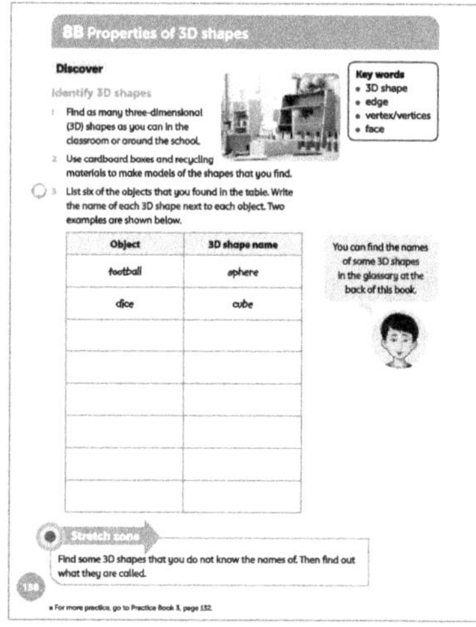

Differentiation

Supporting: Ask questions to support students. Remind them of the names of the shapes they find and model the language.

Consolidating: Ask students to describe the properties of the shapes they find.

Extending: Ask students to define shapes by their properties and to find other examples of the shapes around the classroom.

Stretch zone: *Find some 3D shapes that you do not know the names of. Then find out what they are called.*

Check that students have found the correct names for any shapes that they didn't know. *Where will you look for this information?* Remind them about the glossary at the back of the Student Book, or they could look at a mathematics dictionary, either in the classroom or online.

 **Reflection time**

Collect all the models together at the front of the class. Ask one group to come to the front and to classify the shapes, using suitable sorting criteria. Students should explain the criteria they use. These models can be added to the display table. When the shapes have been sorted, ask another group to come to the front and classify them using a different set of criteria.

Practice Book: Students complete Practice Book page 132. They can do this directly after the main activity, as homework, or as the focus of a separate mathematics session to help students consolidate their learning and build fluency.

Students read clues (properties of 3D shapes) and they have to work out what the shape is, based on the clues given. They write the name of the shape. There is always only one shape that fits each set of clues.

Differentiated outcomes	
All students	should identify some shapes with support.
Most students	will independently identify the shapes accurately and answer questions about their properties.
Some students	may make the shapes accurately and describe their properties.

Answers

Student Book page 158

Check that students have correctly identified the shape names of the objects they have chosen.

Practice Book page 132

1 cube

2 cylinder

3 sphere

4 cone

5 triangular prism

8B Properties of 3D shapes

Explore Student Book page 159 • Practice Book page 133

Specific learning focus

Global skills

- **Creative skills**: exploring
- **Interpersonal skills:** team work

Key vocabulary

- 3D shape, edge, vertex/vertices, face

Resources

- art straws and modelling clay to make models
- a set of 3D shapes from the previous lesson (or use the shapes from the display table)

Language support

Ask students to describe the shapes they are making while they make them. Refer them to the display table for the names and correct vocabulary. Alternatively, ask them to give you instructions to make a 3D shape.

 **Introductory activity**

Select one of the 3D shapes from the previous lesson, but do not show it to students. (You could hide it behind your back or put it in a feely bag.) Students try to guess your shape by taking it in turns to ask you questions, to which you can only answer 'Yes' or 'No'. Repeat this activity several times, with different 3D shapes. As a progression, students could assume the teacher role.

 **Main activity**

Show students the art straws and clay for making 3D shapes from your resources. Look together at page 159 of the Student Book. If you have access to an IWB you could use this. Relate the straws and clay to the picture of the shapes made up of straws and modelling clay on page 159 of the Student Book. Model for the students how to make a cube. *How many straws do we need for the edges of the cube? How many balls of clay do we need for the vertices?* Build the cube and count the edges and vertices together. Students work in groups to make the objects from the previous lesson using art straws and clay, as explained in the Student Book. As students make the models, ask relevant questions such as: *How many vertices does your shape have? What shape is each face of your 3D shape?* Students respond to your questions, using the correct vocabulary. Refer them back to the model sentence from the previous lesson as they answer. For example, they might reply: *My shape has 8 vertices. Each face is a square. It is a cube.*

Differentiation

Supporting: Ask questions to support the students. Remind them of the names of the shapes they make and model the language.

Consolidating: Ask students to describe the properties of the shapes they make.

Extending: Ask students to define shapes by their properties and to find other examples of the shapes around the classroom.

Stretch zone: *Use the shapes you have made to help you to explain the difference between a prism and a pyramid.*

Check that students can explain the difference.

Reflection time

Ask groups to read out their description of the difference between a prism and a pyramid. As a class, decide on the best definition and write this up as a poster to be displayed for the rest of the unit.

Practice Book: Students complete Practice Book page 133. They can do this directly after the main activity, as homework, or as the focus of a separate mathematics session to help students consolidate their learning and build fluency.

Students find 3D shapes in the classroom, at home or outside in the local area. They draw the objects or take photographs and stick them into their books. They could also find images in magazines and then cut them out.

Differentiated outcomes	
All students	should make and name some shapes with support.
Most students	will independently identify the shapes they are making accurately and answer questions about their properties.
Some students	may make a wide range of shapes accurately and describe their properties.

Answers

Student Book page 159

2   Answers will vary but might include:

Object	Shape	Number of edges	Number of vertices	Shape of faces
book	cuboid	12	8	all rectangles
chocolate box	triangular prism	9	6	2 triangles 3 rectangles
water bottle	cylinder	2	0	2 circles 1 rectangular curved surface
roof of a building	square-based pyramid	8	5	1 square 4 triangles

3

	Even number of faces	Not even number of faces
Even number of vertices	cube cuboid (×2) triangular-based pyramid	triangular prism pentagonal prism
Not even number of vertices		square-based pyramid

Practice Book page 133

Check that students have chosen the correct pictures to match the different shape names.

8C Drawing symmetrical shapes

Specific learning focus

- Draw and complete 2D shapes with reflective symmetry and draw reflections of shapes.

Global skills

- **Creative skills:** exploring
- **Interpersonal skills:** communication

Key vocabulary

- pentagon, reflective symmetry, line of symmetry

Resources

- sheets of A4 paper, scissors
- mini whiteboards and markers

Language support

Ask students to name and describe the shape that they think will be made before unfolding the paper. Encourage them to describe what they are visualising using the vocabulary of shape (including vertex and vertices, sides, and right angles).

 ## Introductory activity

Take a sheet of A4 paper and fold it in half across the width. Make one diagonal cut all the way across the paper and ask students to visualise (imagine) the shape that you will get when you open the paper.

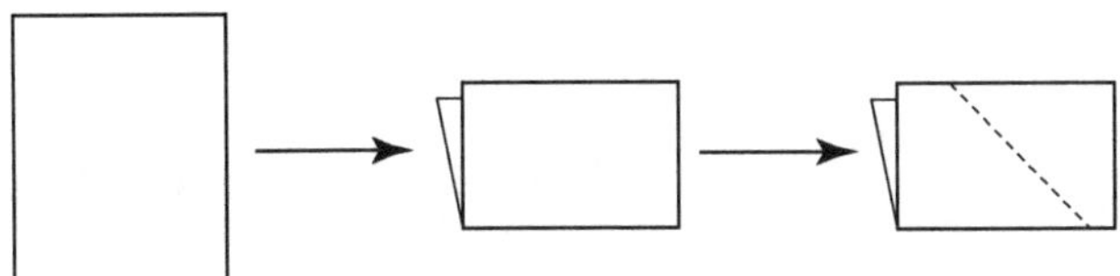

They should sketch this shape on their whiteboards. Open the paper and see which students sketched the correct shape.

Repeat this activity with new pieces of folded paper, asking individual students to make the cuts each time to create different shapes. Other students should predict what shape will be created. With each new piece of paper and shape, students can make one or more cuts after folding the paper, but they must leave some of the fold intact, so they can unfold the shape. Do students get better at visualising the shape as the activity progresses?

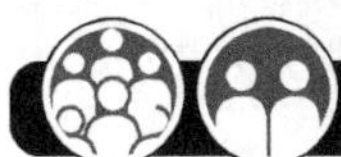 ## Main activity

Discuss what the shapes made in the introduction all have in common: they all have a line of symmetry. Ask students to identify other shapes and objects in the room that have a line of symmetry (e.g. their desk, their chair, the whiteboard, a window, the door). Ask students to sketch some of these shapes and draw in the line of symmetry. *Do any of these shapes or objects have more than one line of symmetry? Look for a vertical line and a horizontal line of symmetry.*

Students should work in pairs on the activity in the Student Book on page 160. Pairs decide between themselves where each cut will go. One student makes the cuts for the first shape and then they take turns.

Ask questions to prompt students, for example: *How many sides does a pentagon have? So how many cuts will you make?*

Ask students to test whether their shapes are symmetrical by folding them along the fold line.

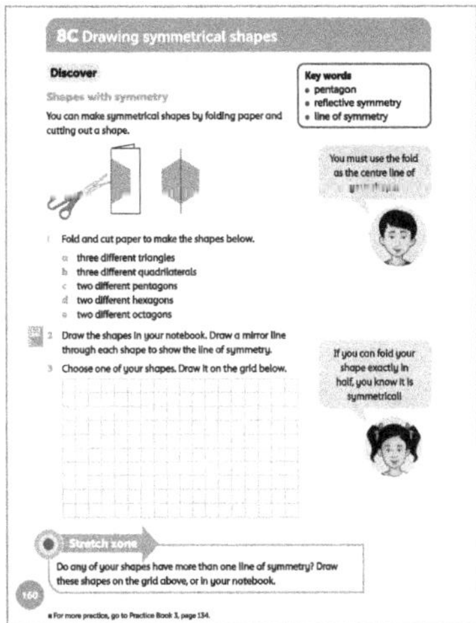

Differentiation

Supporting: Ask students to describe the shape that they think will be made before unfolding the paper.

Consolidating: Ask students to describe how they decided on the cuts they needed for each shape.

Extending: Ask students to find all the possible shapes they can by making just one cut.

Stretch zone: *Do any of your shapes have more than one line of symmetry? Draw these shapes on the grid above, or in your notebook.*

Check that students have recognised lines of symmetry correctly, particularly in any regular shapes, where there will be more than two lines of symmetry.

 ## Reflection time

Ask students to come and draw on the board a shape they made using any number of cuts. Other students should try to work out where the line of symmetry is and what the original shape was. Repeat this for three different shapes.

Practice Book: Students complete Practice Book page 134. They can do this directly after the main activity, as homework, or as the focus of a separate mathematics session to help students consolidate their learning and build fluency.

Students draw shapes that have a line of symmetry. They must follow specific instructions when drawing the shapes, such as side length. *If a square has sides of 4 cm, how long will the horizontal line be either side of the vertical line of symmetry?*

Differentiated outcomes	
All students	should create a range of symmetrical shapes by cutting.
Most students	will predict what shape will be made by the cut.
Some students	may accurately visualise the shapes they are making and describe their properties before they unfold the paper.

Answers

Student Book page 160

Check that students have drawn the required shapes and that they each have a line of symmetry. Ask students to explain how they can tell whether a shape has a line of symmetry or not.

Practice Book page 134

Check that students have drawn the required shapes to a reasonable degree of accuracy: lines should be within 0.5 cm of the correct length and right angles should be approximately correct.

8C Drawing symmetrical shapes

Explore   Student Book page 161 · Practice Book page 135

Specific learning focus

- Draw and complete 2D shapes with reflective symmetry, and draw reflections of shapes.

Global skills

- **Creative skills:** exploring

Key vocabulary

- reflective symmetry, line of symmetry, mirror line, mirror image, pattern, reflection

Resources

- lollipop sticks in jar – each lollipop stick has a different student's name on it
- coloured markers for the whiteboard
- mini whiteboards and markers

Language support

As students are completing the picture of the face, ask questions, such as: *How can you work out where to draw the rest of the mouth? Can you point to the reflection of the ear?* Encourage students to repeat the key mathematics vocabulary in these questions.

 **Introductory activity**

Draw a large square grid on the board with a **mirror line** down the middle. Take a lollipop stick from the jar to select a student to come to the front of the class. The student should shade in a square on one side of the mirror line. Select another student to shade in the square on the other side of the mirror line that is the same distance away. Make sure that you use the vocabulary 'mirror line' and **'mirror image'**. Repeat until several students have had a turn. Tell the class that the completed **pattern** has 'line symmetry': the pattern on one side of the mirror line is a mirror image or **'reflection'** of the pattern on the other side. Remind students that the reflection must be exactly the same distance away from the mirror line as the original pattern.

 **Main activity**

Direct students to look at the 'nose' on the robot's face in question 1 on page 161 of the Student Book. *How is the position of the nose different from the position of the eye?* Students can identify, perhaps with your support, that the nose in the picture is touching the mirror line, rather than being a distance away from it, as is the eye. *Can you see another part of the face that touches the mirror line?* Use the examples of the robot's nose and mouth and help students to visualise how the complete 'shape' would look as if it is on folded paper.

Students should work individually on the activity on page 161 of the Student Book to complete a symmetrical picture and then create their own symmetrical pictures with line symmetry.

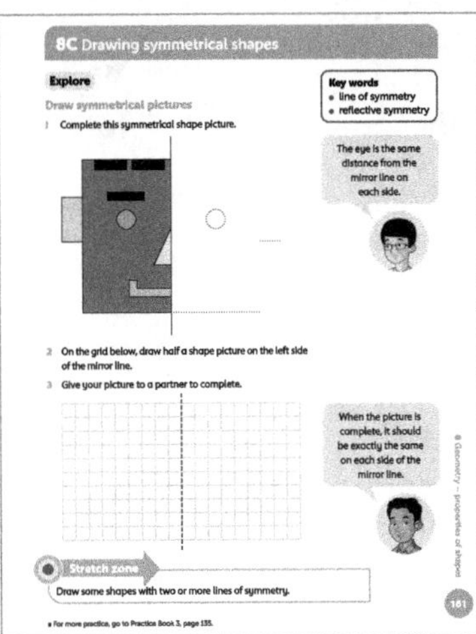

Differentiation

Supporting: Ask students to show you symmetrical shapes in the classroom.

Consolidating: Ask students to explain how they are designing symmetrical patterns. How do they know that they are symmetrical?

Extending: Ask students to find shapes with more than one line of symmetry.

Stretch zone: *Draw some shapes with two or more lines of symmetry.*

Check that students have found shapes that have two lines of symmetry. Have they made a connection with the number of lines of symmetry on regular shapes?

 **Reflection time**

Ask students to draw a picture on their whiteboards of a house that has line symmetry. Select one student to come to the front. They should draw the mirror line and half of the house. *Which features of the house are touching the mirror line? Which are a distance away from it?* The rest of the class should draw the whole house on their whiteboards. Repeat this in pairs with each student drawing half a picture, then swapping whiteboards and the partner completing the drawing. Ask some students to describe what they did to work out how to complete the picture.

Practice Book: Students complete Practice Book page 135. They can do this directly after the main activity, as homework, or as the focus of a separate mathematics session to help students consolidate their learning and build fluency.

Students draw their own symmetrical patterns on squared paper. *Where will you draw the line of symmetry? Does the line of symmetry always need to be vertical or could it also be horizontal?* Encourage more-confident students to draw one pattern with a vertical line of symmetry and one with a horizontal line of symmetry.

Differentiated outcomes	
All students	should understand the line of symmetry as the mirror line.
Most students	will accurately draw pictures and patterns using a line of symmetry.
Some students	may notice shapes with more than one line of symmetry.

Answers

Student Book page 161

Check that students have correctly completed the first symmetrical picture and then drawn the mirror image of their partner's picture. While they are working, ask students to explain how they work out how to complete the pattern on the right-hand side each time.

Practice Book page 135

Check that students have followed the instructions for each pattern and have included the required shapes.

8D Right angles

Specific learning focus

- Use a set square to draw a right angle.
- Compare angles with a right angle and recognise that a straight line is equivalent to two right angles.

Global skills

- **Creative skills**: investigating
- **Interpersonal skills:** team work

Key vocabulary

- set square, right angle

Resources

- straws that bend, or card strips with paper fasteners to make right-angle measurer
- piece of rope or thick cord
- mini whiteboards and markers
- set squares

Language support

Encourage students to tell you which objects make right angles and which objects make angles that are more or less than a right angle. They can do this with the same object, for example: 'These scissors are making a right angle. Now the angle is less than a right angle. Now it is more than a right angle'. Write these model sentences on the board.

 **Introductory activity**

Ask three students to come to the front. Two students should hold opposite ends of the rope, and the third student should hold the middle of the rope. Ask the three students to make a right angle (by making an L-shape with the rope) and observe how they do this.

Ask them to repeat this, but only allow one student to move. This must only be one of the students holding the end of the rope. Explain to students that they need to start with the two ends of the rope touching, one on top of the other and then gradually move one end away, creating an angle where the middle of the rope is.

This models a right angle as a quarter turn. Ask the student who moved the rope end to move all the way through another right angle. This shows that two right angles are equivalent to a straight line. You could also ask them to continue the turn until they are back where they started, to model 'a full turn'.

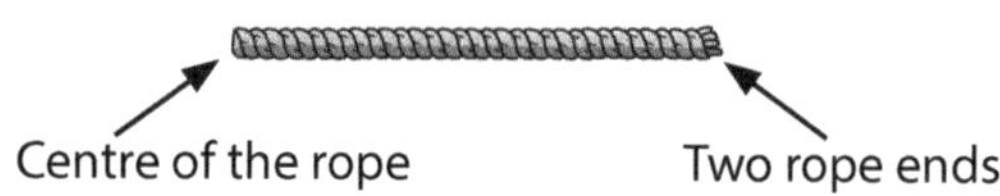

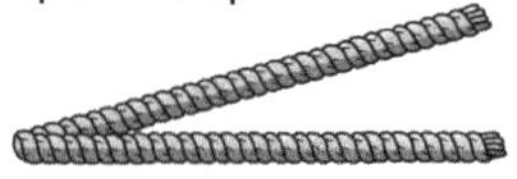

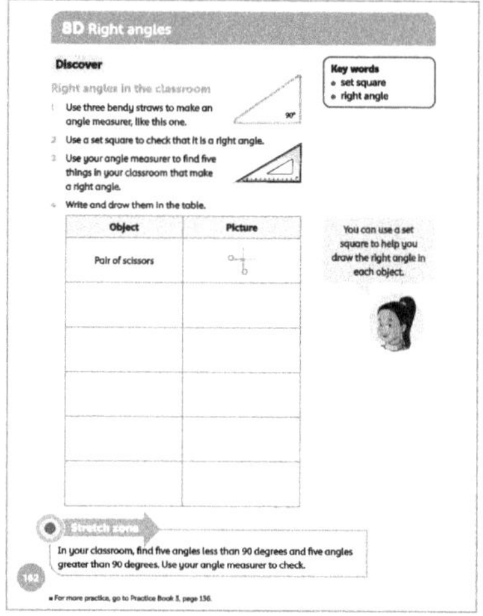 **Main activity**

Students make a right-angle measurer in pairs using bendy straws by following the instructions on page 162 of the Student Book. You can model this with bendy straws. If you prefer, students can make a right-angle measurer by folding a piece of paper. Model this by folding a piece of paper in half, then folding in half again along the first fold to create a right angle.

Students then complete the activity in the Student Book on page 162. They should explore the classroom and find objects that make right angles. Encourage students to find things that turn (e.g. doors, cupboard doors, windows, or the corners of books) as well as right angles in shapes (e.g. the vertices of their book or their desk). They should use a **set square** to draw these images in the table in the Student Book.

Differentiation

Supporting: Ask students to show you right angles around the classroom.

Consolidating: Ask students to make angles with objects that turn. Ask them how they know when to stop turning the object.

Extending: Ask students to estimate angles that are larger or smaller than right angles.

Stretch zone: *In your classroom, find five angles less than 90 degrees and five angles greater than 90 degrees. Use your angle measurer to check.*

Check that students have used their angle measurer accurately to test whether the angles they chose are greater or less than 90 degrees.

 **Reflection time**

Ask pairs of students to give you one example of an object they found and write it on the board. Repeat this until you have at least one object from each pair. Ask students to classify all these objects in some way using their mini whiteboards. For example, they may sort into 'Things that turn' and 'Things that do not turn', or perhaps 'Objects that are moveable' and 'Objects that are fixed'. When they have classified the objects, take feedback from pairs on all the different ways they have found of sorting the objects.

Practice Book: Students complete Practice Book page 136. They can do this directly after the main activity, as homework, or as the focus of a separate mathematics session to help students consolidate their learning and build fluency.

Students draw or take photographs of the objects and then stick them in their books. Alternatively, they may choose to look in magazines for some images. They can cut these out and stick them in their books. *Can you tell me one thing that turns to make a right angle? Can you tell me one object that includes a shape with a right angle?*

Differentiated outcomes	
All students	should recognise right angles.
Most students	will make right angles by turning objects through a quarter turn.
Some students	may identify and check angles that are greater or less than 90 degrees.

Student Book page 162

While they are working, ask students to explain how they can identify right angles. Check that the objects that students have chosen do contain a right angle.

Practice Book page 136

Check that students' drawings or photographs show right angles.

Stretch zone: The definition may include the following. A right angle is an angle of 90°. A right angle is an angle where two perpendicular lines meet.

8D Right angles

Explore Student Book page 163 • Practice Book page 137

Specific learning focus

- Compare angles with a right angle and recognise that a straight line is equivalent to two right angles.

Global skills

- **Creative skills:** investigating
- **Self-development skills:** reflecting on learning
- **Interpersonal skills:** communication

Key vocabulary

- right angle, straight line

Resources

- none required

Language support

Support students with the vocabulary associated with right angles. Make a display of some right angles along with angles smaller than a right angle, and larger than a right angle. Label them appropriately using the words and numbers, for example 'an angle larger than a right angle is more than 90 degrees'.

 **Introductory activity**

Draw several angles on the board. Ask students whether they see any right angles. Ask how they know that they are right angles.

Remind students how they can make a simple right-angle measurer using bendy straws or by folding a piece of paper in half, then folding in half again along the first fold line. Give them some time to do that, then ask a few students to come to the board to check the angles you drew to see which are right angles.

 **Main activity**

Draw a right angle on the board, like an L-shape. Point to the angle in the 'L'. *What do we call this angle?* When they have identified it as a right angle, introduce the right-angle symbol by drawing it on the 'L'. Then ask them to mark all the right angles from the introductory activity with the right-angle symbol. Direct them to look at page 163 in the Student Book and point out the start of question 1, which shows an example of the right-angle symbol being used in a triangle.

Students should work in pairs on the activity in the Student Book on page 163. Students can both check for right angles and each mark them in their own book, using the right-angle symbol. They then see whether they agree for each of the other angles before marking them accordingly with coloured dots. As students are working, ask questions to prompt them. *Is that angle greater than or less than 90 degrees? How do you know? How can you check?*

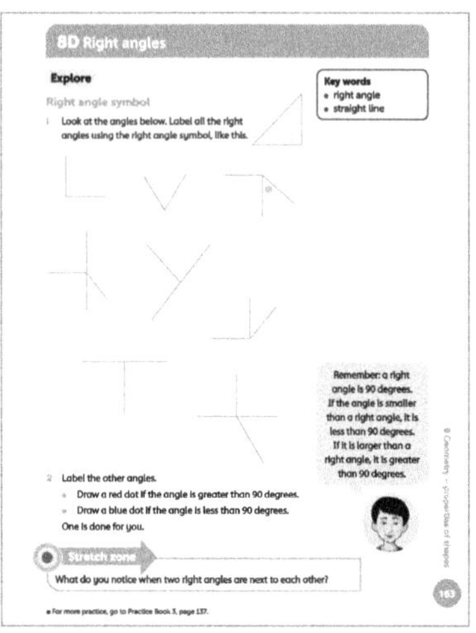

Differentiation

Supporting: Help students to find right angles on each drawing using an angle measurer.

Consolidating: Ask students to mark all right angles with the symbol after they have checked with the angle measurer.

Extending: Ask students to estimate which angles are larger or smaller than right angles before marking them with the appropriate symbol.

Stretch zone: *What do you notice when two right angles are next to each other?*

Check that students understand that two right angles make a **straight line**. *Can you explain why this is?*

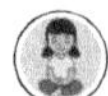 **Reflection time**

Ask students to share their results, indicating which angles on page 163 of the Student Book were right angles, and which were greater than or less than a right angle, and how they knew. *Did you all agree?* Ask students to point out where they can see two right angles together that form a straight line. Did they spot them all? (There are four on page 163.)

Practice Book: Students complete Practice Book page 137. They can do this directly after the main activity, as homework, or as the focus of a separate mathematics session to help students consolidate their learning and build fluency.

Students draw a pattern made out of straight lines. They then label the angles formed by the lines as either 'a right angle', 'an angle less than a right angle' or 'an angle greater than a right angle'. Refer them to the key at the top of the page, which will help them.

Differentiated outcomes	
All students	should identify and label right angles using a measurer with support.
Most students	will identify and label right angles using a measurer.
Some students	may estimate the size of angles greater than or less than a right angle and check with a right-angle measurer before labelling with the symbol.

Answers

Student Book page 163

Check that students have marked all the angles correctly.

Practice Book page 137

Check that students have drawn a suitable pattern and have marked all the angles correctly.

Stretch zone:

An angle less than a right angle is an acute angle.

An angle greater than a right angle (but less than 180 degrees) is an obtuse angle.

8E Lines

Discover
Student Book page 164 • Practice Book page 138

Specific learning focus
- Recognise perpendicular and parallel lines in 2D shapes, drawings and the environment.

Global skills
- **Creative skills**: exploring
- **Self-development skills**: reflecting on learning

Key vocabulary
- parallel, perpendicular

Resources
- mini whiteboards and markers
- rulers, coloured pencils

Language support
Check that students understand the definitions of parallel and perpendicular lines. Model the language needed by saying the following, for example.

- A parallel line is …
- A perpendicular line is …

Make a class poster of the definitions with examples to display.

 **Introductory activity**

*What are **parallel** lines?* Agree that these are two lines that never meet. They are always the same distance apart. Ask students to show what parallel lines look like using their arms.

Ask students to draw pairs of parallel lines in different orientations on their whiteboards. *Can you see any pairs of parallel lines in the classroom?* Ask students each to make a list on their whiteboards of any objects they can see that contain parallel lines (e.g. book covers, window frame, bookshelves). Give them two minutes to do this. Then ask them to swap lists with a partner. Their partner's task is to find and point to the parallel lines on the objects in the list. Pairs can then report their findings to the rest of the class.

 **Main activity**

*What are **perpendicular** lines?* Agree that these are two lines that meet at right angles. Repeat the activity above for perpendicular lines.

Ask students to sketch a square, a rectangle and other quadrilaterals, regular and irregular pentagons and hexagons. Then ask them to identify those with parallel sides and perpendicular sides. Take feedback. Invite students to draw some of these shapes on the board and to point out the two types of lines.

As students work through the activity in the Student Book on page 164, ask questions such as:

- *Can you explain what parallel lines are?*
- *Can you explain in a different way?*
- *Can you explain what perpendicular lines are?*
- *Is there another way?*
- *Where can we see parallel lines in real life?*
- *What about perpendicular lines?*

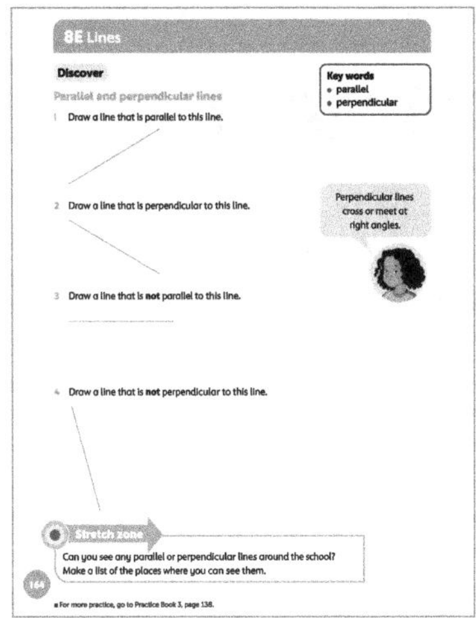

Differentiation
Supporting: Ask students to show you examples of parallel and perpendicular lines.

Consolidating: Ask students to define parallel and perpendicular lines.

Extending: Ask students to tell you which 2D shapes contain parallel and perpendicular lines.

Stretch zone: *Can you see any parallel or perpendicular lines around the school? Make a list of the places where you can see them.*

Check that students have recognised parallel and perpendicular lines correctly.

 **Reflection time**

Ask students to share examples of where they found parallel and perpendicular lines. In each case, can they explain how they know that the lines are either parallel or perpendicular? Ask students to use the internet, or to look in books and magazines to find of pictures of places or buildings that show parallel and perpendicular lines. These can be added to a classroom display.

Practice Book: Students complete Practice Book page 138. They can do this directly after the main activity, as homework, or as the focus of a separate mathematics session to help students consolidate their learning and build fluency.

This activity is best carried out at home, or the research can be done at home and then the activity completed in school. Students look around their home for examples of parallel and perpendicular lines.

Differentiated outcomes	
All students	should identify parallel and perpendicular lines in the environment and use them in their drawings.
Most students	will be able to define parallel and perpendicular lines.
Some students	may use parallel and perpendicular lines to classify a wide range of shapes.

Answers

Student Book page 164

1   Check that students have correctly drawn a line parallel to the one given.

2   Check that students have correctly drawn a line perpendicular to the one given.

3   Check that students have correctly drawn a line not parallel to the one given.

4   Check that students have correctly drawn a line not perpendicular to the one given.

Practice Book page 138

Check that students have drawn appropriate examples of parallel and perpendicular lines they have found around the classroom or their home.

8E Lines

Explore   Student Book page 165 · Practice Book page 139

Specific learning focus

- Recognise perpendicular and parallel lines in 2D shapes, drawings and the environment.

Global skills

- **Creative skills**: exploring
- **Interpersonal skills:** communication

Key vocabulary

- parallel, perpendicular, parallelogram, oblong, rhombus

Resources

- mini whiteboards and markers
- rulers
- coloured pencils

Language support

Check students' understanding of the terms 'parallel' and 'perpendicular' by asking questions, for example: *How do you know that these are parallel lines? How do you know that these are perpendicular? Can you give me another example of where you might see parallel lines? And perpendicular lines?* For additional support, refer students to the poster you made in the previous lesson.

 Introductory activity

Ask pairs of students to draw on their whiteboards different quadrilaterals that have one or two pairs of parallel sides. *Can you name any of these shapes with parallel sides?* You can introduce students to a parallelogram as 'a quadrilateral that has two pairs of parallel sides'. Ask them to identify shapes they have drawn that match this description. Draw some **parallelograms** on the board, including squares, **oblongs** and **rhombuses**. Ask students to show you some of the shapes they have sketched.

 Main activity

Do any triangles have parallel lines? Establish that this isn't possible. Ask students to explain why. (Two parallel lines don't meet. Only shapes with more than three sides can have parallel sides.) Ask students to draw quadrilaterals and triangles with perpendicular sides. Encourage them to be creative with their shapes and not just to draw squares and rectangles. Invite individual students to draw their examples on the board and name them.

Explain the activity on page 165 of the Student Book. First, students draw a picture or a pattern that includes six pairs of parallel lines and six pairs of perpendicular lines. They then have to identify all the pairs of parallel and perpendicular lines in a given picture. They should discuss this picture with a partner and check that they have both found all the sets of lines.

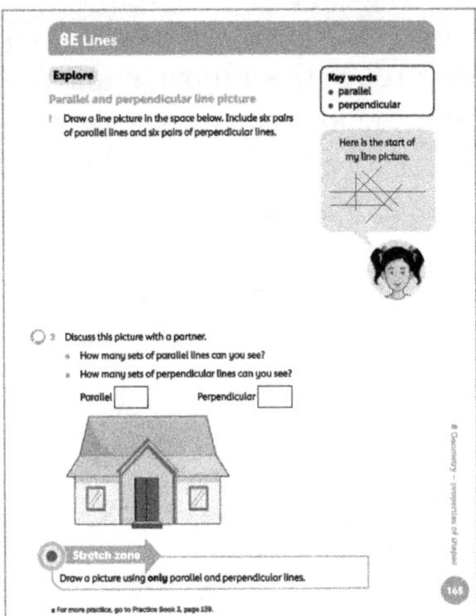

Differentiation

Supporting: Ask students to show you examples of parallel and perpendicular lines in the classroom.

Consolidating: Ask students to define parallel and perpendicular lines.

Extending: Ask students to tell you which 2D shapes contain parallel and perpendicular lines.

Stretch zone: *Draw a picture using only parallel and perpendicular lines.*

Check that students have only used parallel and perpendicular lines in their drawing.

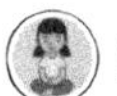 ## Reflection time

Take feedback from the activity. Discuss how many pairs of parallel lines and sets of perpendicular lines students could find. Give each student a piece of plain paper. Ask them to draw a simple pattern using quadrilaterals and triangles that includes at least one set of parallel or perpendicular lines.

Create a class pattern on the board, with students taking it in turns to add either a quadrilateral with parallel or perpendicular sides, a triangle with perpendicular sides, a set of parallel lines or a set of perpendicular lines.

Practice Book: Students complete Practice Book page 139. They can do this directly after the main activity, as homework, or as the focus of a separate mathematics session to help students consolidate their learning and build fluency.

Students find horizontal, vertical, parallel and perpendicular lines in a photograph of the Taj Mahal. They mark the lines on the photograph using different-coloured pens.

Differentiated outcomes	
All students	should identify parallel and perpendicular lines in the drawing.
Most students	will be able to define parallel and perpendicular lines.
Some students	may use parallel and perpendicular lines to classify a wide range of shapes.

Answers

Student Book page 165

1   Students draw their own patterns. Check that they have included six pairs of parallel and six pairs of perpendicular lines.

2   Expect students to find at least ten pairs of parallel lines, and at least eight pairs of perpendicular lines.

Practice Book page 139

Check that students have drawn and coloured the lines in the photograph correctly.

Stretch zone: Answers will vary but students may mention some of the 3D shapes that make up the building (e.g. cuboids). They may mention the columns in front of the building as being cylinders. They may also mention the symmetry of the building.

8 Geometry – properties of shapes

Connect Student Book page 166

Big idea

- I can use the properties of shapes, such as the number of sides, symmetry or the number of faces, edges and vertices, to describe and sort them.

Global skills

- **Creative skills**: exploring
- **Real-world skills**: presenting information
- **Interpersonal skills**: team work

Key vocabulary

- right angle, parallel, perpendicular

Resources

- none needed; digital cameras optional

Language support

Help students to identify the shapes they find and use the correct vocabulary to describe them. Encourage them to describe the properties as well as name the shapes.

 **Introductory activity**

Revise some of the shape names and properties, including symmetry, angles and lines vocabulary. Look together at any posters you have created during the unit. Ask students to find examples of shapes, angles or lines in the classroom.

 **Main activity**

Look together at page 166 of the Student Book. If you have access to an IWB you could use this. Explain that the students will go on a shape walk, either around the school grounds or in the local environment. As they look at buildings, street furniture and objects, they should be looking for shapes they recognise and can name, as well as recognising the properties of shapes and lines (e.g. parallel lines). Use the photographs in the Student Book on page 166 to point out some shapes, lines and angles that students might see around the school.

While on the walk, students should take photographs of the shapes or draw them for follow-up in the classroom. Students can work in small groups using one camera per group. You may find that some groups have similar pictures, but these can be discussed afterwards.

Once back in the classroom, students complete the activity on page 166 of the Student Book. They write about what they found on their walk and should include pictures and details of all the shapes or objects they found. Encourage them to use the correct vocabulary when describing each shape or object, such as: 'The building was a cuboid', or 'The lines on the road were parallel'.

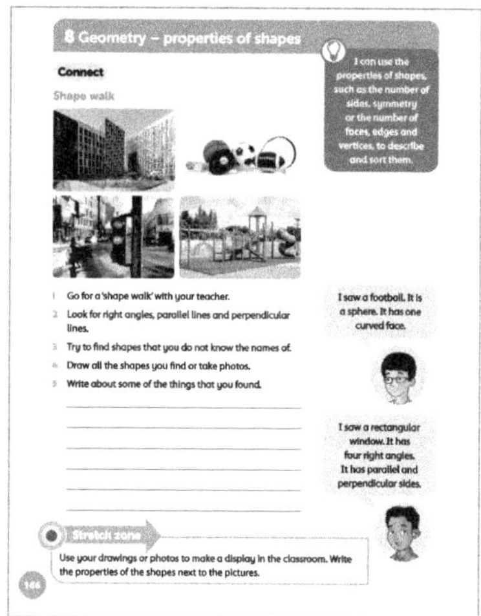

Differentiation

Students should work in mixed-attainment groups so that less-confident learners hear and see appropriate vocabulary and strategies being used.

Stretch zone: *Use your drawings or photographs to make or add to a display in the classroom. Write the properties of the shapes next to the pictures.*

Check that students include properties that reference the number of sides, vertices, edges and faces, as well as right angles and parallel and perpendicular sides.

 **Reflection time**

Bring together the results from the various groups to find out about the shapes they found in the environment. Talk with students about the properties of the shapes and clarify any misconceptions or unknown names of shapes.

Differentiated outcomes	
All students	should name some shapes they see.
Most students	will name the shapes and describe some of the properties of the shapes.
Some students	may describe a shape and its properties in detail.

Answers

Student Book page 166

Check that the shapes are accurately drawn and named.

8 Geometry – properties of shapes

Global skills

- **Self-development skills:** reflecting on learning

Student Book

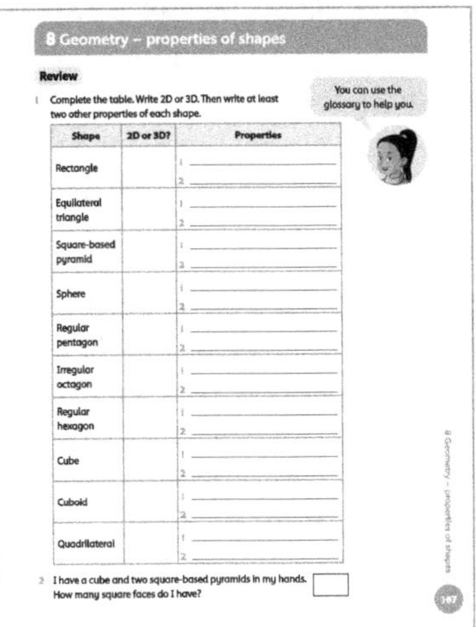

With young children, assessment activities are most effective when carried out as an everyday classroom activity. Students should have access to posters with the names and some properties of shapes, so they can refer to this to support them. You can also make available models of 2D and 3D shapes so that students can handle the physical shapes to explore the properties. They can also refer to the glossary at the back of the Student Book.

Extend by asking questions about the names and properties of different shapes, or ask students to make up their own questions about shapes.

Answers

Student Book page 167

1	rectangle	2D
	equilateral triangle	2D
	square-based pyramid	3D
	sphere	3D
	regular pentagon	2D
	irregular octagon	2D
	regular hexagon	2D
	cube	3D
	cuboid	3D
	quadrilateral	2D

Check that students can correctly distinguish between 2D and 3D shapes. Check that students have included properties relating to each shape's sides, angles, faces, edges, vertices and lines of symmetry, including the number of sides, faces, edges, vertices and lines of symmetry.

2 8 square faces

Practice Book

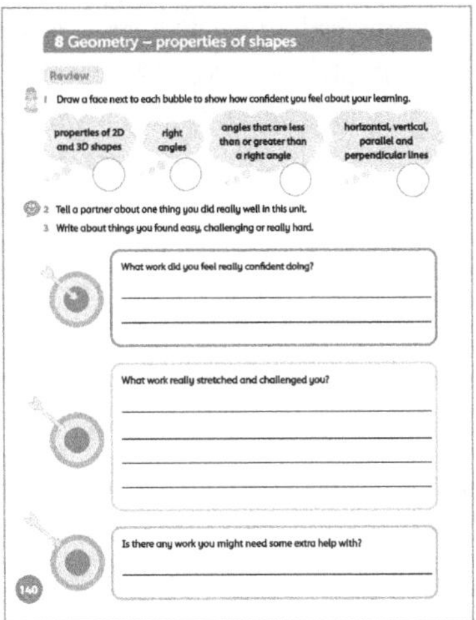

It is appropriate to complete this Practice Book review as a whole-class discussion. You may choose to keep a record of the class discussion or a copy of the review page for your own records. The review provides an opportunity for students to reflect on their learning from the unit, to discuss any areas of mathematics that they feel went particularly well, and any areas that they feel less confident about. Ensure that all students have a copy of the Student Book as a reminder of the areas of mathematics that they have worked on in this unit.

Allow students plenty of time for discussion before asking them to complete the Practice Book page individually, and then, if appropriate, to share their responses with the rest of the class. If students complete this self-assessment at home, encourage them to discuss this with adults. Make a note of areas that students still feel unsure about. As they work, remind them to use the displays. Reassure them that they will become used to the vocabulary as they encounter shapes regularly.

Additional material

There are additional end-of-unit assessments available on the *Oxford Owl for School* website.

9 Geometry – position and direction

Overview

Big idea

The big idea in this unit is that geometry enables us to understand our physical world, through describing it and representing it. Students need to develop a sense of space, which is also called a spatial awareness. This plays an important role in other areas of mathematics and across the curriculum, for example in geography, science and art. Students need to be able to locate and describe the position of objects and their movements, including their own movements. They need the relevant mathematical language to describe position and movement so that they can discuss and visualise.

Early positional language usually focuses on fixed positions. Words and phrases such as 'before', 'after', 'behind', 'in front of', 'on top of' and 'below' are good starting points but need to be developed to include measures and turns. Position and movement games will help students to develop their understanding. These skills play an important role in representing and solving problems in all areas of mathematics and in the real world.

Young students will have discovered through experience that objects, including themselves, can be moved and turned. They will have played with toys, jigsaw puzzles, building blocks and much more, often having to turn the object to place it where they want. It is important that students see angles as a movement (rotation), a measure of turn rather than a fixed property. We can use the movement of a door or window to show that angle is a measurement of the amount of turn. In this way, students see right angles as quarter turns, two right angles as a half turn, and so on.

Look out for

- **Students who only see angles in shapes and not in turns.** Provide lots of opportunities to practise making quarter turns, both clockwise and anti-clockwise, and describing them as right angles.

- **Students who confuse the language of angles.** For example, because they see right angles, they may think that there are 'left' angles. Reinforce the fact that right angles can be seen in many different orientations.

- **Students who confuse clockwise and anti-clockwise.** Model how these terms relate to the movement of the hands on a clock. So, if students visualise the direction the clock hands move, they are visualising clockwise. The opposite direction is therefore anti-clockwise.

Possible misconceptions

- **Students think the order of the coordinates does not matter and so use a mixture of letters and numbers and numbers and letters.** Model for students the principle of reading along the horizontal axis first before reading up along the vertical axis, using reminders such as 'in the house then up the stairs'.

- **Students may think a right angle turn is always from a starting position of north, rather than a right angle turn from the direction being followed.** Guide students to turn maps around to face the same way as they are looking to help with turning right or left in the correct direction.

Key vocabulary

- along, up, column, row, clockwise, anti-clockwise, turn, right angle
- position, above, below, top, bottom, above, below, outside, inside, opposite, middle, in between, in front, behind, front, back, left, right, angle, right angle turn
- to the left of, on top of
- full turn, half turn, quarter turn
- 90 degrees

Coverage in lessons

Learning objective	E	9A	9B	C	R
Identify right angles, recognise that two right angles make a half turn, three make three quarters of a turn and four a complete turn.			✓	✓	
Interpret and create descriptions of position, direction and movement.	✓	✓		✓	✓

9 Geometry – position and direction

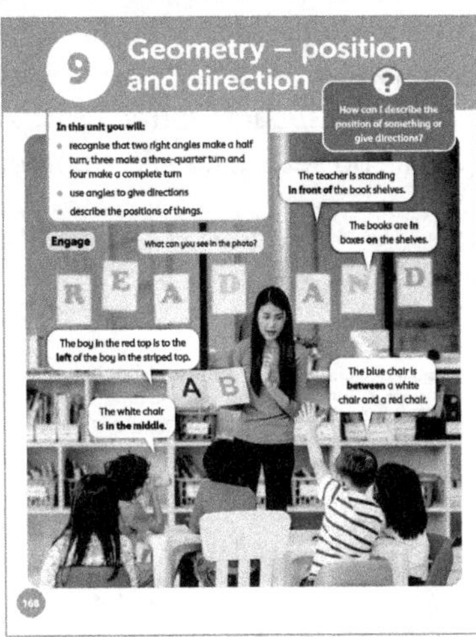

Big question

- How can I describe the position of something or give directions?

Global skills

- **Creative skills:** exploring
- **Interpersonal skills:** communication

Key vocabulary

- position, above, below, top, bottom, outside, inside, opposite, middle, in between, in front, behind, front, back, left, right,
- to the left of, on top of

Resources

- magazines and newspapers

Language support

Use the key vocabulary in a sentence to model the correct use to students. For example, model descriptive sentences about **position**, such as 'the square is to the right of the circle', 'the chair in under the table', 'Archie is sitting opposite Eva'. Students can add words related to position and turns to a class poster and add to it through the unit.

 **Introductory activity**

Look together at page 168 of the Student Book. If you have access to an IWB you could use this. Ask individual students to read out the statements in the speech bubbles, while you point to the objects or images each statement is referring to. Point to the key vocabulary in bold. *This tells us the position of things.* In pairs, students

should look at the image and discuss the relative positions of other objects and people in the picture. Take feedback from each pair and write their sentences on the board. Underline the key vocabulary that they use, which should include:

- position, above, below, top, bottom
- outside, inside
- in front, behind, front, back
- opposite, middle, between
- left, right.

 **Main activity**

In small groups, students should cut out images from magazines and newspapers, to make posters. They then label the images using the vocabulary above to show the relative position of the objects. For example, the football is **to the left of** the goal, the water bottle is **on top of** the table.

Differentiation

Supporting: Ask students questions using key vocabulary, for example: *What is **above** ....?* Encourage them to use some vocabulary of position to describe pictures.

Consolidating: Give students a list of key words and ask them to try to use all the words and a range of vocabulary of position to describe pictures and create posters.

Extending: Students can create their own posters using the key words. Ask them questions that encourage them to use an extensive range of vocabulary of position to describe pictures and create posters.

 **Reflection time**

Each group should present their poster to the whole class. They should read sentences that include the key vocabulary. They can use the examples in the Student Book on page 168. The class should vote for the posters that they want to display for the whole of the unit to help them remember the key vocabulary. Students should complete the glossary at the back of the Student Book using the posters to help them.

9A Using a grid to describe position

Discover 1 Student Book page 169

Specific learning focus

- Find and describe the position of a square on a grid of squares where the rows and columns are labelled.
- Use the language of position, direction and movement.

Global skills

- **Creative skills:** exploring
- **Interpersonal skills:** communication

Key vocabulary

- along, up, column, row

Resources

- lollipop sticks in a jar – each lollipop stick has a different student's name on it
- mini whiteboards and markers

Language support

Provide some simple sentence starters for students to use when giving directions from one place to another, for example: 'First move …, then …'.

Introductory activity

Before the lesson begins, draw a 5 × 5 grid on the board. Label the columns A, B, C, D and E and the rows 1, 2, 3, 4 and 5. Draw five objects in selected squares and then cover each square with a small sheet of paper. Students must try to guess which square the objects are hidden behind by giving you the coordinates of a square. Emphasise that they should say the letter and then the number, so they become used to the order of coordinates. Select individual students by taking lollipop sticks out of the jar. The game is finished when all the objects have been found.

Main activity

Look together at page 169 of the Student Book. If you have access to an IWB you could use this. Ask students to look at the positions of the teacher's desk (C8) and the window (H4). Select one student to give you directions from one of the objects to the other, using the vocabulary from the key words box. For example, to get from the window to the teacher's desk, they might say: 'First move 4 squares **up column** H and then move

5 squares left **along row** 8.' Write these directions on the board and leave them there for students to refer to while they complete the rest of the activity on page 169 of the Student Book, in pairs. Students take it in turns to say how they would get from one object on the grid to another. When their partner agrees this is correct, they can both complete the sentences in their Student Books. While students are working, listen to their directions and check that they are giving correct instructions. Check that they are counting squares correctly: they should not count the starting square for each moment up/down or right/left.

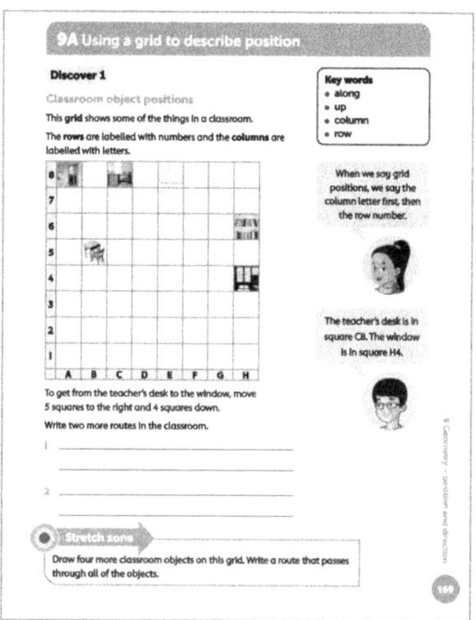

Differentiation

Supporting: Remind students of the order of coordinates.

Consolidating: Students should ask you questions about their grids.

Extending: Students could create their own grids using pictures of plans.

Stretch zone: *Draw four more classroom objects on this grid. Write a route that passes through all of the objects.*

Check that students have written the routes correctly, using the correct vocabulary.

Reflection time

During the main activity, you should prepare a plan of the classroom on the board. Start by drawing a 10 × 10 grid, with columns labelled A–J and rows labelled 1–10. Draw four objects from your classroom on the plan, in the approximate places where they are found in reality. Select an individual student to come to the front and draw another object in approximately the right position. Encourage other students to support them. They may disagree on the position a particular object has been placed. If so, they should give clear instructions, such as 'It should be 2 squares further to the right'. When there are 10 objects on the grid, ask individual students to say one object and then describe the relative position of another object from the first object. Individually, students should write on their own whiteboard which object is being described.

Practice Book: There is no Practice Book page for this lesson.

Differentiated outcomes	
All students	should remember the order of the coordinates with support.
Most students	will remember the order of the coordinates and give directions from one object to another.
Some students	may describe relative positions accurately.

9A Using a grid to describe position

Discover 2 Student Book page 170

Specific learning focus

- Find and describe the position of a square on a grid of squares where the rows and columns are labelled.
- Use the language of position, direction and movement.

Global skills

- **Creative skills:** problem solving
- **Real-world skills:** presenting information
- **Interpersonal skills:** communication

Key vocabulary

- along, up, column, row

Resources

- lollipop sticks in a jar – each lollipop stick has a different student's name on it

Language support

Provide some simple sentence starters for students to use when giving directions from one place to another, for example: 'First move …, then …'.

 **Introductory activity**

Remind students how the grid system can be used to locate objects by specifying their position as being in a row and column, each of which has its own label. In the previous lesson, the horizontal axis was labelled with letters denoting each column of the grid and the vertical axis was numbered for each row.

Draw a grid on the board marked with some objects of your choice spread across the grid (e.g. 2D shapes or coloured dots). Ask students to give you the square reference for each item in turn. You could also ask, *What object is in square F7?* for example.

Student Book page 169

Check that students' directions match the positions of their objects on the grid.

Main activity

Using your grid from the introductory activity, ask students to provide directions to get from one object to another, by describing how many squares to move along to the left or right, or up and down.

Students should work in pairs on the activity in the Student Book on page 170. Remind them of the classroom plan you drew in the previous lesson. They can use ideas from that plan, or they can use some ideas of their own. Explain to students that their plan does not have to be a precise plan of their classroom; it could be their 'ideal classroom'. Refer them to the first speech bubble in the Student Book, which gives further ideas for objects. Also refer students to the second speech bubble so they are clear that each object needs to fit into one square. Students can take it in turns to say how they would get from one object on the grid to another. When their partner agrees that their route is correct, they both complete the sentences in their Student Books.

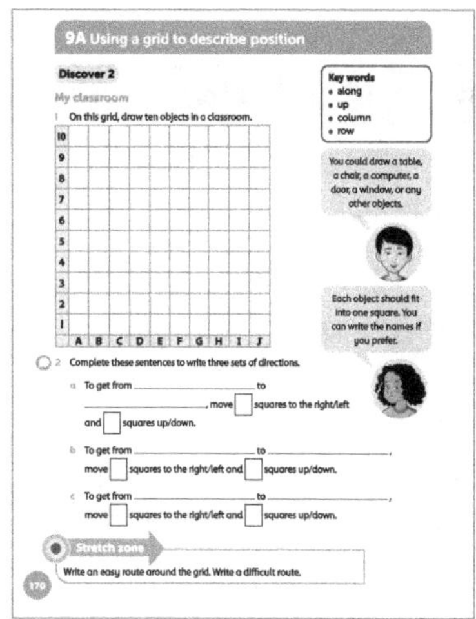

Differentiation

Supporting: Remind students of the order of coordinates.

Consolidating: Students should ask you questions about their grids.

Extending: Students could create their own grids with fictional maps.

Stretch zone: *Write an easy route around the grid. Write a difficult route.*

Check that students have written the routes correctly.

Reflection time

Ask students to share something about their grids. They can describe how they placed their objects and share one of the routes they used to move between objects.

Practice Book: There is no Practice Book page for this lesson.

Differentiated outcomes	
All students	should remember the order of the coordinates with support.
Most students	will remember the order of the coordinates and give directions from one object to another.
Some students	may be able to describe relative positions accurately.

9A Using a grid to describe position

Explore 1 Student Book page 171 • Practice Book page 141

Specific learning focus

- Find and describe the position of a square on a grid of squares where the rows and columns are labelled.
- Use the language of position, direction and movement.

Global skills

- **Creative skills:** problem solving
- **Real-world skills:** interpreting information
- **Interpersonal skills:** communication

Key vocabulary

- along, up, column, row

Resources

- none needed

Language support

Ask students to read their sentences aloud to you and point to the objects in the grid while they do so. They should use the key vocabulary to relate their instructions to movement and position on the grid.

Answers

Student Book page 170

Students grids will vary and so will the directions between grid squares. Check them for accuracy.

If possible, do this activity outside. If this is not possible, the lesson can be easily adapted for the classroom. Before the lesson, draw a large grid on the playground and label it as in the example in the Student Book on page 171. Take the class outside. Ask everyone to stand in a line facing the grid. Ask them to turn through 90 degrees clockwise. Alternate between saying 90 degrees and a quarter turn so that students realise that this vocabulary has the same meaning.

Ask one student to pick a square in the grid for you to stand in. Move to that square. Ask students to take it in turns to give you an instruction, for example: 'Walk 4 steps forward and turn a quarter turn anti-clockwise.' They should try to get you back to your starting point with five instructions. After each move, students should tell you which square you are standing in. If necessary, you can give the first set of instructions, in order to model the use of language for students.

Students should then work in pairs and give each other instructions. They should try to get each other back to the starting point in exactly five moves.

Students work individually on the activity in the Student Book on page 171. Once they have written the grid positions for the animals, they should work with a partner to write sentences to describe the positions, using the writing frame in the stretch zone. They could write, for example: 'The bear is in grid position H6, which is above the eagle' or 'The rhino is in grid position C9, which is two squares to the left of the elephant in grid position E9.'

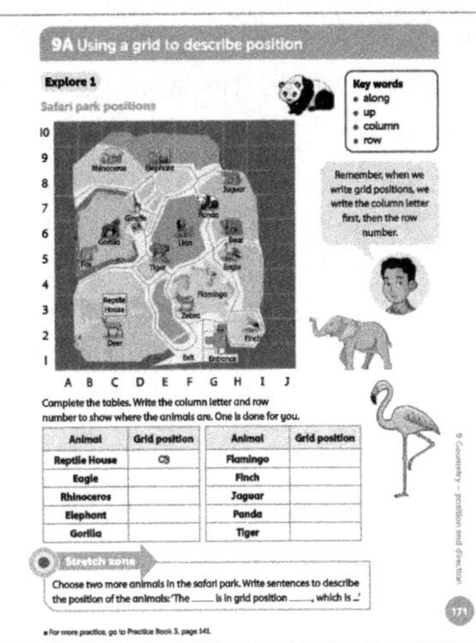

Differentiation

Supporting: You can help write the answers for students as they describe the grid locations. Point to the grid with your finger to help them read the grid reference.

Consolidating: Ask students to use all the key vocabulary when they write sentences about the positions of the animals.

Extending: Ask students to describe journeys from one grid location to another.

Stretch zone: *Choose two more animals in the safari park. Write sentences to describe the position of the animals: 'The ___ is in grid position ___, which is ...'*

Check that students have located the animals correctly and completed the sentences appropriately.

Reflection time

Ask a student to read you one of the sentences without telling you which animal they are describing. The rest of the class can check that they are correct by finding the animal on their grid.

Practice Book: Students complete Practice Book page 141. They can do this directly after the main activity, as homework, or as the focus of a separate mathematics session to help students consolidate their learning and build fluency.

Students draw objects from their bedroom on a grid and then give directions from one object to another. *Try and spread your objects out on the grid so that it is clear how to get from one to another.*

Differentiated outcomes	
All students	should locate grid squares correctly with some help.
Most students	will locate grid squares and complete sentences to describe the locations.
Some students	may locate grid squares and describe journeys between grid squares.

Answers

Student Book page 171

	Column	Row
Reptile House	C	3
Eagle	H	5
Rhinoceros	C	9
Elephant	E	9
Gorilla	C	6
Flamingo	G	4
Finch	I	2
Jaguar	H	8
Panda	G	7
Tiger	E	5

Practice Book page 141

Students' answers will vary. Check that they have written the sentences correctly to match their grid.

9A Using a grid to describe position

Explore 2   Student Book page 172

Specific learning focus

- Find and describe the position of a square on a grid of squares where the rows and columns are labelled.
- Use the language of position, direction and movement.

Global skills

- **Creative skills:** exploring
- **Real-world skills:** presenting information
- **Interpersonal skills:** communication

Key vocabulary

- along, up, column, row, half turn, quarter turn

Resources

- none needed

Language support

Ask students to read their sentences aloud to you and point to the objects in the grid while they do so. They should use the key vocabulary to relate their instructions to movement and position on the grid.

 Introductory activity

Revise the language of position and movement with the whole class. Students should stand up and move according to the instructions you give using 'quarter turn', 'half turn', 'clockwise' and 'anti-clockwise'.

Then choose two students. One will give directions to the other, who will move around the classroom before returning to their seat. The directions should include directions using '**quarter turn**', 'half turn', 'clockwise' and 'anti-clockwise', and moves forward or backwards by a number of steps.

Choose other pairs of students to repeat the activity.

 Main activity

Look together at page 171 of the Student Book. If you have access to an IWB you could use this. Briefly review the safari park activity. Explain that, using the ideas from the previous lesson, students will design their own safari park on a grid. They should think about which animals and birds they wish to have in their park, then draw the enclosures on the grid. Students work individually on the activity in the Student Book on page 172.

Once they have written the grid positions for the animals, they can complete the safari park 'map' with paths, lakes and so on. They should then work with a partner to make up some sentences to describe the positions. Refer students to the first speech bubble, which reminds them to write the labels for the axes of the grid. Explain that the second speech bubble gives an example of how to write directions from one animal to another, using grid references.

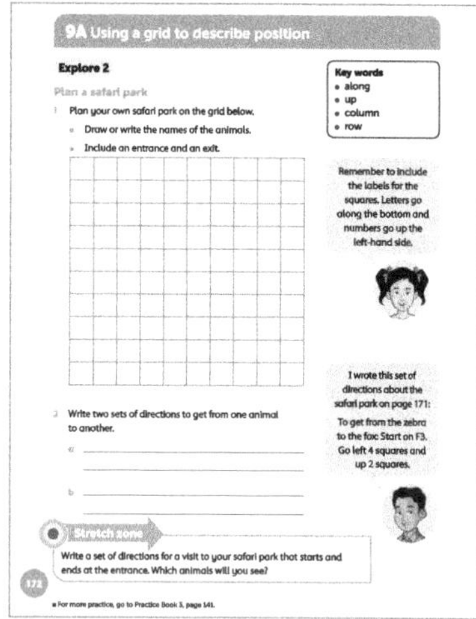

Differentiation

Supporting: You can help to write things down for students as they add animals and birds to the grid. Point to the grid with your finger to help them read the grid reference.

Consolidating: Ask students to use all the key vocabulary of position and direction.

Extending: Ask students to describe journeys from one grid location to another.

Stretch zone: *Write a set of directions for a visit to your safari park that starts and ends at the entrance. Which animals will you see?*

Check that students have devised a set of directions appropriately.

 Reflection time

Ask students to read you one of the sets of directions they have written to move from one area of the safari park to another. Ask them to make the same journey using a different set of directions. Can they find the shortest possible journey? Can they find a journey that passes through exactly 10 squares of the grid?

Practice Book: There is no Practice Book page for this lesson.

<table>
<tr><th colspan="2">Differentiated outcomes</th></tr>
<tr><td>All students</td><td>should locate grid squares correctly with some help.</td></tr>
<tr><td>Most students</td><td>will locate grid squares and complete sentences to describe the locations.</td></tr>
<tr><td>Some students</td><td>may locate grid squares and describe journeys between grid squares.</td></tr>
</table>

9B Using right angles to give directions

Discover Student Book page 173 · Practice Book page 142

Specific learning focus

- Identify right angles, recognise that two right angles make a half-turn, three make three quarters of a turn and four a complete turn.

Global skills

- **Creative skills:** problem solving
- **Real-world skills:** interpreting information
- **Interpersonal skills:** communication

Key vocabulary

- clockwise, anti-clockwise, turn, right angle, half turn, right angle turn

Resources

- Resource sheet 9.1

Language support

Support students in reading the instructions aloud. As students say the instructions, trace with your finger the route in the Student Book on page 173. This will help students to see when they give you a wrong instruction.

Introductory activity

Ask students to all stand up and face the front of the class. Ask them to turn one **right angle clockwise**. Then ask them to **turn** one right angle **anti-clockwise**. *What do you notice?* Now ask the students on one side of the room to turn two right angles clockwise, and those on the other side to turn two right angles anti-clockwise. *What do you notice? What single turn is the same as two right angles?* (a **half-turn**)

Answers

Student Book page 172

Check that students have drawn a suitable safari park on their grid and completed the journeys between animals using correct vocabulary of position.

Discuss how many clockwise **right angle turns** would be the same as one anti-clockwise right angle turn and vice versa.

Main activity

Draw a 10 × 10 grid on the board or, if you have access to an IWB, use a pre-drawn one on the IWB. You could use Resource sheet 9.1 . Explain that this activity is about drawing a route and giving directions to follow it. Ask one student to come to the front. They should choose a starting square and write 'Start' in this square. They should write 'Finish' in another square. Students then take it in turns to give directions for a journey within the grid from the start square to the finish square. Provide students with a writing frame to help them focus on right angles, such as: *Make ___ right-angle turn(s) clockwise/anti-clockwise. Move forward ____ squares.*

The student at the front should follow the instructions and draw the route they followed. Tell students that they must always start facing the top of the grid and can only move in the direction they are facing.

Students should work in pairs on the activity in the Student Book on page 173. One of the pair will draw the journey and the other will write down the description of the route. Students should use one square on the grid to represent one step. They then swap roles: the student who wrote the description follows their own instructions to check that the route is drawn correctly, while the other student makes any necessary corrections to the directions.

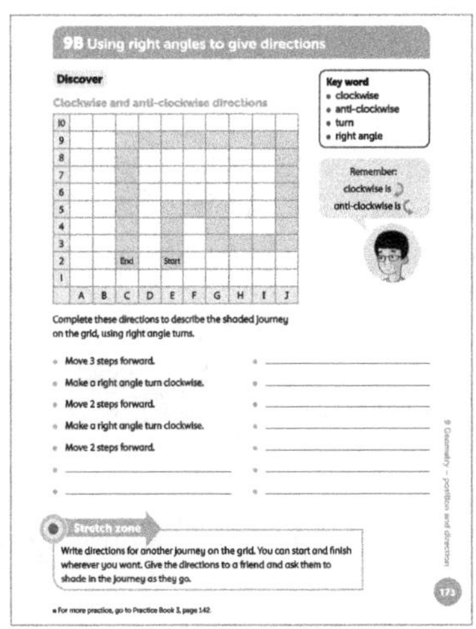

Differentiation

Challenge students to draw the most complicated journeys that they can. This will vary from student to student.

Stretch zone: *Write directions for another journey on the grid. You can start and finish wherever you want. Give the directions to a friend and ask them to shade in the journey as they go.*

Check that students have correctly chosen directions and communicated them accurately.

 **Reflection time**

Clear a space at the front of the classroom and ask pairs to come forward and model the journey they have drawn in their books. One student should read the instructions and the other should follow those instructions.

Practice Book: Students complete Practice Book page 142. They can do this directly after the main activity, as homework, or as the focus of a separate mathematics session to help students consolidate their learning and build fluency.

Students draw a path on a grid and then write the instructions to follow the path through a maze. *Remember you can only move in the direction you are facing so you will need to include clockwise and anti-clockwise turns in your instructions.* There are example instructions for students to refer to.

Differentiated outcomes	
All students	should give instructions for simple journeys.
Most students	will give instructions for more complex journeys.
Some students	may be able to give instructions for very complex journeys.

Answers

Student Book page 173

Turn anti-clockwise through a right angle, or clockwise through 3 right angles.

Take 3 steps forward

Turn anti-clockwise through a right angle, or clockwise through 3 right angles.

Take 6 steps forward

Turn anti-clockwise through a right angle, or clockwise through 3 right angles.

Take 7 steps forward

Turn anti-clockwise through a right angle, or clockwise through 3 right angles.

Take 7 steps forward

Practice Book page 142

Check that students' directions match the route drawn on the grid.

9B Using right angles to give directions

Explore   Student Book page 174 • Practice Book page 143

Specific learning focus

- Identify right angles, recognise that two right angles make a half-turn, three make three quarters of a turn and four a complete turn

Global skills

- **Creative skills:** exploring
- **Real-world skills:** presenting information

Key vocabulary

- clockwise, anti-clockwise, turn, right angle

Resources

- compass (and screen image of compass)

Language support

Support students in writing their instructions and reading them aloud. As they say the instructions, trace with your finger the route they have designed. This will help students to see whether they give you any wrong instructions.

 **Introductory activity**

Display a compass on the board. Give students some time to look at it and discuss what they can see with a partner. Take feedback using question prompts such as: *What can you see? What do you think the letters on the compass mean? Have you seen compasses being used anywhere?* Ensure that students know the compass shows north, south, east and west.

Ask students questions relating to right angle turns to compass direction, for example: *If we are facing north and turn one right angle clockwise, which direction will we be facing?*

Ask students to think of their journey from home to school. Ask them to describe their journey to school to a partner, listing the stages of the journey using language such as 'walk for approximately __ steps/minutes', 'turn one right angle turn clockwise'.

Students should work on the activity in the Student Book on page 174. They should think of the names of buildings you might find in a town, then draw the journey that visits all the buildings and write down the description of the route using the language of right angle turns.

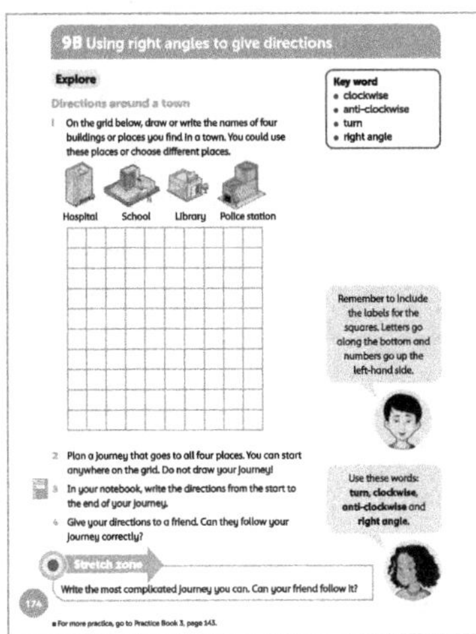

Differentiation

Challenge students to draw the most complicated journeys that they can. This will vary between students. Some may use cardinal directions as well, such as: 'You are facing north. Move forward 20 steps, then make one right angle turn so you are facing east …'.

Stretch zone: *Write the most complicated journey you can. Can your friend follow it?*

Check that students have correctly chosen directions and communicated them accurately.

 ## Reflection time

Choose three students to describe the journey they have written in their books. As they do so, other students could draw the buildings and journey on blank grids. When all three students have finished, ask: *Which journey was the easiest to follow and why? Which journey was the most difficult to follow and why? How do the three journeys that were shared compare with your own? Were they easier or more difficult? Why?*

Practice Book: Students complete Practice Book page 143. They can do this directly after the main activity, as homework, or as the focus of a separate mathematics session to help students consolidate their learning and build fluency.

Students write directions for a journey around a room. *Remember you can only move in the direction you are facing so you will need to include clockwise and anti-clockwise turns in your instructions. The Practice Book also gives a list of vocabulary to use, such as quarter turn, half turn, three-quarter turn.*

Differentiated outcomes	
All students	should give instructions for simple journeys.
Most students	will give instructions for more complex journeys.
Some students	may give instructions for complex journeys that may include cardinal directions.

Student Book page 174

Students will draw their own maps and plan their own journeys. Check them for accuracy.

Practice Book page 143

Students will draw their own room plans and plan their own journeys. Check them for accuracy.

9 Geometry – position and direction

Connect Student Book page 175

Big idea

I can describe journeys and the position of objects using mathematical language such as up, down, left, right, right angle turn, clockwise and anti-clockwise.

Global skills

- **Creative skills:** investigating
- **Real-world skills:** presenting information
- **Interpersonal skills:** team work

Key vocabulary

- clockwise, anti-clockwise, full turn, half turn, quarter turn, right angle

Resources

- map of school grounds (or students can make their own)
- measuring tapes and metre rules
- compasses
- squared paper

Language support

Help students with using the correct language in their sets of directions for their maths trails. Encourage them to use the vocabulary of turns that they have learned, focusing on right angles, clockwise and anti-clockwise, and cardinal directions.

 **Introductory activity**

Draw students' attention to the class poster started in the Engage lesson and added to during the unit. Reflect on the vocabulary they have learned. Ask them for examples of how it can be used in giving directions.

 **Main activity**

Ask students to look at the activity in the Student Book on page 175. Explain that they are going to make maths trails in the school grounds. Tell them that this will be a set of instructions to follow that will lead from one part of the school to another.

Students can work in pairs or small groups and should plan their trail carefully, using metre rules, tape measures and compasses. They might like to find some locations around the school that have shapes, numbers and so

on before planning a route from one to another. They can include steps forward, right angle turns and cardinal directions in their routes but should remember that different people have different step lengths.

Once their trails are completed, students should 'test' them and make any adjustment if they need to. They can then swap trails with another group and try to follow each other's trail.

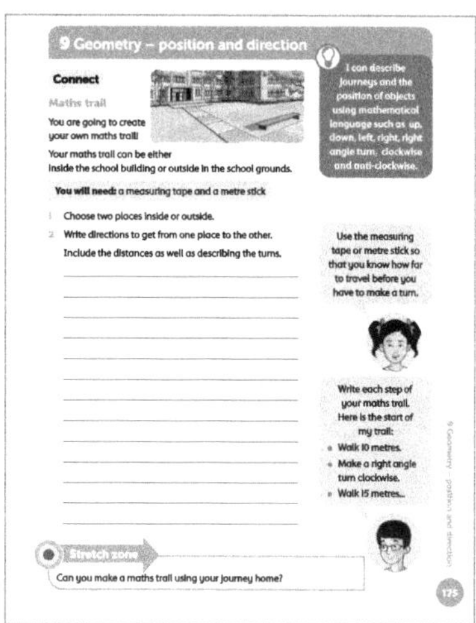

Differentiation

Students should work in mixed-attainment groups for this activity so that less-confident students hear others modelling the correct use of vocabulary. This also means that higher-attaining students have to explain their thinking to others, which consolidates their understanding.

Stretch zone: *Can you make a maths trail using your journey home?*

Check that students have used appropriate directions.

 **Reflection time**

Ask each group to report on how they devised their trail, and how successful their instructions were when they tested the trail. If they followed another group's trail, they can make suggestions for improvements so that they can share the trail with other students in the school.

Differentiated outcomes	
All students	should create a maths trail using simple directions such as turn left, right and turn at a right angle.
Most students	will create a maths trail using more complex directions, for example, clockwise and anti-clockwise.
Some students	will give directions using a wide range of vocabulary including cardinal directions.

9 Geometry – position and direction

Review   Student Book page 176 · Practice Book page 144

Global skills

- **Self-development skills:** reflecting on learning

Student Book

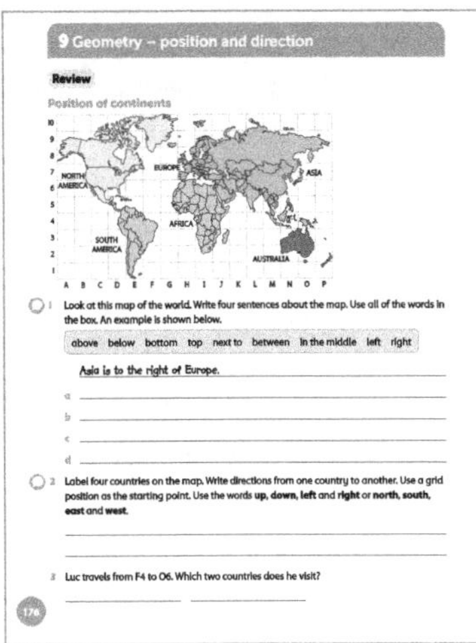

With young children, assessment activities are most effective when carried out as an everyday classroom activity. Students should have available a list of the key vocabulary from the unit to support them, perhaps referring to the class poster.

Watch as students complete the activities that require locations or routes to be identified. Look together at page 176 of the Student Book. If you have access to an IWB you could use this. Check that students understand the language of position and direction such as up, down, left, right and other vocabulary associated with describing position or moving between positions.

Answers

Student Book page 176

1 Check that students have written accurate sentences about the world map.

2 Check that students have used correct language to describe directions between the countries they have selected.

3 Brazil and Japan

Practice Book

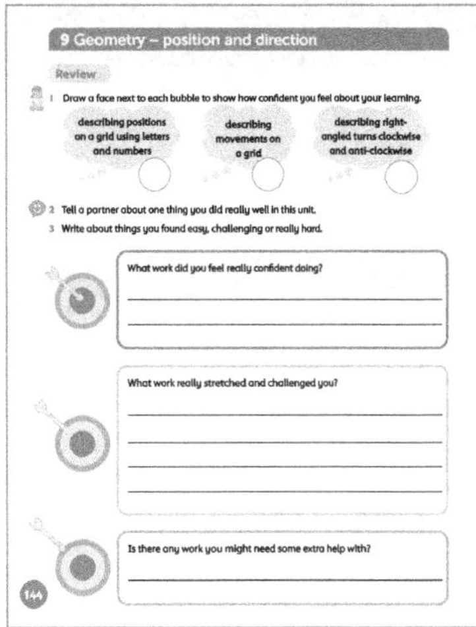

It is appropriate to complete this Practice Book review as a whole-class discussion. You may choose to keep a record of the class discussion or a copy of the review page for your own records. The review provides an opportunity for students to reflect on their learning from the unit, to discuss any areas of mathematics that they feel went particularly well, and any areas that they feel less confident about. Ensure that all students have a copy of the Student Book as a reminder of the areas of mathematics that they have worked on in this unit.

Allow students plenty of time for discussion before asking them to complete the Practice Book page individually, and then, if appropriate, to share their responses with the rest of the class. If students complete this self-assessment at home, encourage them to discuss this with adults. Make a note of areas that students still feel unsure about. These might include, for example, clockwise and anti-clockwise, or which way to turn right on a map if the map is not facing the same way as they are facing.

Additional material

There are additional end-of-unit assessments available on the *Oxford Owl for School* website.

10 Statistics

Big idea

The big idea that underpins data handling at all levels is that data handling is a cycle. This process is as follows.

1 Define the problem and identify the key question.

2 Collect data that can answer the question.

3 Process and represent the data so it can be interpreted and analysed.

4 Interpret the data.

Students should be engaged in defining their own questions at all stages so that they come to understand the different processes involved in data handling.

In this unit students are introduced to bar charts, Venn diagrams and Carroll diagrams as ways of representing data, and to tally charts and frequency tables as ways of recording data. Several of the activities run together in order to ensure that students understand how the stages of the data-handling cycle follow on from each other.

Look out for

- **Students who see each stage of the data-handling process as separate. This means that they do not 'own' the data, and so cannot make inappropriate decisions about how to represent the data. They may also interpret the data in a fairly shallow way as they are trying to answer a question that is in the teacher's head rather than a question that they have become interested in themselves.** To help with this, during the unit students will set up a role-play shop scenario to help them in collecting and processing data themselves, allowing them to be part of the complete data-handling cycle.

- **Students who do not record data accurately in tables because they are using tally marks incorrectly.** Model the tally method for students, showing them how to make the 5-bar gate, crossing four tallies with a fifth one to make counting easier.

Possible misconceptions

- **Students confuse the lines and the spaces when plotting data on bar charts.** Help students by modelling how to label the axes for a bar chart correctly: on the y-axis labelling the lines that cross the axis and not labelling the spaces between the lines; and on the x-axis labelling the spaces where the bar will go, not the lines that the bar goes between.

- **Students assume that one picture in a pictogram always represents one of something instead of recognising that the pictures may represent multiple data.** Give students examples of pictograms where the picture represents more than one item of data and help them to interpret the pictograms by using a key.

Key vocabulary

- count, tally, sort
- graph, pictogram
- list, chart, bar chart, tally chart
- table, frequency table
- Carroll diagram, Venn diagram, criteria
- label, title, axis, axes
- most popular, most common, least popular, least common
- criteria, sort

Coverage in lessons

Learning objective	E	10A	10B	C	R
Interpret and present data using bar charts, pictograms and tables.	✓	✓		✓	✓
Solve one-step and two-step questions using information presented in bar charts and pictograms and tables.		✓			✓
Continue to interpret data presented in many contexts and representations.			✓		✓

10 Statistics

Engage Student Book page 177

Big question

- How can I collect, organise and interpret information to help me find things out?

Global skills

- **Creative skills:** exploring
- **Real-world skills:** interpreting information

Key vocabulary

- tally, frequency table, pictogram, bar chart

Resources

- none needed

Language support

Use the key phrases in the Student Book on page 177. The important vocabulary includes 'favourite' (*'What is your favourite?'*), and 'organise' (*'How do the shops organise their items?'*). You should also introduce the word 'frequency' to describe how often something occurs.

Introductory activity

If possible, plan to take students on a visit to a local shopping mall at the start of this unit.

In the classroom, introduce the unit by telling students that they will be exploring real-life data. As a prompt, look together at page 177 of the Student Book. If you have access to an IWB you could use this.

In pairs, students should list all the different shops, then the restaurants or cafés, that they visit. Collect all this data, then organise a vote to find out the five most popular shops and restaurants or cafés.

Main activity

Ask students how they could represent the data about the five favourite shops and restaurants or cafés from the introductory activity. Students should be able to say '**pictogram**' and 'block diagram'. Create a whole-class pictogram of this data and then ask how useful this is. Ask students to come up with questions about the data that they can answer using the pictogram.

Differentiation

As this is a discussion-based activity, it is expected that all students will take turns to speak and express their ideas, although some students will be able to write questions about the data.

Reflection time

Ask students to share some of the questions they came up with, during the main activity, about the five favourite shops and restaurants or cafés. *How easy is it to answer these questions using the pictogram? Could there be a different way of representing the data that could make it clearer?*

Tell students that later in the unit they will be setting up an imaginary shop. At this point, they should discuss what type of shop they will want to set up. *Does the information in the class pictogram help you to decide what would be a good type of shop to set up?*

10A Bar charts, frequency tables, pictograms

Discover 1
Student Book page 178 · Practice Book page 145

Specific learning focus
- Answer a real-life question by collecting, organising and interpreting data.
- Draw and interpret pictograms.

Global skills
- **Creative skills:** problem solving
- **Real-world skills:** presenting information

Key vocabulary
- data, key, pictogram, tally

Resources
- mini whiteboards and markers
- class list of names

Language support
Make sure that the vocabulary of 'tally chart' and 'pictogram' is displayed. It would be helpful to make large versions of the definitions of these terms to display, and to add to these throughout the unit.

 **Introductory activity**

Take students outside or to a large open space. Ask them to line up in order of birthday. The person with the birthday nearest to 1 January should be on the left and the person with the birthday nearest to 31 December should be on the right. When they are all lined up, ask them to form lines to represent a 'human pictogram', classifying birthdays by month. Remind students that a pictogram is a chart that shows **data** using pictures or images.

Demonstrate how you can count the number of students with a birthday in a particular month, and keep track of the count, by using a **tally** that shows 'sticks'; after four sticks are drawn there is a strike-through for the fifth. Repeat for the other months and ask students to keep a tally of the numbers on their mini whiteboards. Remind them that this is called a tally chart.

 **Main activity**

In the classroom, demonstrate how the information can be represented in a pictogram in the Student Book on page 178. They should each create their own pictogram, using simple 'stick images', or emoticon-type faces for the images. Discuss what each picture represents – students will probably be using 1 picture = 1 person at this stage – and say that they should add this to the **key** at the bottom of their pictogram. Some students might use 1 picture = 2 (or 5) people. If not, mention that this could be used. Refer students to the first speech bubble and ask them to add a title as well as a key to their pictogram.

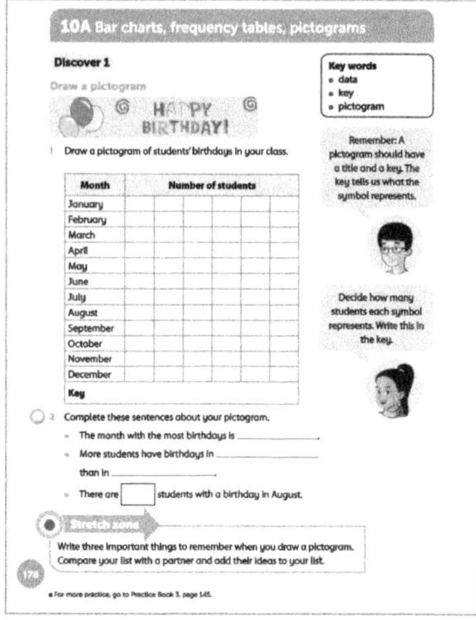

Differentiation

Supporting: Support students by modelling how to use the collected data to create the pictogram.

Consolidating: Ask students whether they can think of any other questions they could ask or answer about the pictogram.

Extending: Ask students to support those who are less confident so they will consolidate their understanding by forming questions and helping others to find the answers on the pictogram.

Stretch zone: *Write three important things to remember when you draw a pictogram.*

Compare your list with a partner and add their ideas to your list.

Listen to students' responses and check for accuracy, using correct vocabulary.

 **Reflection time**

Take feedback from students to check that they all have the correct data in the form of the pictogram. Model how you use this pictogram to interpret the data by asking questions, for example: *Which month had the most birthdays? Which month had the fewest birthdays? How many more birthdays are there in _______ than _______?* Where some students were able to write more questions, they can be asked to share them with the class.

Practice Book: Students complete Practice Book page 145. They can do this directly after the main activity, as homework, or as the focus of a separate mathematics session to help students consolidate their learning and build fluency.

Students draw a pictogram to represent the different-coloured cars that passed a house in one hour. Remind them to write a title and complete a key for their pictogram.

Differentiated outcomes	
All students	should create pictograms about class birthdays with support.
Most students	will accurately create pictograms independently.
Some students	may be able to model creating pictograms and interpreting them for others.

Answers

Student Book page 178

Answers will vary according to students' birthdays and names. Check that that pictogram and questions have been completed correctly.

Practice Book page 145

Check that students have drawn the correct pictogram for the car-colour data in the table.

Stretch zone: Answers will vary but could include these.

There were 5 more black cars than grey cars.

There were 3 more white cars than silver cars.

10A Bar charts, frequency tables, pictograms

Discover 2 Student Book page 179 • Practice Book page 146

Specific learning focus

- Answer a real-life question by collecting, organising and interpreting data.
- Use tally charts, frequency tables and bar charts.

Global skills

- **Creative skills:** problem solving
- **Real-world skills:** presenting information
- **Self-development skills:** reflecting on learning
- **Interpersonal skills:** communication

Key vocabulary

- data, frequency table, bar chart, axes

Resources

- class list of names and birthdays

Language support

Make sure that the vocabulary of 'tally chart', 'frequency table' and '**bar chart**' is displayed. It would be helpful to make large versions of the definitions of these terms to display throughout the unit.

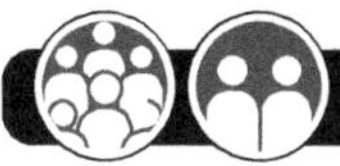

Introductory activity

Give each student a list of names of everyone in the class and their birthday month. Use the data collected in the previous lesson to remind students how you can count the number of students that have a birthday in a

particular month and keep track of the count by tallying: using groups of up to four 'sticks' and then a strike-through for the fifth. Write the names of the months on the board and complete a class tally chart of birthdays.

Demonstrate how the information in the tally chart can be turned into a frequency table by writing the numbers for each tally in a new column.

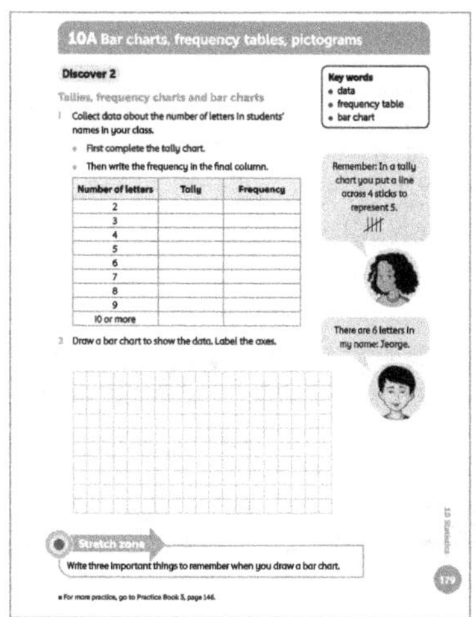

Main activity

Show how the information can be shown as a bar chart where each bar is the height of the number of students with a birthday in a particular month. Show students how to construct a bar chart. *First draw the **axes**. Write the months along the bottom (the horizontal axis) and the numbers on the vertical axis. Complete the bar chart by drawing the bars for each month. Don't forget to add a title to the bar chart.*

Students should complete the tally chart and frequency table in the Student Book on page 179. They then complete a bar chart with the information they have collected. Students work in pairs, with one partner counting the number of letters for each name and the other completing the tally chart. They count the tally to produce a frequency table. They then discuss the bar chart and both draw this in their own books.

Differentiation

Supporting: Help students collect data to create tally charts and model 'bundling' in fives.

Consolidating: Ask students to describe how they used data to create bar charts.

Extending: Ask students to support other less-confident students so they will consolidate their understanding.

Stretch zone: *Write three important things to remember when you draw a bar chart.*

Ask students to compare their list with a partner's list. Students add each other's ideas to their lists.

Listen to students' responses and check for accuracy using correct vocabulary.

 **Reflection time**

Take feedback from students to check that they all have the correct data, and then focus on interpreting the data using questions such as: *How many students have two letters in their name? How many have seven? What is the most frequent number of letters in all your names? What is the least frequent?*

Practice Book: Students complete Practice Book page 146. They can do this directly after the main activity, as homework, or as the focus of a separate mathematics session to help students consolidate their learning and build fluency.

Students consolidate their learning from the main activity, using a page from their favourite book to generate the words, then counting the number of letters in each word. They then complete a tally and frequency chart to show the data.

Differentiated outcomes	
All students	should create a tally chart about the number of letters in names with support.
Most students	will create tally charts and frequency tables accurately and independently, and bar charts with support.
Some students	may model creating bar charts for others.

Answers

Student Book page 179

Answers will vary according to students' names. Check that each type of chart and table has been completed correctly for its type.

Practice Book page 146

Check that students have tallied correctly (in groups of five, with four 'sticks' and the fifth striking the others through) and that they have counted them up to write the correct frequency for each number of letters.

10A Bar charts, frequency tables, pictograms

Explore 1   Student Book page 180 · Practice Book page 147

Specific learning focus

- Answer a real-life question by collecting, organising and interpreting data.
- Use tally charts to record and interpret data.

Global skills

- **Creative skills:** investigating
- **Real-world skills:** presenting information
- **Interpersonal skills:** communication

Key vocabulary

- data, frequency table, tally chart, axis, axes

Resources

- none needed

Language support

Refer students to the tally, frequency and bar charts that were produced in the previous lesson to remind them of the key vocabulary.

 **Introductory activity**

Look together at page 177 of the Student Book to remind students of the shops discussed in the Engage lesson. If you have access to an IWB you could use this. Introduce the main activity by explaining to students that they need to imagine that they are setting up their own shop. Use the results of the class vote in the Engage lesson and use this type of shop as an example of a shop they are going to set up. For example, if they voted a clothes shop as their favourite kind of shop in the Engage lesson, use this to model the main activity.

If you are setting up a clothes shop, what sort of information might you need to know before setting up the shop? Have a class discussion and agree on information such as the age group students are appealing to, whether they will sell men's, women's and children's clothes, or specialise in one type of clothing. Will they sell sportswear, casual clothes or formal wear? Based on who they are appealing to, what days of the week and at what times are these people likely to visit the shop?

Main activity

Look together at page 180 of the Student Book. If you have access to an IWB you could use this. Explain that, in groups, they must first decide on the type of shop they would like to set up and then think of some good questions that would give them information they might need to know before setting up their shop, as modelled in the introductory activity. *What sort of shop would you like to set up? What information would be good to know before you decide what to sell?*

In their small groups, students discuss the possible questions they might ask. This could be finding out favourite items to sell, comparing prices, thinking about when people are likely to visit and so on. Take feedback and list all the different questions on the board. Students should decide in their groups to explore the question they are most interested in. They then complete a tally chart to collect information to answer their question. For example, if they want to find out what clothes people are most likely to buy, they could list different items of clothing or type of clothes: sports clothes, beachwear, casual clothes, smart clothes/formal wear, designer clothes and so on. They then collect data in their **tally chart** by asking other students in the class.

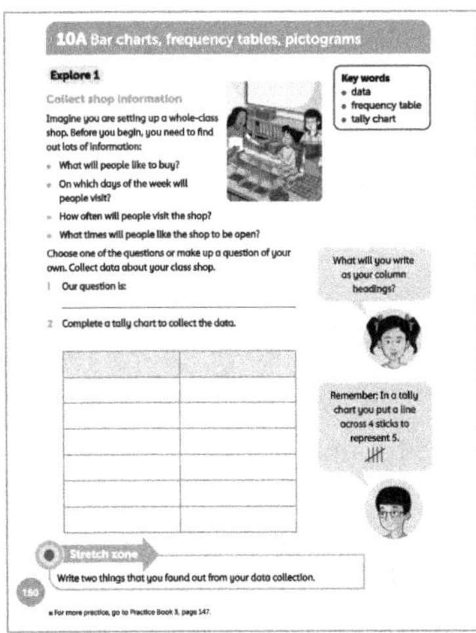

Differentiation

All members of the small group should work together to collect the data.

Stretch zone: *Write two things that you found out from your data collection.*

Check that students are interpreting the data correctly.

 **Reflection time**

Each group should share their findings with the whole class. Ask for feedback on the findings using the 'two stars and a wish' strategy (students tell one another two things they really liked about each other's tally charts, and one thing they would change).

Practice Book: Students complete Practice Book page 147. They can do this directly after the main activity, as homework, or as the focus of a separate mathematics session to help students consolidate their learning and build fluency.

Students complete a bar chart based on the data they collected on the numbers of letters in words from a page of their favourite book. *What must you remember to include on your bar chart?* (They must include a title, and labels for both axes.) You may need to discuss this activity first with students, pointing out any vocabulary that is new to them, such as 'y-**axis**' and 'axes'. You may also want to discuss the scale they could use. This will depend on the number of words that were on the page they chose. Alternatively, this activity could be completed once students have completed in the Explore 2 lesson.

Differentiated outcomes	
All students	The work should be distributed between the group with the more-confident students supporting the less-confident ones.

Answers

Student Book page 180

Answers will vary according to students' questions and the data collected. Check that the tally chart has been completed correctly.

Practice Book page 147

Answers will vary according to the data collected for the activity on page 146 of the Practice Book. Check that the height of each bar in the chart corresponds to the information in the completed table on page 146, and that an appropriate scale is used.

Stretch zone: Check that students have interpreted the information in the bar chart correctly. They should be able to say which length of word appears the greatest number of times on the page and which appears the fewest number of times on the page. They may also make comparisons between two or more lengths of words.

10A Bar charts, frequency tables, pictograms

Explore 2
Student Book page 181 • Practice Book page 148

Specific learning focus
- Answer a real-life question by collecting, organising and interpreting data.
- Use bar charts to record and interpret data.

Global skills
- **Creative skills:** investigating
- **Real-world skills:** presenting information
- **Interpersonal skills:** communication

Key vocabulary
- bar chart, frequency table, axis/axes

Resources
- none needed

Language support
Use the bar charts that were produced in the Discover 2 lesson to remind students of the key vocabulary. Introduce the vocabulary of 'axis' and 'axes' when you ask them to 'label the bar chart'.

Introductory activity

Students should have already decided on the questions they will ask about what they are going to sell in their shop, and collected data using a tally chart.

Discuss the data they have collected and ask them about the information they found out from their tally charts. *Was anything surprising? Did you find anything you thought was interesting? How could you present this information other than in a tally chart?* If no one mentions it, explain that a frequency chart and a bar chart are both useful ways of presenting this sort of data.

Main activity

Look together at page 181 of the Student Book. If you have access to an IWB, you could use this. Explain to students that they now need to construct a frequency table and then a bar chart from the data they collected in the Explore 1 lesson. Discuss how they will construct their bar charts. Model this for students with step-by-step instructions and discuss the vocabulary such as 'axes' and 'axis'. *What will you write on each of the axes? What scale will you use on the vertical axis?* and so on. Students should work individually or in pairs to complete the charts.

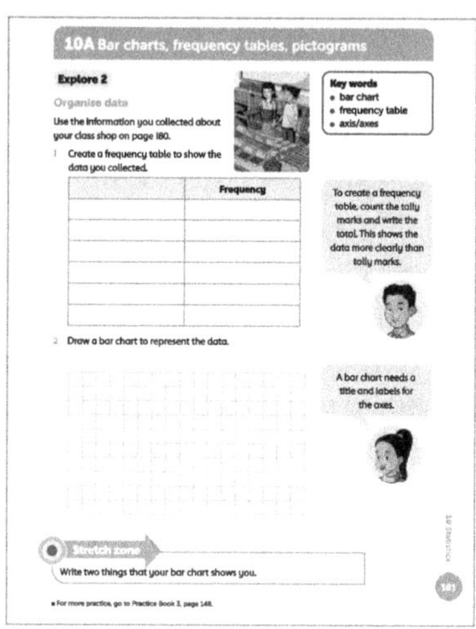

Differentiation
Supporting: Help students to create their frequency table and bar chart.

Consolidating: Ask students to describe how they used their data to create the frequency table and bar chart.

Extending: Challenge students to interpret their bar chart to answer the questions they had devised in the Explore 1 lesson.

Stretch zone: *Write two things that your bar chart shows you.*

Check that students can interpret the bar chart correctly.

 Reflection time

Students should share their findings with the whole class. Choose several students to do this and ask for feedback on the findings using the 'two stars and a wish' strategy (students tell one another two things they really liked about the bar charts, and one thing they would change).

Practice Book: Students can complete Practice Book page 148. They can do this directly after the main activity, as homework, or as the focus of a separate mathematics session to help students consolidate their learning and build fluency.

Students start a new data-handling cycle in their Practice Book, based on finding out the most popular types of exercise. *Who are you going to ask? What question will you ask them? How will you collect this data?*

Differentiated outcomes	
All students	should create a frequency table and bar chart with support.
Most students	will create frequency tables and bar charts for their shopping data.
Some students	may model creating frequency tables and bar charts for others and begin to interpret their own charts.

Student Book page 181

Answers will vary according to the data collected. Check that each type of chart has been completed correctly for its type and the data.

Practice Book page 148

Answers will vary according to the data collected. Check that the frequency table has been completed correctly for its type and the data, and that students have used it correctly to identify three facts about the data.

10A Bar charts, frequency tables, pictograms

Explore 3 Student Book page 182 • Practice Book page 149

Specific learning focus

- Interpret data on a pictogram.

Global skills

- **Creative skills:** exploring
- **Real-world skills:** interpreting information

Key vocabulary

- pictogram, key, interpret

Resources

- none needed

Language support

Use the pictogram in the Student Book on page 182 to remind students of the key vocabulary. Reinforce the vocabulary of 'frequency' when you ask students to draw a pictogram from the frequency table made previously.

 Introductory activity

Discuss with students how a key works on a pictogram. Conduct a quick show of hands to gather data on, for example, students' favourite season of the year. Suppose there are 2 students who prefer winter, 4 for spring, 12 for summer and 7 for autumn.

Choose a symbol to represent the numbers of students, perhaps a smiley face like an emoji. Explain that each face could represent one student, so there would be 2 faces for winter, 4 for spring, 12 summer and 7 autumn. Or, each face might represent 2 students, so there would only be one smiley face to represent the two students who prefer winter. Ask students how many faces there would be for each of the other seasons.

 Main activity

Direct students to look at the pictogram on page 182 of the Student Book. Just from looking at the pictogram, ask students whether they can tell you what information it is showing. They should be able to tell from the heading that the pictogram is about lunch choices.

Now ask them to focus on the key at the bottom of the pictogram. How many students does each blue spot represent? (2) Refer students to the speech bubble if they need a hint at this stage. Now ask students to **interpret** the data by looking at the number of spots against each lunch choice. For example, for vegetable lasagne, there are $2\frac{1}{2}$ blue spots, so that makes 5 students who prefer the lasagne.

By counting the blue spots and using the key, students answer the questions on page 182. Once finished, they can try making up their own questions about the data in the pictogram.

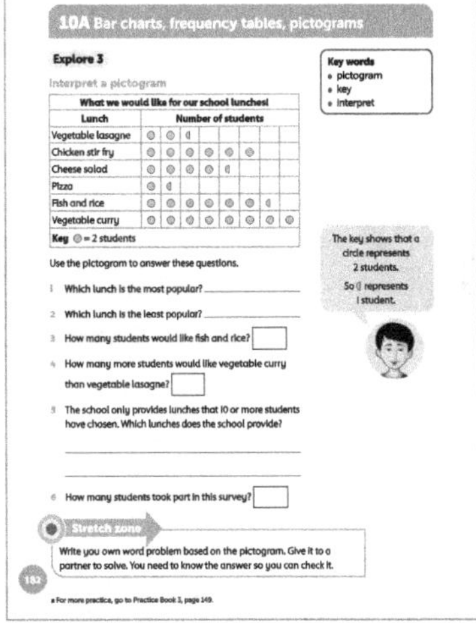

Differentiation

Supporting: Ask students to read the data in the pictogram to see which lunches were most and least popular.

Consolidating: Ask students to tell you everything they can about the pictogram.

Extending: Ask students to support other less-confident students to interpret the data so that they will consolidate their understanding.

Stretch zone: *Write your own word problem based on the pictogram. Give it to a partner to solve. You need to know the answer so you can check it.*

Check that students write appropriate word problems for the data in the pictogram and can give the correct answer.

 ### Reflection time

Ask students to share what they found out from the data. *How did you work out how many students preferred each lunch option? Can you use the pictogram to put the lunches in order of popularity? How many students were asked about their lunch preferences?*

Practice Book: Students complete Practice Book page 149. They can do this directly after the main activity, as homework, or as the focus of a separate mathematics session to help students consolidate their learning and build fluency.

Students continue with their exercise survey as part of their data-handling cycle. This time they construct a pictogram to represent the data collected on the previous page and then write some sentences about their pictogram.

Differentiated outcomes	
All students	should recognise what the pictogram is showing.
Most students	will be able to interpret the data using the key.
Some students	may ask deeper questions about the data.

Answers

Student Book page 182

1 Vegetable curry

2 Pizza

3 13

4 11

5 Chicken stir fry, fish and rice, vegetable curry

6 58

Practice Book page 149

Students draw their own pictogram based on data they collected. Check that their pictogram and their sentences match their data.

Check that they have used the terms 'fewer' and 'more' correctly to describe the data in their pictogram. For example, their sentences might include: 'More people like swimming than cycling' or 'Fewer people like to run than to walk.'

10A Bar charts, frequency tables, pictograms

Explore 4   Student Book page 183 · Practice Book page 150

Specific learning focus
- Use bar charts to interpret data.

Global skills
- **Creative skills:** exploring
- **Real-world skills:** interpreting information

Key vocabulary
- bar chart, data, interpret

Resources
- none needed

Language support

Use the bar charts that were produced in previous lessons to remind students of the key vocabulary. Remind students of the vocabulary of 'axis' and 'axes'.

 ### Introductory activity

Choose a bar chart that has been completed previously and display it for the class to see. If you have access to an IWB you could use this. Discuss with students the key features of bar charts, as they have explored previously. Ask them to describe the axes and how they are labelled. Remind them that the bars need to be separate from each other as the data represents distinct categories.

 ### Main activity

Look together at the bar chart on page 183 of the Student Book. If you have access to an IWB you could use this. From looking at the bar chart, ask students whether they can tell you what information it is showing. They should be able to tell from the heading that the bar chart is about car sales in a given month.

Ask them to focus on the axes. *How many BMW cars were sold?* (40, reading the height of the bar for BMW against the numbers on the axis.) Now ask students to interpret the data by answering the questions on page 183 of the Student Book. Then they can try making up their own questions about the data in the bar chart.

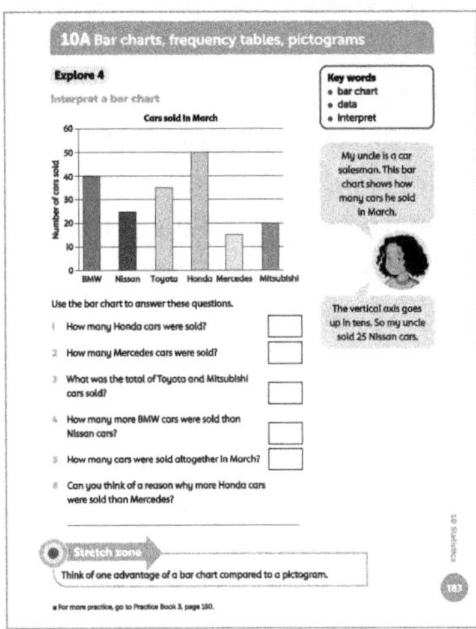

Differentiation

Supporting: Ask students to tell you which make of cars were sold least and most.

Consolidating: Ask students to tell you everything that they can about the bar chart.

Extending: Ask students to support other less-confident students to interpret the data so that they will consolidate their understanding.

Stretch zone: *Think of one advantage of a bar chart compared to a pictogram.*

Check that students can suggest an appropriate advantage. For example, they may suggest reading the numbers off the axis directly instead of having to count pictures.

 **Reflection time**

Ask students to share what they found about the data. How did they work out how many of each car were sold? Can they use the bar chart to put the cars in order of numbers sold?' How many cars were sold altogether? Encourage students to think of their own questions about the data.

Practice Book: Students can complete Practice Book page 150. They can do this directly after the main activity, as homework, or as the focus of a separate mathematics session to help students consolidate their learning and build fluency.

Students continue with their exercise survey as part of their data-handling cycle. This time they construct a bar chart to represent the data collected previously and then write some sentences about their pictogram. *What do you always need to remember to include on your bar chart?*

Differentiated outcomes	
All students	should recognise what the bar chart is showing.
Most students	will be able to interpret the data using the axes.
Some students	may ask deeper questions about the data.

Answers

Student Book page 183

1 50

2 15

3 55

4 15

5 185

6 It could be because Honda cars are cheaper than cars made by Mercedes.

Practice Book page 150

Students draw their own bar chart based on data they collected. Check that their bar chart matches their data and that it includes a title and correctly labelled axes. Also check that they have written sensible sentences about their bar chart.

10B Venn diagrams and Carroll diagrams

Discover 1 — Student Book page 184

Specific learning focus

- Present and interpret data using many contexts and representations (e.g. Carroll diagrams).

Global skills

- **Creative skills:** problem solving
- **Real-world skills:** presenting and interpreting information
- **Interpersonal skills:** communication

Key vocabulary

- Carroll diagram, criteria, sort

Resources

- none needed

Language support

Use display paper to model the Carroll diagrams so that students can refer to these to support them throughout these lessons. When students are deciding how they want to classify the objects, key vocabulary includes 'criteria'.

Introductory activity

Remind students of the birth dates activity from the Discover 1 lesson in Unit 10A. Ask each student to tell you the date in the month they were born and write these numbers on the board. Just use the dates – so write 8 for 8 June. Do not repeat any numbers.

When you have the list, draw a **Carroll diagram** on the board:

	Even	Not even
Less than 15		
Not less than 15		

Remind students that for Carroll diagrams we always use the convention of 'criterion' and '**not** criterion' rather than writing 'even' and 'odd'.

Students should come to the front of the class and decide where their birth date would go in the Carroll

diagram. Group students with the same birth date together so if two or three were all born on the 8th of a month, they can help each other decide where the number 8 goes.

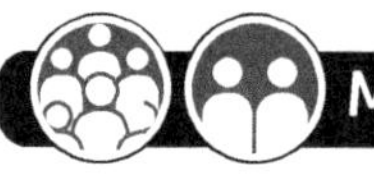

Main activity

Remind students of the work they did in Unit 10A on setting up a shop and deciding what to sell in their shop. Explain that they are going to use items from their shop in this lesson. (If they didn't decide as part of their survey in Unit 10A what items they would sell, they can use data from another group.) Look together at page 184 of the Student Book. If you have access to an IWB you could use this. Refer students to the Think back, which shows how various items have been sorted, based on their colour and whether we can eat them.

Students work through the activities in the Student Book on page 184. They work in pairs to support each other in thinking of different ways to **sort** the items from their shop in Unit 10A. They should take it in turns to think of **criteria** for sorting. At this stage, they should only use two criteria. While students work on this, notice pairs who have classified in an interesting or unusual way, and who could share this in reflection time. Students complete two different Carroll diagrams, sorting the items in two different ways.

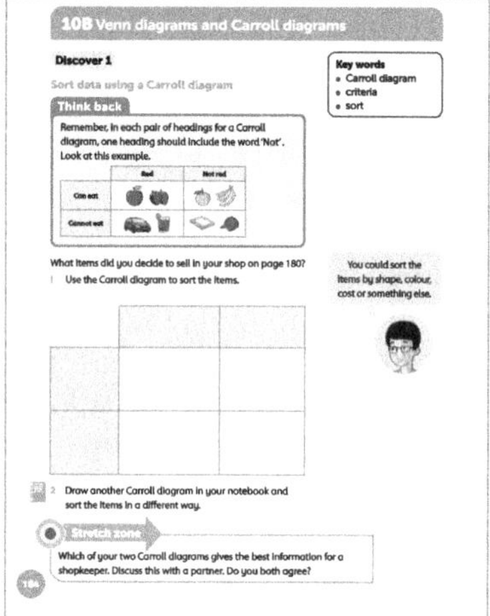

Differentiation

Supporting: You can help students by writing the items to be sorted into the Carroll diagram – they should tell you where to write each item.

Consolidating: Ask students to explain their thinking as they complete the Carroll diagrams.

Extending: Encourage students to be creative in the criteria they use to classify objects.

Stretch zone: *Which of your two Carroll diagrams gives the best information for a shopkeeper? Discuss this with a partner. Do you both agree?*

Check that students draw the diagrams correctly according to the sorting criteria they choose and can explain why their choice of diagram is best.

 ## Reflection time

Ask pairs to share the criteria they used to sort the objects. After each presentation, students should discuss what this result tells them about their choices of criteria and whether this would help them as a shopkeeper.

Practice Book: There is no Practice Book page for this lesson.

Differentiated outcomes	
All students	should complete Carroll diagrams with support.
Most students	will complete Carroll diagrams independently.
Some students	may think of creative ways to classify objects.

10B Venn diagrams and Carroll diagrams

Discover 2 Student Book page 185

Specific learning focus

- Present and interpret data using many contexts and representations (e.g. Venn diagrams).

Global skills

- **Creative skills:** exploring
- **Real-world skills:** presenting information
- **Self-development skills:** reflecting on learning
- **Interpersonal skills:** communication

Key vocabulary

- Venn diagram, criteria, data

Resources

- none needed

Language support

Use display paper to model the Venn diagrams so that students can refer to these to support them. When students are deciding how they want to classify the objects, key vocabulary includes 'criteria'.

Introductory activity

Ask students to think of different animals and make a short list of about ten animals. Now ask them to think of anything they can that has four legs, and again make

Answers

Student Book page 184

Answers will vary according to the items chosen and the sorting criteria. Check that the Carroll diagrams have been labelled correctly and that the items have been placed in the correct places on them.

a list of about ten. Now compare the two lists – is there anything that has appeared on both lists? Look together at page 185 of the Student Book. If you have access to an IWB you could use this. Refer students to the Think back to give them an example of a Venn diagram with these same sorting criteria.

Remind students that for **Venn diagrams** we always use the convention of having two distinct criteria, with a circle for each and an overlap for items that have both criteria. Draw a blank Venn diagram on the board and label the circles for 'animals' and 'things with four legs'.

Students should come to the front of the class one at a time and decide where each item from the lists should go in the Venn diagram.

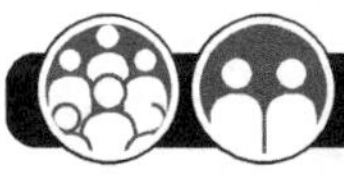 Main activity

Write a list of the numbers from 1–20 on the board. Draw a Venn diagram that sorts numbers into 'Even numbers' and 'Multiples of 3'. Remind students that they place numbers outside the circles if they do not fit in either circle, and in the intersection if they fit in both.

Students work through the activities in the Student Book on page 185. They work in pairs to support each other in thinking of different ways to sort the items from their shops in Unit 10A. They should take it in turns to think of criteria for sorting. At this stage, they should only use two criteria. While students work on this, notice pairs who have classified in an interesting or unusual way and who could share this during reflection time. Students complete two different Venn diagrams, so they are sorting the items in two different ways.

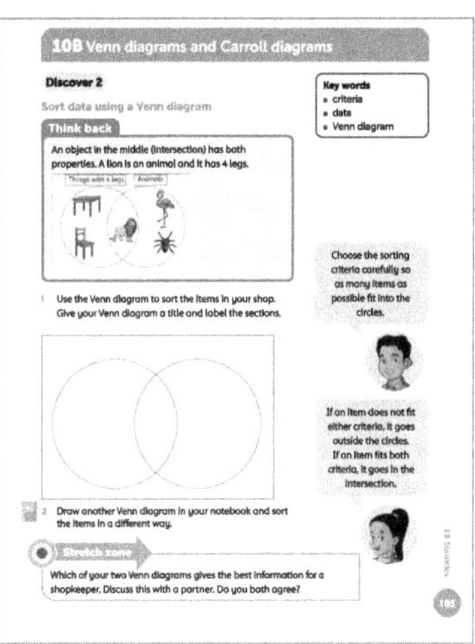

Differentiation

Supporting: You can help students by writing the items to be sorted into the Venn diagram – they should tell you where to write each item. They can draw the items if they prefer.

Consolidating: Ask students to explain their thinking as they complete their Venn diagrams.

Extending: Encourage students to be creative in the criteria they use to classify objects.

Stretch zone: *Which of your two Venn diagrams gives the best information for a shopkeeper? Discuss this with a partner. Do you both agree?*

Check that students draw the diagrams correctly according to the sorting criteria they choose and then can explain why their choice of diagram is best.

Reflection time

Ask pairs to present to the class the criteria they used to sort the objects. After each presentation, students should discuss what this result tells them about their choice of criteria and whether this would help them as a shopkeeper.

Practice Book: There is no Practice Book page for this lesson.

Differentiated outcomes	
All students	should complete Venn diagrams with support.
Most students	will complete Venn diagrams independently.
Some students	will think of creative ways to classify objects.

Answers

Student Book page 185

Answers will vary according to the items chosen and the sorting criteria. Check that the Venn diagrams have been labelled correctly and that the items have been placed in the correct places on them.

10B Venn diagrams and Carroll diagrams

Explore Student Book pages 186–187 • Practice Book page 151

Specific learning focus

- Present and interpret data using many contexts and representations (e.g. Carrol and Venn diagrams).

Global skills

- **Creative skills:** problem solving
- **Real-world skills:** presenting and interpreting information

Key vocabulary

- Carroll diagram, Venn diagram, criteria

Resources

- none needed

Language support

Ask students closed questions to help them remember the properties of numbers and then shapes: *Is it odd? Is it a multiple of 3?* and so on; *Is it regular? How many sides does it have?*

Introductory activity

Draw a blank Carroll diagram on the board. Do not label it. Add the numbers 1–15, starting with 1, and then 2 and so on. The completed diagram should be as follows (but do not write in the criteria until the end of the activity):

	Multiple of 3	Not multiple of 3
Less than 10	3, 6, 9	1, 2, 4, 5, 7, 8
Not less than 10	12, 15	10, 11, 13, 14

As you write the numbers in the diagram, students should try to guess what criteria you are using. Take several guesses at each stage to reinforce the properties of the numbers you are writing.

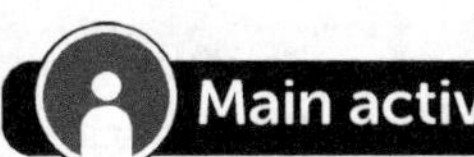

Main activity

Students should individually complete questions 1–4 in the Student Book on pages 186–187. Remind students to focus on whether each item has a criterion or not. There are three Carroll diagrams for students to complete. If students finish quickly, they can choose one of the sets to sort again, using a Venn diagram and choosing their own criteria.

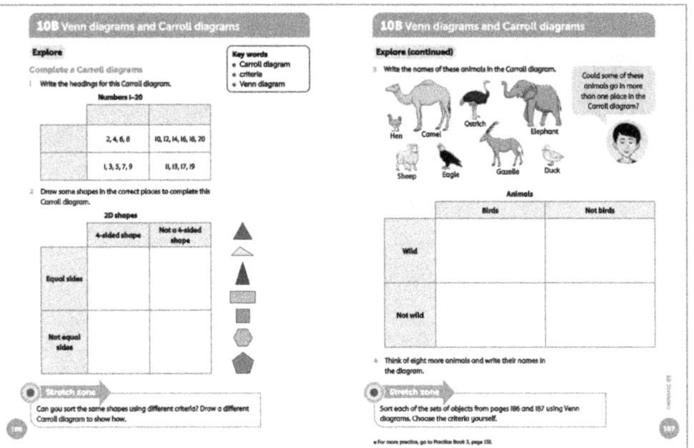

Differentiation

Supporting: You can help write the items to be sorted. Students should tell you where to write each item.

Consolidating: Ask students to explain their thinking as they complete the diagrams.

Extending: Encourage students to be creative in the criteria they use to classify objects.

Stretch zone: (end of page 186) *Can you sort the same shapes using different criteria. Draw a different Carroll diagram to show how.*

Sort each of the sets of objects from pages 186 and 187 using Venn diagrams. Choose the criteria yourself.

Check that students have completed the Venn diagrams correctly. They can sort all three sets of objects using three separate Venn diagrams. *Which objects were better sorted using a Carroll diagram, and which were better sorted using a Venn diagram? Why?*

 Reflection time

Take feedback from individuals about how they completed the Carroll diagram in question 3. Make a large copy of the diagram on the board and ask students to come and add the animals they discussed. *Which animals did you find difficult to place? Were there animals that could have been placed in more than one of the cells? Would a Venn diagram be better to sort these animals? Why?*

Practice Book: Students complete Practice Book page 151. They can do this directly after the main activity, as homework, or as the focus of a separate mathematics session to help students consolidate their learning and build fluency.

Students sort food items, first using a Carroll diagram and then using a Venn diagram. They choose their own sorting criteria for the Venn diagram. *Which diagram was more suitable for sorting food items?*

Differentiated outcomes	
All students	should complete Venn diagrams and Carroll diagrams with support.
Most students	will complete Venn diagrams and Carroll diagrams independently.
Some students	may think of creative ways to classify objects.

Answers

Student Book pages 186–187

1 Horizontal headings: 'Numbers less than 10' and 'Numbers not less than 10'

Vertical headings: 'Even numbers' and 'Not even numbers'

2

	4-sided shape	Not a 4-sided shape
Equal sides	square	equilateral triangle regular pentagon regular hexagon
Not equal sides	rectangle	isosceles triangle scalene triangle

3 Answers may vary depending on students' reasoning but could include:

	Birds	Not birds
Wild	ostrich eagle duck	elephant gazelle
Not wild	hen	sheep camel

4 Check that the animals have been placed correctly.

Practice Book page 151

Answers will vary according to the data collected. Check that the completed Venn diagrams correspond to the information in the completed Carroll diagrams.

Stretch zone:

Same: both diagrams sort data using two or more criteria. Both provide a visual representation of the sorted data.

Different: a Venn diagram allows data to be sorted into more than one section, using the intersection. In a Carroll diagram data is sorted into one section and cannot be in more than one section.

10 Statistics

Connect — Student Book pages 188–189

Big idea

- I can collect data using tally charts and frequency charts. I can organise data using pictograms, bar charts, Carroll and Venn diagrams. I can interpret information in charts to help me find out things.

Global skills

- **Real-world skills:** research, presenting information, interpreting information
- **Interpersonal skills:** communication, teamwork
- **Self-development skills:** reflecting on learning

Key vocabulary

- tally chart, frequency table, pictogram, bar chart, criteria, Carroll diagram, Venn diagram

Resources

- access to the internet or to reference books
- Top Trumps® cards (optional) – available from toptrumps.comx

Language support

You can help write the answers for students who may need support in writing. This will help you judge whether they understand the mathematics.

Introductory activity

What is shown on Top Trumps cards? (They show facts about specific things, for example, footballers, cars, superheroes, animals, Disney characters and so on.) Explain that students are going to use the facts on a set of Top Trumps cards to generate some data about a particular topic. Play a game of Top Trumps so students become familiar with the cards.

Main activity

In their groups, students should either select a pack of Top Trumps cards, or design their own Top Trumps game, first using the internet or reference books to research facts about their chosen topic. They could choose to look at the length or mass of various animals, or the height of some of the world's most famous buildings, or the top speed of some sports cars, for example. They then use the data they find out to complete the table in the Student Book on page 188. Look together at this page. If you have an IWB you could use this. Refer students to the part-completed table, so students can see how to complete it. Talk through this example before students complete their own table for their chosen topic. You may also choose to talk through how to complete a table for some of the specific topics chosen by students.

On page 189 of the Student Book, students choose to create a bar chart, a Venn diagram, or a Carroll diagram based on the data in their table. This activity may need to be spread over more than one lesson to ensure that students have time to engage fully with it.

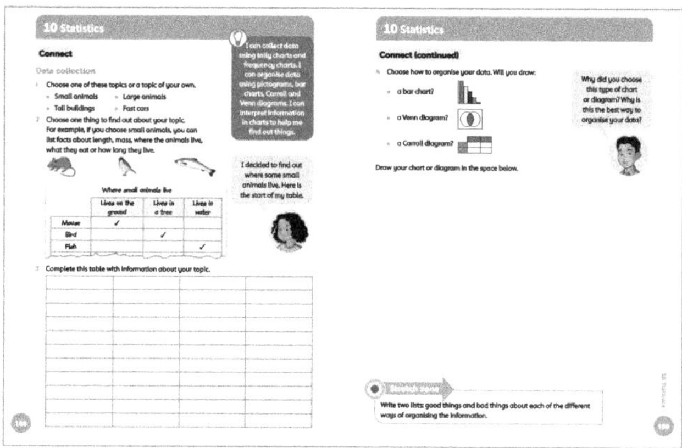

Differentiation

Supporting: Help students complete the table of information. They should tell you where to write each item.

Consolidating: Ask students to explain their thinking as they complete the table, for example, *Why did you choose those headings for your table? Where will you write that information?*

Extending: Challenge students to be creative in the criteria they use for their information table.

Stretch zone: *Write two lists: good things and bad things about each of the different ways of organising the information.*

Check that students have written useful lists about the different types of diagram.

Reflection time

Choose students to share their table and chart with the class and say something about the information in it. Take feedback from the rest of the class. *Do they think each group chose the best way of organising the data from the tables?* Allow students to ask questions about each table or chart that is shared.

Differentiated outcomes	
All students	should complete an information table with support.
Most students	will complete an information table independently.
Some students	may think of creative ways to classify objects in their information table.

10 Statistics

Review   Student Book page 190 • Practice Book page 152

Student Book

- Global skills
- **Self-development skills:** reflecting on learning

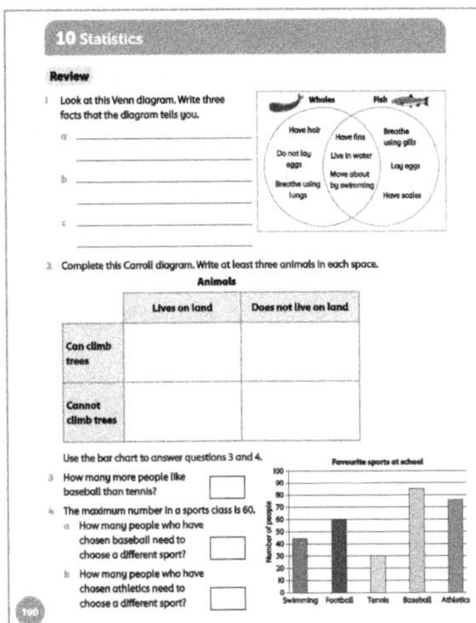

With young children, assessment activities are most effective when carried out as an everyday classroom activity. Students should have examples of different types of chart available so they can refer to these to support them.

Watch as students complete the charts and interpret the data from these on page 190 of the Student Book. Listen to check that they understand the names of the different types of chart and understand how to read them.

Answers

Student Book page 190

1 Answers will vary according to the set of facts chosen.

2 Check that students have completed the Carroll Diagram correctly with their chosen animals.

3 55

4 **a** 25

   b 15

Practice Book

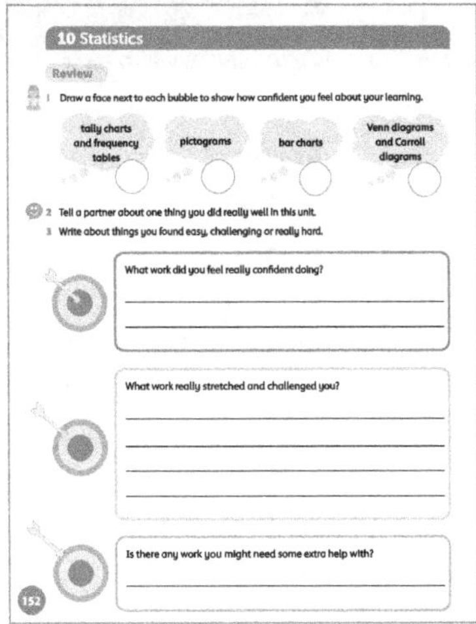

It is appropriate to complete this Practice Book review as a whole-class discussion. You may choose to keep a record of the class discussion or a copy of the review page for your own records. The review provides an opportunity for students to reflect on their learning from the unit, to discuss any areas of mathematics that they feel went particularly well, and any areas that they feel less confident about. Ensure that all students have a copy of the Student Book as a reminder of the areas of mathematics that they have worked on in this unit.

Allow students plenty of time for discussion before asking them to complete the Practice Book page individually, and then, if appropriate, to share their responses with the rest of the class. If students complete this self-assessment at home, encourage them to discuss this with adults. Make a note of areas that students still feel unsure about, for example the difference between tally charts and frequency tables, or the difference between Carroll and Venn diagrams.

Additional material

There are additional end-of-unit assessments available on the *Oxford Owl for School* website.

acute	a type of angle that is less than 90°
a.m.	stands for *ante meridiem*, which is Latin for 'before midday' (noon). It is used to show times after 12 midnight but before 12 noon. For example, 9:30 a.m. is half past nine in the morning
angle	an amount of turn measured in degrees
approximate	a number that is near enough the exact answer. Similar words are 'nearly', 'about' and 'near enough'. The mathematical symbol for an approximate answer is ≈
axis /axes	the axes of a graph or chart are the horizontal and vertical lines that make it. The horizontal line is the x-axis and the vertical line is the y-axis
bar chart	a graph that uses bars to show information. The bars are all the same thickness and can be horizontal or vertical
base-10	the everyday number system, where each digit of a number is between 0 and 9
calendar	shows time divided into years, months, weeks, and days
Carroll diagram	a table or diagram that sorts things into different categories, using 'X' or 'Not X' sorting rules
century	100 years
chart	a graph or table that shows information in an orderly way. Examples are a bar chart and a frequency table
column addition / column subtraction	a type of written calculation that uses columns for ones, tens, hundreds, etc. You always start by adding or subtracting the ones first and you transfer amounts from one column to the next
complement	the amount that you need to add to a number to reach a target number. For example, the complement to 100 for 47 is 53
date	tells us when something happens, usually the day, month and year
decimal	a decimal number has a decimal point to separate its tens and ones from its tenths and hundredths
diagonal	a straight line that joins one vertex of a polygon with an opposite vertex
digital time	time shown using numerals only. It often uses the 24-hour clock. For example, 14:30 is half past two in the afternoon
division	sharing things equally or grouping into sets of the same size. The mathematical symbol for division is ÷
equivalent	equal in size
estimate	make a reasonable guess to get an approximate answer
frequency table	frequency is how often something happens. Tally marks are often used to show the frequency in a frequency table
greater than (>)	a number or fraction is greater than another when its value is higher. For example, 3 > 2
hexagonal	describes any polygon that has six sides
horizontal	a line in the same direction as the horizon (or ground). A table has a horizontal top
hundreds	groups of 100. In a place-value grid, a number in the hundreds column represents that number of hundreds
integer	a whole number, positive or negative
inverse	opposite. For example, addition and subtraction are inverse operations, one operation undoes the other operation
kilometre (km)	a metric unit of length used to measure long distances. 1000 metres = 1 kilometre
less than (<)	a number or fraction is less than another when its value is lower. For example, 2 < 3
mile	an imperial unit used to measure long distances. 1 mile = approximately 1600 metres
multiple	the result of multiplying a number by another number. For example, the multiples of 4 are 4, 8, 12, 16 and so on

<table>
<tr><td>multiplication</td><td>adding the same number again and again. The mathematical symbol for multiplication is ×</td></tr>
<tr><td>non-unit fraction</td><td>a fraction with a numerator that is not 1</td></tr>
<tr><td>note</td><td>a piece of money made from paper</td></tr>
<tr><td>obtuse</td><td>an angle that measures between 90° and 180°</td></tr>
<tr><td>octagonal</td><td>describes any polygon that has eight sides</td></tr>
<tr><td>ones</td><td>groups of 1. In a place-value grid, a number in the ones column represents that number of ones</td></tr>
<tr><td>parallel</td><td>parallel lines are straight lines that always stay the same distance apart and never meet</td></tr>
<tr><td>partition</td><td>split a number into separate parts (for example, ones, tens, hundreds) to make it easier to work with</td></tr>
<tr><td>pentagonal</td><td>describes any polygon that has five sides</td></tr>
<tr><td>perpendicular</td><td>if two lines are perpendicular they meet to make a 90° angle</td></tr>
<tr><td>plan</td><td>a diagram showing where things are. Plans often have a scale because they cannot be drawn life size</td></tr>
<tr><td>p.m.</td><td>stands for post meridiem, which is Latin for 'after midday' (noon). It is used to show times after 12 noon but before 12 midnight. For example, 4:15 p.m. is quarter past four in the afternoon</td></tr>
<tr><td>prism</td><td>a solid shape with matching ends that are polygons. Cubes and cuboids are special types of prism</td></tr>
<tr><td>product</td><td>the answer you get by multiplying numbers together</td></tr>
<tr><td>quadrilateral</td><td>any polygon that has four sides and four angles. The four angles add up to 360°</td></tr>
<tr><td>remainder</td><td>what is left over after you share something, or the amount left over when one quantity cannot be exactly divided by another</td></tr>
<tr><td>right angle</td><td>an angle that measures 90°; it is a quarter of a whole turn</td></tr>
<tr><td>right-angled triangle</td><td>a triangle that contains one right angle</td></tr>
<tr><td>Roman numerals</td><td>numbers that were used in Ancient Rome. Letters from the Latin alphabet represent numbers. For example, V = 5</td></tr>
<tr><td>round</td><td>write a number as an approximate number by adjusting it up or down to the nearest 10 or 100</td></tr>
<tr><td>semi-circle</td><td>half of a circle. The straight side is a diameter of the circle</td></tr>
<tr><td>tens</td><td>groups of 10. In a place-value grid, a number in the tens column represents that number of tens</td></tr>
<tr><td>tenth</td><td>one of ten equal parts of a shape, quantity or number</td></tr>
<tr><td>third</td><td>one of three equal parts of a shape or number. It can be written as $\frac{1}{3}$</td></tr>
<tr><td>value</td><td>what something is worth</td></tr>
<tr><td>Venn diagram</td><td>a sorting diagram that uses circles to show sets of things that have different properties. Anything in the overlapping part of two circles has both properties</td></tr>
<tr><td>vertex / vertices</td><td>a point at which two or more lines meet in an object or shape</td></tr>
<tr><td>vertical</td><td>a line that goes up and down, at right angles to a horizontal line</td></tr>
</table>